What Your Colleagues Are Saying . . .

"With this new book, Karp, Fennell, Kobett, Andrews, Suh, and Knighten have delivered an innovative strengths-based intervention approach that focuses on priming rather than remediation. The structures they have designed around co-planning, assessment, and accessible tasks can provide much-needed clarity to the complex task of math intervention work. This work takes the idea of building from student strengths and makes it tangible. I am so excited to see schools and districts implement this work!"

Rachel Lambert
Associate Professor of Special Education and Mathematics
Santa Barbara, CA

"What a great resource for supporting interventions! One of the key features of this book is that it provides a detailed process for planning interventions that will assist students in accessing grade-level content. The tasks provided are especially helpful in using rich tasks that engage students as they make sense of mathematics."

Barbara J. Dougherty
Mathematics Instructional Coach
Key West, FL

"This book transforms mathematics intervention! It skillfully weaves together research-based practices of mathematics education and special education to empower educators and their students to be active doers of mathematics. The innovative, strengths-based approach positions all students to be successful by promoting student agency, educator collaboration, and a solid mathematical foundation. This is a must-have resource for all as it offers a new way of thinking about mathematics intervention to drive much-needed change."

Dawn Pilotti
K–12 Faculty Chair of Mathematics,
McRae Family Foundation, Currey Ingram Academy
Brentwood, TN

"*Proactive Mathematics Interventions* is a game changer! The authors masterfully combine research-based frameworks and instruction to create a coherent picture on how to effectively implement interventions. This must-read book provides clear examples, detailed descriptions, and numerous resources."

Joleigh Honey
Founder & Consultant, JHoneyMath
Former K–12 STEM Coordinator, Utah State Board of Education
Salt Lake, UT

"This book reminds us that effective intervention isn't about fixing kids. It's about fixing systems. With deep respect for teachers and students, the authors share a proactive, strengths-based approach full of practical strategies that help students engage, reason, and see themselves as capable, confident math learners. It's a must-have for every educator."

Zak Champagne
Chief Content Officer, Flynn Education
Olympia, WA

"*Proactive Mathematics Interventions* advocates for rethinking our traditional deficit approaches to intervention and provides a vision for a strengths-based approach that uncovers and builds upon what students know and can do in time for them to shine within their classrooms. In addition to advocating for a priming approach, this resource supports educators to enact this instructional shift with tasks and activities that revisit and strengthen foundational understandings while proactively positioning every student to thrive with grade-level content."

Nicole Rigelman
Professor of Mathematics Education and Education Program Officer,
Portland State University and The Math Learning Center
Portland, OR

"As mathematics specialists, we're always looking for ways to enhance our support for both teachers and students. *Proactive Mathematics Interventions, Grades 2–5* offers precisely that shift—moving us from a reactive stance to a proactive one within our mathematics and MTSS frameworks. This resource champions the development of a more robust system of support, one centered on a truly transformational approach to mathematics intervention. This book emphasizes a strengths-based perspective, where we leverage the insights gained from formative assessment to guide students in 'doing the math' through thoughtfully chosen, purposeful tasks. This book is an invaluable tool to lead the way in fostering a more proactive and effective mathematics learning environment."

Spencer Jamieson
Elementary Mathematics Specialist
Fairfax, VA

"This book will have a permanent spot on your professional bookshelf, but it will hardly ever actually be on the bookshelf. It makes the path for shifting from reactive interventions to proactive interventions clear and the authors provide a comprehensive plan to get there. The task resources are robust and the critical discussion of how to transform a system will impact students for years."

Karla Bandemer
Grades 3–5 Math Teacher Leader,
Lincoln Public Schools
Lincoln, NE

"*Proactive Mathematics Interventions* shifts the focus from remediation to readiness with a powerful 'priming' approach. Designed for grades 2–5, it offers timely, targeted strategies grounded in math progressions. With practical tasks and clear guidance, this resource equips teachers, coaches, and leaders to support all learners—especially those who struggle—before they fall behind. A must-have for proactive, meaningful math instruction."

Hampden-Wilbraham Regional School District Math Team
Wilbraham, MA

Proactive Mathematics Interventions Grades 2–5

Priming for Success Through Engaging Tasks and Purposeful Design

Karen S. Karp

Francis (Skip) Fennell

Beth McCord Kobett

Delise R. Andrews

Jennifer Suh

Latrenda Knighten

CORWIN

FOR INFORMATION:

Corwin

A SAGE Company

2455 Teller Road

Thousand Oaks, California 91320

(800) 233-9936

www.corwin.com

SAGE Publications Ltd.

1 Oliver's Yard

55 City Road

London EC1Y 1SP

United Kingdom

SAGE Publications India Pvt. Ltd.

Unit No 323–333, Third Floor, F-Block

International Trade Tower Nehru Place

New Delhi 110 019

India

SAGE Publications Asia-Pacific Pte. Ltd.

18 Cross Street #10–10/11/12

China Square Central

Singapore 048423

Vice President and Editorial
  Director: Monica Eckman

Associate Director
  and Publisher, STEM: Erin Null

Senior Editorial Assistant: Nyle De Leon

Production Editor: Tori Mirsadjadi

Copy Editor: Kim Husband

Typesetter: C&M Digitals (P) Ltd.

Proofreader: Jennifer Grubba

Cover Designer: Rose Storey

Marketing Manager: Margaret O'Connor

Copyright © 2026 by Corwin Press, Inc.

All rights reserved. Except as permitted by U.S. copyright law, no part of this work may be reproduced or distributed in any form or by any means, or stored in a database or retrieval system, without permission in writing from the publisher.

When forms and sample documents appearing in this work are intended for reproduction, they will be marked as such. Reproduction of their use is authorized for educational use by educators, local school sites, and/or noncommercial or nonprofit entities that have purchased the book.

All third-party trademarks referenced or depicted herein are included solely for the purpose of illustration and are the property of their respective owners. Reference to these trademarks in no way indicates any relationship with, or endorsement by, the trademark owner.

No AI training. Without in any way limiting the author's and publisher's exclusive rights under copyright, any use of this publication to "train" generative artificial intelligence (AI) or for other AI uses is expressly prohibited. The publisher reserves all rights to license uses of this publication for generative AI training or other AI uses.

Printed and bound by CPI Group (UK) Ltd, Croydon, CR0 4YY

Library of Congress Cataloging-in-Publication Data

Names: Karp, Karen S. author | Fennell, Francis M., 1944- author | Kobett, Beth McCord author | Andrews, Delise author | Suh, Jennifer M., 1971- author | Knighten, Latrenda author

Title: Proactive mathematics interventions, grades 2–5: priming for success through engaging tasks and purposeful design / Karen S. Karp, Francis (Skip) Fennell, Beth McCord Kobett, Delise R. Andrews, Jennifer Suh, Latrenda Duretta Knighten.

Description: Thousand Oaks, California: Corwin, [2026] | Series: Corwin mathematics series | Includes bibliographical references and index.

Identifiers: LCCN 2025019012 | ISBN 9781071973677 paperback | ISBN 9781071936450 epub | ISBN 9781071936597 pdf

Subjects: LCSH: Mathematics—Study and teaching (Elementary)—United States | Learning disabled children—Education (Elementary)—United States | Multi-tiered systems of support (Education)—United States | Response to intervention (Learning disabled children)—United States | Priming (Psychology)

Classification: LCC QA135.6 .K366 2026
LC record available at https://lccn.loc.gov/2025019012

This book is printed on acid-free paper.

25 26 27 28 29 10 9 8 7 6 5 4 3 2 1

DISCLAIMER: This book may direct you to access third-party content via web links, QR codes, or other scannable technologies, which are provided for your reference by the author(s). Corwin makes no guarantee that such third-party content will be available for your use and encourages you to review the terms and conditions of such third-party content. Corwin takes no responsibility and assumes no liability for your use of any third-party content, nor does Corwin approve, sponsor, endorse, verify, or certify such third-party content.

Contents

Preface	xi
What Is This Book About?	xi
Who Is This Book For?	xiii
How Does This Book Work?	xiv
More on Priming	xvi
How Do We Use Certain Language in This Book?	xix
Why Do We Choose Certain Terms?	xix
How Does Strengths-Based Language Change Mindsets?	xix
How Do We Define Terms Throughout the Text?	xix
Acknowledgments	xxi
About the Authors	xxiii

Part 1: Introduction to Priming   1

Chapter 1: Introduction. Proactive Intervention: Fixing Structures, Not Children   3

So What's the Problem?	3
What's Being Done?	5
What Needs to Change?	6
What Is the Research Base?	8
Recommendation 1: Systematic Instruction	9
Recommendation 2: Mathematical Language	9
Recommendation 3: Representations	9
Recommendation 4: Number Lines	10
Recommendation 5: Word Problems	10
Recommendation 6: Timed Activities	10
When We Talk About Intervention, What Do We Mean?	12
What Do We Mean by Tutoring?	14
What Are Some Key Elements of the Intervention Instruction?	14
Employing Multiple Representations	14
Collaborating Across the Team	16
Using a Preemptive and Proactive Approach	17
What Should We Think About Next?	20
Reflection Opportunities for Chapter 1	20

Chapter 2: Characteristics of Effective Strengths-Based Interventions   21

What Is Strengths-Based Teaching and Learning?	21
What Is the Strengths-Spotting Process?	22
Four Steps to Spotting Strengths	22
Example of the Process Using a Student's Representation	27
How Do We Build Rapport and Relationships?	29
What Does It Mean to Be a Warm Demander?	30
How Can We Integrate Culturally Responsive Mathematics Teaching as We Engage in Intervention Activities?	31
How Can We Leverage Family Involvement in Supporting Math Proactive Intervention?	32
How Can We Help Students Avoid or Overcome Math Anxiety?	33
Reflection Opportunities for Chapter 2	33

Chapter 3: Formative Assessment: Targeting and Monitoring Interventions — 35
    What Is Formative Assessment, and Why Is It Important? — 36
    What's the Role of Summative Assessments? — 36
    And What About Feedback? — 37
    How Can We Focus on Formative Assessment in Intervention Settings? — 38
        Observation Tools — 38
        Interview Tools — 41
        Show Me Tool — 45
    What Is the Importance of Task Selection in Assessment? — 48
    How Can We Connect Assessment and Instruction? — 49
    Reflection Opportunities for Chapter 3 — 50

Chapter 4: Your Turn: A Proactive Preventative Mathematics Intervention/Tutoring Model in Action — 53
    Considering Elements of Your Mathematics Intervention Program — 54
    How Is Our Approach Different? — 54
        What Does Planning for Priming Look Like? — 54
        What Does a Planning Conversation Sound Like? — 60
    How Do We Connect Strengths-Based Teaching and Formative Assessment? — 62
    How Can We Communicate Student Progress? — 67
    What Are Other Considerations We Need to Think About? — 68
        Staffing — 68
        Time and Location — 68
        Program Size — 68
        Instructional Materials — 69
        Instruction and Assessment — 69
    Where Do You Begin Priming Intervention Lessons? — 69
    What Can We Learn From Listening to the Voices of Others Trying Priming in Their Schools? — 70
    And Now, It's Your Turn! — 72
    Reflection Opportunities for Chapter 4 — 72

Part 2: Putting Priming Into Action — 74

Chapter 5 Intervention Tasks: — 79
    Task 1: Mystery Number Riddles — 80
    Task 2: Come on Down! — 88
    Task 3: Trading Places — 95
    Task 4: Adding It Up — 105
    Task 5: What's the Difference? — 114
    Task 6: Targeted Sum — 128
    Task 7: Targeted Differences — 134
    Task 8: Valuable Digits — 140
    Task 9: Picture This — 149
    Task 10: Balancing Act — 158
    Task 11: Escape Room — 166
    Task 12: Things That Come in Groups — 170
    Task 13: Garden Spaces — 180
    Task 14: Build It Bingo — 187
    Task 15: Rearrange It — 197
    Task 16: Multiplication Chains — 209
    Task 17: Speed Estimator — 220
    Task 18: Magic Pot Patterns — 234

Task 19: Fraction Frenzy	241
Task 20: Cake Time	250
Task 21: Line 'Em Up	260
Task 22: To Be or Not to Be: Equivalent Fractions	270
Task 23: Fraction Equivalence Through Quilts and Fringes	276
Task 24: Sport Stats	284
Task 25: Fraction Zap	290
Task 26: Fraction Sums	295
Task 27: Fraction Multiplication: Fundraiser	304
Task 28: Fraction Multiplication: Farmland	308
Task 29: Fraction Races	313
Task 30: Shape Sort	320
Task 31: Alike and Different	325
Task 32: Shape Shifting	331
Task 33: Shapes: Build It!	337
Task 34: What's My Rule?	343
Task 35: Counting Coins	352
Task 36: Smart About Money	359
Task 37: Measure Up!	365
Task 38: Comparing With Measurement	369
Task 39: On the Edge	373
Task 40: Field Day	377
Task 41: Broken Clock	382
Task 42: Tip and Plot	388
Task 43: Picture Perfect	396
References	409
Index	415

Visit the companion website at
https://companion.corwin.com/courses/
ProactiveMathIntervention
for appendices and downloadable printables.

Companion Website Contents

Part 1: Introduction to Priming

- **Figure 1.2 (and Figure 4.1)** Planning Model: General Education and Special Education Teacher Teams
- **Figure 2.1** Math Strengths Checklist
- **Figure 3.2** Observations: Small Group Implementation and Recording Tool (blank template version)
- **Figure 3.3** Observations: Student Mathematics Strengths Log (blank template version)
- **Figure 3.4** Interview Planning Tool
- **Figure 3.5** Interview: Individual Interview Prompt Recording Sheet (blank template version)
- **Figure 3.6** Show Me: Show Me Recording Sheet (blank template version)
- **Figure 4.2** The Planning for Priming Intervention Planning Tool (blank template version)
- **Figure 4.3** Monitoring Student Strengths Tool
- **Figure 4.4** Progress Monitoring Tool—Priming Task
- **Figure 4.5** Intervention Updates

Part 2: Putting Priming Into Action

- Task 1: Mystery Number Riddle Slides
  Task 1: Hundreds Chart
  Task 1: Place Value Mat
- Task 2: Place Value Column Labels
  Task 2: Teacher Support for Come on Down!
- Task 3: Place Value Mat
  Task 3: Directions for Making a Jumbo Place Value Chart
  Task 3: Place Value Cards
  Task 3: Equivalent or Not? Activity Pages
- Task 4: Place Value Mat
  Task 4: Directions for Making a Jumbo Place Value Chart
  Task 4: Place Value Cards
  Task 4: Adding It Up! Addition Mat
- Task 5: Place Value Mat
  Task 5: Directions for Making a Jumbo Place Value Chart
  Task 5: Place Value Cards
  Task 5: What's the Difference? Subtraction: Take From Mat
  Task 5: What's the Difference? Subtraction: Compare Mat
- Task 6: 0–9 Digit Cards
  Task 6: Target Sum Challenge Cards
- Task 7: 0–9 Digit Cards
  Task 7: Target Differences Challenge Cards
- Task 8: Valuable Digits Game Board
  Task 8: 0–9 Digit Cards
- Task 9: Math Story Cards (Additive Situations: no questions)
  Task 9: Math Story Cards (Additive Situations: with question)
  Task 9: Math Story Cards (Multiplicative Situations: no question)
  Task 9: Math Story Cards (Multiplicative Situations: with question)
  Task 9: Math Representations Game Cards
- Task 10: Pattern Block Cutouts
  Task 10: Balancing Act Game Board
  Task 10: Balancing Act Game Cards and Relationship Cards
  Task 10: Balancing Expressions
  Task 10: Balancing Shapes
  Task 10: Is It True?
- Task 11: Student Task Sets—Animals vs. Humans; Insect Data
  Task 11: Escape Room Teacher Recording Sheet and Codes
  Task 11: Translation Task Template
- Task 12: Group Images
  Task 12: Representing Equal Groups Recording Sheet
  Task 12: Boxes and Candy Recording Sheet
  Task 12: Translation Task Template
- Task 13: 1-centimeter grid paper
  Task 13: Garden Area Game Board
- Task 14: 1-centimeter grid paper
  Task 14: 1-inch grid paper
  Task 14: Bingo Cards
  Task 14: Equation Cards (multiplication)
  Task 14: Equation Cards (multiplication with variables)
  Task 14: Equation Cards (division)
  Task 14: Equation Cards (division with variables)
- Task 15: 1-centimeter grid paper
  Task 15: 1-inch grid paper

- Task 15: Place Value Cards
  Task 15: Array Cards
  Task 15: Associative Expression Mat
- Task 16: 1-centimeter grid paper
  Task 16: 1-inch grid paper
  Task 16: Multiplication Chain Number Cards
  Task 16: Multiplication Chain Expression Frame
  Task 16: Multiplication Chains Activity Page
- Task 17: Estimation Game Board A
  Task 17: Array Estimation Game Board B (2 pages)
  Task 17: Area Estimation Game Board B (2 pages)
  Task 17: First Factor Cards
  Task 17: Factor Cards
- Task 18: Input–Output Table Chart
  Task 18: Function Machine
  Task 18: Magic Pot
- Task 19: Unit Fraction Strips (tabloid sized)
  Task 19: Unit Fraction Strips (letter sized)
  Task 19: Unit Fraction Circles
- Task 20: Pattern Block Cutouts
  Task 20: Hexagons Recording Sheet
  Task 20: Comparing Fractions Activity Page
  Task 20: Fraction Compare Cards
- Task 21: Unit Fraction Strips (tabloid sized)
  Task 21: Unit Fraction Strips (letter sized)
  Task 21: Unit Fraction Circles
  Task 21: Line 'Em Up Common Numerator Cards
  Task 21: Line 'Em Up Common Denominator Cards
  Task 21: Line 'Em Up Benchmark Cards
  Task 21: Compare 'Em Cards Set A
  Task 21: Compare 'Em Cards Set B
  Task 21: Compare 'Em Cards Set C
  Task 21: Compare 'Em Cards Set D
  Task 21: Compare 'Em Cards Set E
- Task 22: 1-inch grid paper
  Task 22: Brownie Tray Activity Sheet
  Task 22: Unit Fraction Strips (tabloid sized)
  Task 22: Unit Fraction Strips (letter sized)
  Task 22: Unit Fraction Circles
- Task 23: Unit Fraction Strips (tabloid sized)
  Task 23: Unit Fraction Strips (letter sized)
- Task 24: Decimal Grids
  Task 24: Hundredths Wheel
  Task 24: Fraction–Decimal Game Cards
- Task 25: Fraction Zap!
- Task 26: Unit Fraction Circles
  Task 26: Fraction Sum Games Pages
- Task 27: Unit Fraction Strips (tabloid sized)
  Task 27: Unit Fraction Strips (letter sized)
  Task 27: Unit Fraction Circles
- Task 28: Unit Fraction Strips (tabloid sized)
  Task 28: Unit Fraction Strips (letter sized)
  Task 28: Unit Fraction Circles
- Task 29: Unit Fraction Strips (tabloid sized)
  Task 29: Unit Fraction Strips (letter sized)
  Task 29: Unit Fraction Circles

Task 29: Whole Number Tracks
Task 29: Unit Fraction Tracks
Task 29: Equal Shares Cards
- Task 30: Shapes for Sorting
- Task 31: Whole-group Charts
Task 31: Shapes for Sorting
Task 31: Shape Riddle Cards
- Task 32: Geometry Necklace Cards
Task 32: Questions and Checklist
- Task 33: Pattern Block Cutouts
Task 33: Triangle Outline
Task 33: Tangram Pieces
- Task 34: Polygons Sorting Set Shape Cards
Task 34: Polygons Sorting Set Note Page
Task 34: Polygons Sorting Set Group Labels
Task 34: Attributes Sorting Set Shape Cards
Task 34: Attributes Sorting Set Note Page
Task 34: Attributes Sorting Set Group Labels
Task 34: Triangles Sorting Set Shape Cards
Task 34: Triangles Sorting Set Note Page
Task 34: Triangles Sorting Set Group Labels
Task 34: Group Number Cards
- Task 35: Coin Spinners
Task 35: Counting Coins Recording Sheet
Task 35: Make 25 Cents Game Board
Task 35: Make $1.00 Game Board
Task 35: Hundreds Chart
- Task 36: SMART Slides (to be used when reading the poem for visuals)
Task 36: Is It a Fair Trade? Problem Cards
Task 36: Get Smart Recording Sheet
Task 36: Let's Go Shopping Activity Cards
Task 36: Let's Go Shopping Recording Sheet
Task 36: Counting Tower Challenge Recording Sheet
- Task 37: Measuring Worm Units
Task 37: Cuisenaire Rods
Task 37: Measuring Cuisenaire Rods Recording Sheet
- Task 38: Comparing With Measurement Recording Sheet
- Task 39: 1-centimeter grid paper
Task 39: 1-inch grid paper
- Task 40: Station Directions
Task 40: Unit Spinners
Task 40: Field Day Recording Sheet—Metric
Task 40: Field Day Recording Sheet—Customary
- Task 41: Clock Spinner
Task 41: Student Timeline Game Boards
Task 41: What's the Hour? Cards
Task 41: 1–12 Number Cards
Task 41: Student Paper Clocks

Appendices

- Appendix A – Children's Literature Reference
- Appendix B – Home Connections
- Appendix C – Interventions at a Glance
- Appendix D – Digital Resources

Preface

Welcome! If you have picked up this book, you are one of the important adults working with elementary-aged students who will benefit from supportive intervention opportunities in mathematics. You play a critical role in shaping students' confidence and mathematical abilities, helping them unlock what already exists within them to leverage for mathematical success.

What Is This Book About?

Many educators suggest that they are in need of ideas to create the structures necessary for student success in mathematics learning because they have not seen the improvements in students' performances they wish to see. This book focuses on mathematics intervention and related practices that make multitiered systems of support (MTSS) and its academic component, response to intervention (RtI), effective and successful.

This book also posits that adopting a proactive approach to intervention offers the potential to create a more robust support system that addresses educational and behavioral needs, ensuring every student can thrive. We call this approach *Priming*.

Throughout the book, we will share the voices of educators to communicate their wisdom on how they have implemented these ideas and how student improvement has emerged and expanded on several levels. For example, let's hear from a fifth-grade teacher who, along with many others in her school system, is dedicated to Priming students for success:

> *Our district has been really working hard over the last 2 years to implement an MTSS model and we have a dedicated time for students to receive mathematics intervention, acceleration, and so on. However, math was a bit of a mystery for us. There wasn't a program we could buy or a framework we could adopt that felt good to us or matched what we wanted math instruction to look like. After attending presentations where we heard about the thinking shared in this book, we started to talk in larger teams about the idea of introducing prerequisite skills weeks before new content in our intervention sessions. After more conversation with the authors, we got to work! Priming felt like the answer we were ready to work with. Grade-level teams met this past summer and talked about formative and summative assessments and how we were going to identify the prerequisite skills to Prime.*
>
> *Now, students are working on prerequisite skills, teachers are giving pre-assessments to identify students to Prime and figure out what their prior knowledge and strengths are. We are not over-assessing, there are some students who never get the pre-assessments because we know*

Multitiered systems of support (MTSS) is a preventative framework designed to support student learning. It has two components: academic (response to intervention) and behavioral (positive behavior support).

Priming means to proactively address foundational skills that need strengthening for students to become successful in grade-level mathematics content.

(based on other data) they won't need the intervention, and other students that we know will absolutely need the intervention. We are not progress monitoring constantly and in so doing not taking away valuable instructional time. We are seeing far more success on end-of-unit assessments than we would have ever seen historically from students.

Another fifth-grade teacher added,

By the next week, we were looking at our curriculum in a whole new way. We used the summer to map out prerequisite skills and planned small, targeted sessions to introduce them to students before the upcoming lessons. It wasn't about "fixing kids"—It was about giving them the tools they needed to walk into class ready to succeed.

And it worked. The shifts were subtle at first—students answering questions with more confidence, trying problems they might have avoided before. But those small changes built momentum, and we saw something we hadn't seen in a while: students excited about math. For the first time, it felt like we were ahead of the curve, giving kids what they needed before they even knew they needed it. "This is the way forward," one of us said, and we all cheered.

Unlike other intervention resources, we aim to go beyond defining and explaining existing approaches, models, and tiers for instruction. Instead, we describe a transformative approach to mathematics intervention. We provide a support system for educators working to change the school's instructional structure to serve students better—both those with and without formal diagnoses of disabilities—who are identified as needing intervention or in-school tutoring.

We address the needs of educators and staff who embrace diverse students, including those with learning disabilities, extensive needs, who are twice exceptional (both exceptional abilities and learning disabilities documented by federal or state criteria; Asbell-Clarke, 2023; King, 2022), and without any disabilities who may be challenged with an upcoming topic area or need to build confidence in their ability to succeed in mathematics. Also, our approach is flexible and can be successfully implemented across various instructional settings and intervention formats. Whether in inclusion settings where learners receive Tier 1 instruction and are pulled out for intervention or identified for tutoring and/or in settings with a specific self-contained classroom, we are committed to providing the support needed with this book.

As educators, we know that tailoring and modifying our instruction helps us meet the unique needs of our diverse students regardless of the setting. Whether one is implementing a formal tiered approach, providing before- or after-school tutoring, or working in general education or small-group intervention classrooms, this book advocates for a proactive model of pedagogical support for the optimal mathematical experience. In our view, four principles set our approach apart from traditional intervention practices:

1. **Our approach is proactive**—As mentioned, instead of waiting until after instruction to identify and address student gaps, we employ a practice called *Priming*. This proactive strategy prepares students for *upcoming* instruction, ensuring they have the foundational knowledge to succeed. Our goal is to fix structures, not children.

2. **Our approach is strengths based**—Rather than defaulting to the prevailing model, which often focuses on repairing students' perceived weaknesses, we emphasize using a strengths-based model. This approach, supported by Cohen and Lotan (2014), allows teachers to recognize and build students' competencies, fostering their positive mathematical identities and confidence in their mathematical abilities.

3. **Our approach heavily leverages formative assessment**—By using intervention-focused formative assessment, teachers may directly impact their planning and instruction. This approach provides valuable, student-centered, individualized insights that help monitor student progress and provide timely feedback.

4. **Our approach focuses on the right tasks**—We emphasize "Doing Math" tasks (Kobett et al., 2021), which we'll explore at length in Chapter 1.

Strengths-based approaches focus on what students know and can do rather than on their deficits, "gaps," or "learning loss."

A positive mathematical identity refers to an individual's belief in their ability to understand and succeed in learning mathematics. It involves seeing themselves as capable, competent, and valued within the mathematics community.

With the right tasks and tools, you can enact a truly preventative and proactive approach to intervention. Our aim with this book is to help students enter their classrooms prepared for grade-level learning and full of confidence in their strengths. Additionally, as research suggests (Harbour et al., 2022), this book emphasizes the powerful combination of co-teaching models that pair special education with mathematics education and employ a high-quality series of mathematics tasks for intervention.

Who Is This Book For?

We designed this book with many audiences in mind. First and foremost, it is intended for people working with students in Grades 2 through 5, though the principles apply to any intervention setting. You may wonder why the book focuses on Grades 2 through 5. First, it is because in many, if not most, elementary schools, students aren't typically formally considered "at risk" until second grade, and they often aren't identified with a learning disability until third grade. We know that targeted and proactive interventions for children who need intensive support, delivered in this grade range, can set them up for long-term success.

Second, we know from dynamic development systems theory (Osher et al., 2020) that humans—in this case, children—are expected to be highly variable and have multiple development pathways in a learning landscape. We want to recognize and celebrate this variability based on this complexity and consideration of individuals and their development as a backdrop (Lambert, 2024). By exposing such variability, we may better explain students' mathematical thinking and provide learning opportunities that can ensure success among all students.

With this in mind, you are likely one of many adults who influence this success among children. You may be a general education teacher, a special education teacher, an elementary mathematics specialist, or a coach. You may be an interventionist, a paraprofessional, a mathematics coordinator, a special education coordinator, or a school principal. You may use this book as part of a professional learning experience, as a focus area within a professional learning community (PLC), or as a book club at your elementary school. You may even be a homeschooling parent, caregiver, or tutor working in a school-based setting before and after the regular school day. You may work in higher education as a mathematics teacher educator, or you may prepare special education teachers and hope to use this book in your methods classes or field-based experiences.

Regardless of your role, we have designed this book to be as valuable and implementable as possible. At the heart of it, what matters most is the children who need you. We aim to equip you with the right mindset and tools to help them shine their brightest in mathematics.

How Does This Book Work?

Part 1 includes the first four chapters of this book. These chapters will help us explore what proactive intervention looks like in terms of fixing structures, not children (Chapter 1); moving next into the value and characteristics of effective strength-based interventions (Chapter 2); Chapter 3 discusses formative assessment in the intervention setting; and Chapter 4 shows you what a proactive preventative mathematics intervention/tutoring model looks like in action.

Part 2 of the book comprises the task section. These 43 tasks and the more than 100 instructional activities within those tasks are designed for intervention and tutoring and draw from strong foundational "must-have" mathematics content on essential grade-level (Grades 2–5) understandings.

The tasks go beyond whole-number concepts and related operations, which can be a limitation of many intervention tasks found elsewhere. Why? Although working with numbers and operations is foundational, particularly at the elementary school level, we recognize and value the importance of learning experiences and related connections to numbers and operations. Such connections involve algebraic thinking, geometry, measurement, and data. Our task-based activities will align with the current curriculum standards while ramping students up or *Priming* them. As previously mentioned, to Prime means to address foundational skills that need strengthening for students to become successful in grade-level mathematics content.

The Standards for Mathematical Practice (SMPs) were developed based on the National Council of Teachers of Mathematics (NCTM, 2000) processes of problem-solving, reasoning and proof, communication, representation, and connections and the strands of mathematical proficiency specified in *Adding It Up* (Findell et al., 2001). The SMPs guide mathematics teachers in suggesting instructional considerations related to important mathematics content and elements of student engagement as their mathematical understandings develop (they can also serve as grade-level mathematics objectives). This intervention model prepares students to truly engage in the SMPs as they learn mathematics. Beyond teaching mathematics content through tasks, the intervention regularly exposes them to the practices and processes needed to successfully engage in and do the mathematics they are learning. In *Rethinking Disability and Mathematics*, Lambert (2024, p. 59) described the following features of accessible math tasks:

1. They have a low floor (within reach), which ensures that every student can engage in the problem, even with limited knowledge of the topic.

2. They have a high ceiling (adaptable to higher sophistication), meaning the problem has pedagogical possibilities, allowing extensions into more complex mathematical topics.

3. They are multimodal, meaning they have many options for arriving at solutions.

Each task presented in this book has these features. You can use such features to work with every child at their current learning level. Then, you can find students' strengths and work toward advancing their mathematical understanding to more complex ideas.

We recognize the significant demands on teachers' time and the varied lengths of intervention opportunities or tutoring sessions; thus, the intervention options in this book are ready to use and adaptable. Given the variety of activities *within* each task, you can use the tasks to engage students across multiple grade levels. You can also repeat the tasks with students so that they can practice their growing knowledge of concepts and skills and build confidence.

Across this book, we also provide plenty of the following:

- connections to recent research on the critical components of intervention;
- emphasis on students' strengths and how teachers can capitalize on them;
- effective classroom-based formative assessment techniques and tools adapted for use in intervention settings;
- collaborative planning structures;
- student work samples;
- examples that describe instructional practice and instances of teacher–student and teacher–teacher dialogue;
- end-of-chapter reflection opportunities for use in professional learning; and
- ready-to-use intervention tasks to build concepts and skills for upcoming grade-level lessons, including printable recording pages and other tools to implement the intervention activities.

Finally, we provide viewpoints from teachers who have implemented the Priming Approach, moving from "pulling skills from here and there" to a cohesive strategy in which "previewing what's coming" is the norm. These are teachers who shifted their focus from constantly assessing students to instructing students during intervention and classroom time. Throughout the book, you'll find feedback from educators using this approach, including their success stories and helpful comments. Through this book, we invite you to join us in learning how to bring these ideas to life in your classroom.

More on Priming

As mentioned, Priming is the time teachers spend with students within math interventions, teaching what they need to know to succeed in the *upcoming* instruction. However, this is *not* the same as preteaching the topic. Priming is a way to develop skills and conceptual knowledge to be used across one or more future lessons; Priming is *not* focused on simply covering the content that will occur in the next lesson.

To better explain Priming, imagine entering a room and knowing only that you will learn something new in math class. You don't know exactly what you're going to learn, how you're expected to learn, or even how others will determine that you've learned it. For some, this lack of knowing may be just fine. However, others may want time to "warm up" and acclimate to the topic over time. If you fall into the latter category, you may wish to relate foundational experiences to make the most of this next mathematics learning opportunity. Priming removes such ambiguity and uncertainty; students are supported with the key ingredients for preparing them to understand the particular mathematics focus of a lesson or series of lessons.

Priming is like a carefully considered and organized set of warm-ups, conditioning, and practice activities one may complete before that long run or big game. These activities are targeted to the specific muscles one intends to engage when it is time to *do* the long run or *play* the big game. Similarly, parents prime their children for kindergarten by telling them what they will hear, what they will see, and how to behave while having them listen to stories and count items around the house. In both scenarios, Priming reduces anxiety and creates space for optimal performance—whether learning in a classroom or excelling in a sport. All students do best when they have the necessary preparation for learning.

Let's look at how the brain works to demonstrate the benefits of the Priming model. Cognitive and neuroscience research suggests students learn best through active engagement (Fischer, 2009). To be best prepared, there's no substitute for students actively thinking about and doing mathematics, supported by teachers who can keep them engaged in the content. For quite a while, research in cognitive psychology and neuroscience recognized a change in a person's ability to produce information by providing previous encounters with the phenomena—this is known as Priming (Tulving & Schacter, 1990). Originally, this was sometimes unconscious Priming—but here, we are consciously telling the learners we are preparing them with what they need for success.

Priming helps students identify useful and relevant information from what they already know and mentally organize those ideas to best employ them for upcoming mathematics topics. Taking students down familiar pathways makes learning less complicated; new knowledge connects to prior learning, making the content more accessible in their memory (Wyer, 2007). Unsurprisingly, the background knowledge students bring to new content predicts how easily they will learn it (Fennema et al., 1993; La Paro & Pianta, 2000). As we define it in this book, time spent on stimulating and expanding students' precisely needed prior knowledge is the essence of Priming. Priming requires collaborating educators to locate the logical threshold of knowledge necessary to facilitate understanding mathematics content while leveraging individual students' strengths and needs. More discussion on that process can be found in Chapters 2 and 4.

> *Priming is like a carefully considered and organized set of warm-ups, conditioning, and practice activities one may complete before that long run or big game.*

Priming aims to positively support cognitive processes such as reasoning and decision-making—important components for acquiring mathematics knowledge. Priming can also help to build lasting understandings, such as what representations to use or what schema to employ when solving word problems. Research on cognition suggests that objects and drawings are more easily remembered and expressed than abstract concepts or words (Paivio & Csapo, 1973 (picture superiority); Paivio, 2013). For instance, teachers often encourage students to "make a picture" to help them strategize an approach to a word problem. Translating to a new format facilitates students' remembering the information. However, without precise guidance initially, these images may not fully support the mathematical relationships needed to solve the problem. They can even be time-consuming and counterproductive. You have probably had a moment when you asked a student to represent or show their thinking using a visual model, and you found yourself staring at an illustration of a story where the mathematics was lost in the art. For example, let's consider several visual images created by students with learning differences to solve problems; see Figure I.1. One set refers to a problem about shooting baskets in basketball, trying to reach 500 baskets by the end of the week. Another is of a 900-mile family trip where they are driving a distance each day.

Figure I.1 Comparing Useful (a) and Not as Helpful (b) Semiconcrete Representations

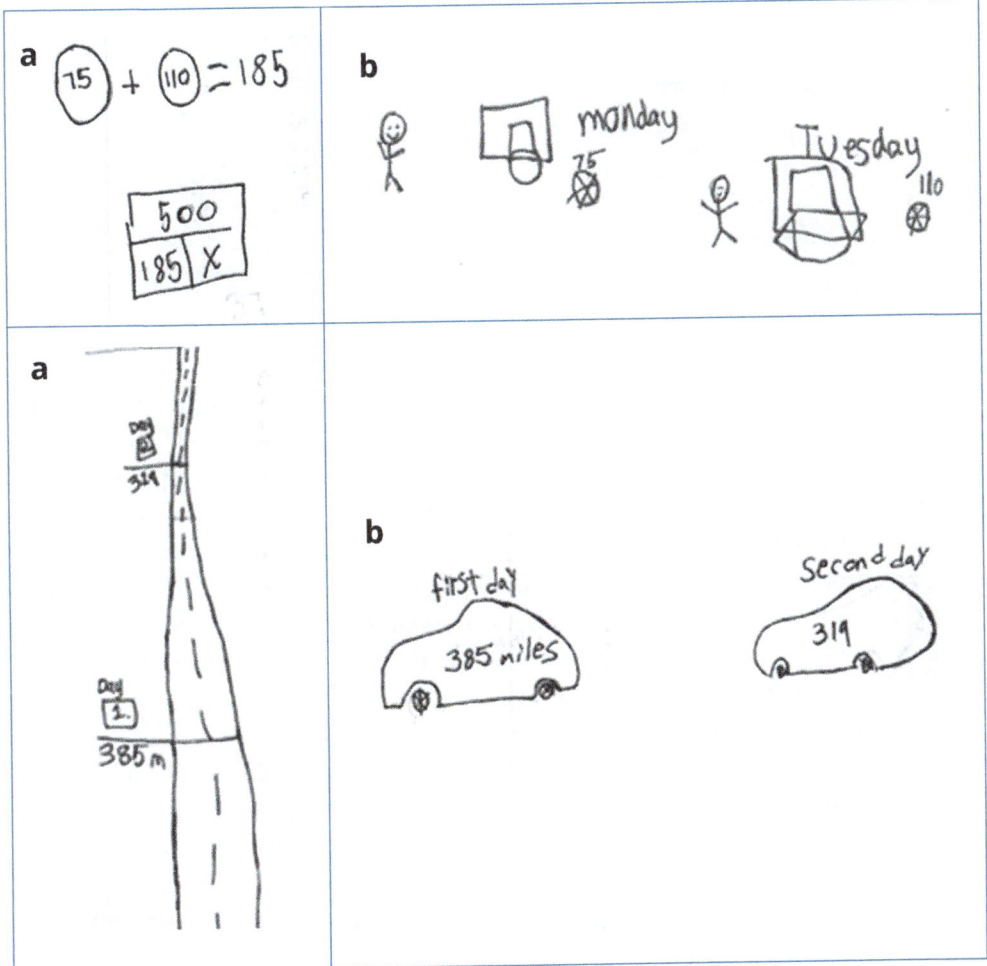

As you can see, image (a) in both cases meaningfully represents the mathematical context the student is trying to understand—in these cases, a representation of the completed parts is presented, providing a potential comparison to the total (whole). The other images shown in part (b) are more like pictures or illustrations of the story that do not show the relationships between the mathematical ideas needed to solve the problem. They are likely being used as thinking tools to organize the information given but may not always lead to a solution strategy.

In intervention settings, students need direct guidance on what drawings and sketches are the most "telling" concerning the context and making the mathematical information meaningful (Scandurra, 2024). Scandurra (2024) found that young students benefit from instruction about math sketches (Figure I.2), which enhanced their ability to visualize and gain number sense. Developing and connecting such mental representations through concrete, semiconcrete, and abstract means are critical to student success (Fuchs et al., 2021). This deliberate practice in making useful representations serves the student well in the short term for the planned new mathematical content and in the long term for their future mathematics studies (see Task 9 in Part 2 of this book for a Priming Activity related to this idea).

Figure I.2 Students' Images of Five

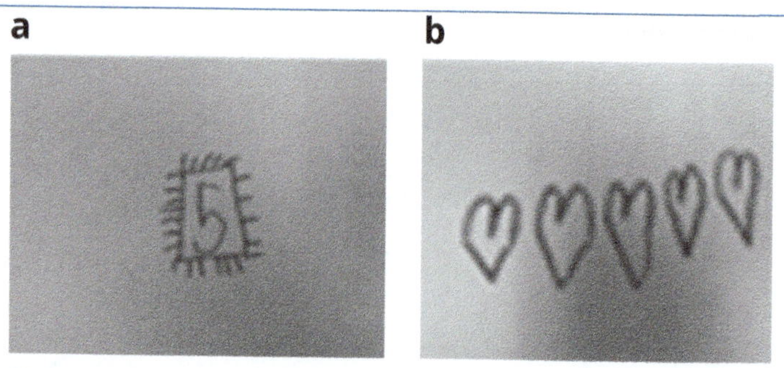

Source: Adapted from unpublished paper by J. Scandurra, 2024. Used with permission.

Figure I.2a displays an image made by students responding to the prompt, "When you see the number five in your mind, what do you see?" (Scandurra, 2024, p. 20). After working with multiple mathematical representations, including 10 frames and other discussions, students drew representations of five, like the hearts on the right in Figure I.2b. This second illustration was more helpful for taking the next steps in understanding numbers and developing number sense. These examples suggest the importance of teachers consistently engaging students in reading, writing, and representing their thinking when doing math.

In summary, Priming is many things: locating the sweet spots for instructional support, digging into such necessary details as how to get students to use multiple representations effectively, and explaining how to prepare students for immediate success with mathematics lessons focusing on new topics. Most importantly, interventions become a preventative point, where interventionists can Prime students for what is coming with a carefully choreographed set of preparatory intervention sessions and related activities. We will explore in more depth exactly how you, as an interventionist, can achieve this goal in the coming chapters.

How Do We Use Certain Language in This Book?

The language teachers use matters. It shapes how they think about students, interact with them, and perceive themselves. In this book, as teachers ourselves, we have made intentional choices about the language and terms we use to describe students, their abilities, and teaching and learning processes. We made these decisions based on a commitment to respect learners' abilities, build inclusivity, and empower all learners and teachers of mathematics.

Why Do We Choose Certain Terms?

When referring to students with disabilities, we use person-first language (e.g., "students with disabilities" rather than "disabled students"). This word choice emphasizes the individual first, not their diagnosis or learning challenge. It aligns with what is currently suggested by the National Council of Teachers of Mathematics and the Council for Exceptional Children Joint Position Statement on Teaching Mathematics to Students With Disabilities (NCTM/Council for Exceptional Children [CEC], 2024). We believe this approach highlights each student's unique identity and reminds us that no single label defines an individual's abilities or potential.

At the same time, we recognize that language evolves. Thus, some individuals or organizations may prefer identity-first language (e.g., "autistic student"). Although this book primarily uses person-first language to maintain consistency, we encourage educators to remain sensitive to individual preferences when working with students and families.

How Does Strengths-Based Language Change Mindsets?

Throughout the book, you'll notice a focus on strengths-based language. Instead of framing students by what they "cannot" do or where they "struggle," we aim to emphasize what students *can* do and how their unique skills and perspectives can serve as valuable assets in their mathematical growth. For instance, rather than describing a student as "behind in math," we may say they are "building foundational skills" or "developing confidence in mathematical reasoning." NCTM (2023) strongly advises against using labels like "high," "medium," and "low" to categorize students' mathematical abilities and instead advocates building on students' strengths. Our book supports this recommendation and focuses on strength-based language.

How Do We Define Terms Throughout the Text?

Instead of providing a separate glossary, we define technical language or specialized terms directly within the text as they appear. This process ensures a "just-in-time" approach, where definitions are provided in context. We hope to make it easier to connect the meaning of a term with its application. By embedding definitions, we hope to maintain the flow of the text while ensuring clarity and accessibility for readers. By making these intentional choices, we hope to foster an environment of respect, equity, and high expectations for all learners. Language is a powerful tool, and we encourage you, as an educator, to use it to inspire, uplift, and connect with your students.

Now, we are ready to move onward. The Preface was intended to prepare you to consider the importance and potential of proactive mathematics intervention opportunities. By embracing proactive, strengths-based approaches to mathematical intervention, you can make a profound difference in the lives of your students. Together, we can build learning environments in which every child feels capable, valued, and prepared to tackle new mathematical challenges. We appreciate your dedication to this critical work—it truly matters. Chapter 1 will delve into all elements of our approach. Let's get started!

Acknowledgments

We would like to acknowledge our families and friends who support us during work on large projects such as the preparation for and writing of this book. We are privileged in that they gave us the space to occasionally postpone other happenings and allowed us to be occupied with all the thinking and activities that go into an effort of this nature. We are most appreciative of the many educators who provided feedback and suggestions to us and for the early adopters of the Priming Approach. This collection of amazing teachers, interventionists, coaches, curriculum specialists, administrators, researchers, and teacher educators includes Isabella Adkins, Lauren Alben, Karla Bandemer, Jay Blackstone, Lisa Curtin, Meaghan Ferrera, Phil Howell, Mistie Parsons, and Sara Wright, with special thanks to Dawn Pilotti for her important contributions. We also celebrate the amazing children with learning differences who contributed their thinking to this book through samples of their problem-solving. We are indebted to their families, who generously gave permission and shared their children's work with us.

We are also very grateful for the continuing cheerleading and support from Erin Null, STEM associate director and publisher at Corwin. She is a champion for equity for students with disabilities and believes that all students deserve the opportunity to learn mathematics by building on strengths instead of highlighting their gaps or "lack ofs." She is patient and gives "just-in-time" feedback to move manuscripts to new heights. We also want to thank Nyle De Leon, whose attention to detail made the numerous figures and resources find just the right places in the book and on the companion website, where readers can easily locate them.

Publisher's Acknowledgments

Corwin gratefully acknowledges the contributions of the following reviewers:

Cheryl Berkuta
Elementary Math Coach & Specialist, Old Bridge Township Public Schools
Monroe Township, NJ

P. Renee Hill-Cunningham
Associate Professor of Mathematics Education, University of Mississippi
Houston, MS

Emily Dwivedi
Baltimore County Public Schools

About the Authors

Karen S. Karp is a mathematics educator who focuses on the intersection of mathematics education and special education. She is a former professor at Johns Hopkins University and at the University of Louisville, where she is professor emerita. Early in her career, she received a Development Award from the Joseph P. Kennedy, Jr. Foundation to support more seamless integration between general education and special education. She is the author or co-author of numerous publications including *Assisting Students Struggling With Mathematics: Intervention in the Elementary Grades* and *Elementary and Middle School Mathematics*. Karen was on the writing team of the NCTM/CEC joint position statement on *Teaching Mathematics to Students with Learning Disabilities*. In 2024, she chaired the Topic Study Group on Teaching Mathematics to Students with Special Needs at the International Congress on Mathematical Education in Australia. She holds teaching/administrative certifications in elementary education, secondary mathematics, K–12 special education, and K–12 educational administration.

Francis (Skip) Fennell is professor of education and graduate and professional studies, emeritus, at McDaniel College in Maryland. He is a former classroom teacher, principal, and supervisor of instruction and past president of the Association of Mathematics Teacher Educators (AMTE), the Research Council on Mathematics Learning (RCML), and the National Council of Teachers of Mathematics (NCTM). He is a recipient of the Mathematics Educator of the Year Award from the Maryland Council of Teachers of Mathematics (MCTM), the Glenn Gilbert National Leadership Award from NCSM: Leadership in Mathematics Education, the Excellence in Leadership and Service in Mathematics Teacher Education Award from AMTE, the James Heddens Distinguished Service Award from RCML, and Lifetime Achievement Awards from both MCTM and NCTM. Skip's many publications, including *Achieving Fluency: Special Education and Mathematics* (NCTM, 2011) and *The Formative 5: In Action* (Corwin, 2024), have been influenced by his classroom experiences and decades-long focus on assessment, number sense, fractions, elementary mathematics specialists, and teacher education.

Beth McCord Kobett serves as professor and dean in the School of Education at Stevenson University, where she works closely with early childhood, elementary, and middle school preservice teachers. She brings experience as a classroom teacher, mathematics specialist, and university supervisor. Beth served on the NCTM board and served as president of the Association of Maryland Mathematics Teacher Educators. Beth has authored 10 mathematics education books and supports professional learning efforts nationwide. She has been honored with awards such as the MCTM Mathematics Educator of the Year and Stevenson's Rose Dawson Award for Excellence in Teaching. Deeply committed to her students, she strives to create a supportive, strengths-based learning environment that fosters curiosity, collaboration, and meaningful growth.

Delise R. Andrews is the 3–5 mathematics coordinator for Lincoln Public Schools in Lincoln, Nebraska. During her career, she has worked in both rural and urban districts and has taught mathematics to students at every age from Kindergarten through the eighth grade, undergraduate mathematics methods and mathematics content courses for preservice teachers, and graduate-level courses for teachers of mathematics. Delise is a recipient of the Presidential Award for Excellence in Mathematics and Science Teaching and is a Robert Noyce Master Teaching Fellow. Delise is an active member of the NCTM, serving as a past member and chair of the Professional Development Services Committee, member of regional conference committees, chair of the St. Louis annual conference committee, Professional Services facilitator, and associate editor for the *Mathematics Teacher: Learning and Teaching K–12* journal. She is a co-author of the Grades K–1 and Grades 4–5 books in the *Classroom-Ready Rich Math Tasks: Engaging Students in Doing Math* series.

Jennifer Suh is a mathematics educator at George Mason University, leading efforts to enhance K–8 math instruction through strength-based formative assessments and bridging activities. In partnership with the Virginia Department of Education, her project, *Bridging for Math Strength*, focuses on using rich mathematics tasks across grade levels, unpacking the learning progression to enhance teaching and learning. Jennifer uses this project for math intervention and multitiered instruction, working closely with special education and general education math teachers to meet diverse learners' needs. She directs the Center for Outreach in Mathematics Professional Learning and Educational Technology (COMPLETE), conducting Lesson Study in schools emphasizing problem-based tasks to promote equitable access to 21st-century skills, including creativity, critical thinking, communication, and collaboration, particularly for diverse student populations.

Latrenda D. Knighten is president (2024–2026) of the National Council of Teachers of Mathematics (NCTM) and a retired mathematics supervisor from Baton Rouge, Louisiana. She has been an educator for more than 30 years, during which she has been a classroom teacher, science specialist, mathematics coach, instructional coach, and mathematics coordinator. An active member of many professional organizations, Latrenda is a past member of the board of directors for the National Council of Teachers of Mathematics, NCSM: Leadership in Mathematics Education, and the Benjamin Banneker Association, Inc. She is also the co-author of two books: *Classroom-Ready Rich Math Tasks, Grades K–1* and *Five to Thrive: Answers to Your Biggest Questions About Teaching Elementary Math, K–5*.

*This book is dedicated to the late Eunice Kennedy Shriver,
whose goal was to meld together the wisdom and energies of general education
and special education to benefit students with disabilities. We honor
her vision with the work provided within these pages.*

PART 1
INTRODUCTION TO PRIMING

Introduction

CHAPTER 1

Proactive Intervention: Fixing Structures, Not Children

As an experienced classroom teacher and school-based math coach, I get it. Many schools and school districts recognize a need to improve the mathematics performance of their students. Providing mathematics intervention opportunities—however defined—has become a response. However, we must get this right! Attention to the goals of intervention programs, which includes engaging students in doing mathematics they are learning, and, importantly, recognizing that such programs do not replace daily mathematics class opportunities is an organizational and instructional priority. Mathematics intervention starts by identifying individual student strengths, and then addresses, daily, the planning to both support and supplement student learning.

—An elementary math coach

> The **National Assessment of Educational Progress** is a congressionally mandated, large-scale assessment administered by the National Center for Education Statistics for Grades 4 and 8. The Nation's Report Card provides results on student performance based on gender, race/ethnicity, public or nonpublic school, teacher experience, and hundreds of other factors (see https://nces.ed.gov/nationsreportcard./).

So What's the Problem?

Let's consider the following: Recent results from the National Assessment of Educational Progress (NAEP, 2019, 2022, 2024) are periodic reminders that clearly indicate that students' scores remain below those prior to the pandemic even with recently improved scores at the fourth-grade level (Table 1.1). Among the many suggestions for finding time to help accelerate and deepen student learning of mathematics are well-designed and carefully implemented and monitored mathematics intervention and tutoring programs.

Table 1.1 Percentages of All Students on the NAEP in the Below Basic Level in Mathematics

	2019	2022	2024
Grade 4	19	25	24
Grade 8	31	38	39

Source: 2019, 2022, and 2024 NAEP Assessment Highlights.

Starting in 2020 and subsequent years, the long-term effects of the COVID-19 pandemic became apparent in communities, schools, and students. Children lacked access to consistent education, indicating the need for reinforcement in their learning. As teachers, we saw the need to change instruction in response by creating new and evolving long-term plans to address *all* our students' learning needs.

During this period, many people identified students with disabilities as the most vulnerable population that would be disproportionately affected (Stelitano et al., 2021b). However, the magnitude of the resulting disparities in student performance at the Below Basic level continues to be of concern (Table 1.1).

Let's also consider the long-term trend (LTT) NAEP scores. Since the 1970s, the NAEP has monitored the academic performance of 9-, 13-, and 17-year-old students over time using what's known as the NAEP LTT assessments. The average scores on the LTT assessments for mathematics for 9-year-olds in 2022 were higher than the earliest assessments in the 1970s but lower compared to the previous assessments in 2020 (see https://qrs.ly/i1gjc7u for more details). The average LTT-NAEP mathematics score for 13-year-olds in 2023 was 5 points higher than in 1973 but 9 points lower than in 2020.

Interestingly, the LTT-NAEP also reported that the percentage of 13-year-old students missing 5 or more days of school monthly has doubled since 2020, which, as is often reported, has become a topic of national concern. It was also reported that mathematics scores declined compared to 2020 for most student groups and students at all reported levels of parental education (NAEP, 2023). The NAEP (2024) also demonstrated that eighth-grade students with disabilities are at a level of achievement that is close to what average fourth graders score, with only 5% reaching the Proficient (4%) or Advanced (1%) levels (NAEP, 2024).

Based on these data, what's our takeaway here? Until recently (NAEP 2024), mathematics scores have decreased on both the main and LTT national assessments. In addition, the survey data provided by the LTT NAEP indicates that student absenteeism has become an issue of concern. The findings suggest the need to control mathematics learning opportunities for all, including interventions and tutoring. To validate our statement about the importance of mathematics learning opportunities, we suggest considering what Tom Kane, an economics professor at Harvard University, has to say. Kane suggested that the current national need for learning support is enormous (Sparks, 2022a). As an economist, he predicted that if this situation is ignored and the drop in scores continues (or remains permanent), an eventual 1.6% decline in income might occur for students when they grow up. Considering all the children in K–12 schools, that equals trillions of dollars in lost income! The long-term impact can be significant! Further, the estimated time it will take for early elementary-aged students to return to grade-level growth progression is 3 years; upper-grade students may require 5 years to regain the needed levels of learning (Sparks, 2022b).

We share these data not to fixate on test scores but to raise awareness of the challenges ahead when increasing students' standards-based understandings, which persist after the lesson as "mental residue" (Dougherty, 2008). So here's our starting line: Recent Main and LTT-NAEP results clearly indicate the critical need to prioritize mathematics teaching that connects mathematics to the lives of their students (Latrenda Knighten as cited in NCTM, 2025). Let's do this!

> *The NAEP (2024) scores also demonstrated that eighth-grade students with disabilities are at a level of achievement that is close to what average fourth graders score, with only 5% reaching the Proficient (4%) or Advanced (1%) levels.*

What's Being Done?

As a result of these realities, at the time of this book's publication, seven U.S. states required by law that students who struggle in learning mathematics get support (Schwartz, 2023). Following long-standing attention to reading practices, this legislation requires students to be identified and monitored through various assessments. They then receive targeted support to direct their learning to meet proficiency levels with grade-level mathematics standards.

In some states, there is also a mandate to support teachers. States must provide in-service teachers with learning opportunities to develop the mathematics knowledge and skills inherent in instruction based on best practices and preservice teachers with teacher-preparation programs that include this information. Some states require having at least one mathematics coach/specialist for every elementary school. Complying with this requirement would cause a significant shift from the current reality: Although expert reading support is available in about half of the elementary schools reporting, mathematics coaching from well-prepared specialists is hard to find, with only 23% of schools reporting that mathematics specialists are available (Korbey, 2024; NCES, 2022).

Often, interventions in mathematics rely on instructional practices that are not demonstrated to be effective. For example, perhaps you have had the experience or expectation that students in math intervention should receive a review of the same content taught in Tier 1 or general education class settings. Nothing is different except that you share this content in a small group rather than the whole class. Will providing the same instruction again help? Another example is only relying on abstract representations without interweaving the development of concepts utilizing concrete and semiconcrete models. If you provide time for students to grapple with the foundational concepts as a starting point, they will be better equipped to grapple with the new grade-level learning and find greater success. For example, students in intervention settings are sometimes taught repeatedly the current mathematics topic the class is learning. Although applicable in some situations, this model rarely addresses the need to find and focus on foundational concepts and related understandings underpinning students' confusion or driving challenges with the grade-level material.

One example is when students find it challenging to compare and order fractions because they do not understand the magnitude (size) of the fractions (e.g., they do not yet recognize that $\frac{3}{4}$ is a number with a value less than 1). So when the topic of adding fractions is presented, the same foundational understandings must come into play. Just as we would not expect learners to add whole numbers such as 3 + 5 if they didn't already grasp the quantities of 3 and 5, they cannot add fractions successfully without knowing the underlying concepts of the magnitude of the fractions they are combining. Interventionists should recognize that learning-trajectory-based instruction can build solid understandings focusing on the progression of ideas.

When students build familiarity with foundational mathematical ideas by talking about, describing, and applying concepts and procedures, this familiarity

> A **learning trajectory** refers to the path or progression of a learner's knowledge and understanding from foundational ideas to more advanced concepts. It describes how an individual moves from their initial level of understanding to a deeper, more sophisticated level of knowledge as they engage with mathematics tasks, experiences, and feedback.

is reinforced by different processes in the brain called recollection memory (Yonelinas et al., 2010). Those processes are the mental muscles learners should be encouraged to develop and regularly flex. These progressions may involve revisiting and strengthening student understanding, connecting concepts, and engaging with more complex ideas or problems.

Granted, a high dosage of interventions each week or double-dosing of tutoring both in school and outside of school in private settings may be options for students (Carr, 2022, as cited in Sparks, 2022b, n.p.), but it's what goes on in those sessions that actually makes the difference. Additionally, the expectations involved with access, cost, and availability of outside tutoring for many students are significant equity issues. States have considered other options supported by their public schools. As a significant percentage of students have been identified as needing extra math instruction (referred to throughout this book as *interventions or tutoring*), such scaled-up implementation can cause complications with resources and capacity, such as space, time, and availability of additional qualified mathematics educators in schools. It's not as much about the role of *who* provides the intervention—classroom teacher, special educator, interventionist, tutor, math specialist, etc.—but *how* the students are taught.

> *Although a learning trajectory is often thought of as a linear progression, it's important to recognize that this path is not always linear.*

What Needs to Change?

In this book, we propose a refreshed vision of proactive intervention opportunities embedded throughout the school day. We also recognize that some schools are using intensive tutoring before and after school hours with learning opportunities designed to be genuinely preventative. While—as a starting point—we focus on students who are identified as having disabilities according to the Individuals with Disabilities Education Act (IDEA, 1997), we recognize that these ideas, assessments, models for instruction, and "*doing* math" tasks (Kobett et al., 2021) can and should be expanded. This expansion should include a broader audience of children with and without disabilities who may not have equal access to the grade-level curriculum (required by law). For example, many students face instructional barriers created by missing mathematics content from previous grades (NCTM/CEC Joint Position Statement, 2024). Mathematics intervention programs and learning opportunities should enhance student learning in MTSS Tier 2 and Tier 3 instruction but also, significantly, in Tier 1 instruction. So we encourage a cohesive and coordinated instructional shift to move all students forward.

However, you may wonder, why not just buy a packaged intervention program or access one delivered electronically? Available commercial mathematics programs (print or electronic) for intervention sessions may only briefly address or even omit key topics, so people have posted on various online platforms that they need recommendations and suggestions for something that works. We find that student success in learning mathematics isn't about something you buy; it's about instructional strategies you try. It isn't just about teaching a commercially available or school district–designed curriculum; it's about understanding children where they are and bringing them to greater success in grade-level content through a structured intervention planning model. Priming is a major shift in instructional practice.

> *Mathematics intervention programs and learning opportunities should enhance student learning in MTSS Tier 2 and Tier 3 instruction but also, significantly, in Tier 1 instruction. So we encourage a cohesive and coordinated instructional shift to move all students forward.*

We suggest starting with a set of common grade-level curriculum materials as a tool to build consistency and predictability across the grades. This way, teachers in subsequent grades can expect and rely on the fact that the content has been taught in a particular way (Karp et al., 2021). You can, of course, also use supplements and modifications. However, be careful in how you do this. Research from 4,414 teachers found that 72% of special education teachers and 41% of general educators reported high levels of modification to the curriculum, including at least 50% or more for students with disabilities (Stelitano et al., 2021a, p. 4). Additionally, more than half of special educators and a third of general educators suggested they created their instructional materials from scratch. Sadly, this use of unaligned resources generally continues over the grades as the content knowledge of any given child gets more and more splintered and unpredictable. The researchers also stated that it is unknown what background and capacity these teachers have to develop well-aligned curricular content, and they recommended more professional learning in this area. Without this expertise, some teachers will mistakenly provide alternative mathematics instruction that reduces students' work—in mathematical reasoning and sense-making, for example—particularly for students with disabilities. Teachers may do so because they are unsure whether students can become engaged in and carry out this level of thinking.

However, as teachers ourselves, we know the students can! We do not advocate for children copying what the teacher presents using a step-by-step approach. Candidly, this book is not intended for someone who wants a script or a "foolproof" guide filled with procedures or quick-fix tricks as *the* recipe for student success. As Peter Liljedahl (2020) suggested, students tend to stop thinking once they start mimicking the teacher: "Mimicking is an addiction that is easily acquired at lower grades and difficult to give up" (p. 30). He added that it's even difficult for parents to think there are other better ways to teach than mimicking. This expectation to mimic causes students to wait for a teacher to explain the lesson over and over.

Proactive Priming interventions in mathematics are encouraged to shift the thinking process by eliciting students' thinking and guiding math discussions in small groups. This goal can be achieved through probing questions and then turning back to the students to hear their ideas. That's engagement! So our focus here is on the importance of a genuine investment in developing children into active thinkers about mathematical ideas by investing in the preparation for your instructional time together. The criterion for doing this work is believing that—whenever possible—you should refuse to accept limitations on what children can learn. That means examining your beliefs about who can access and learn important mathematics. Children must also be exposed to and practice reasoning; hence, you are encouraged to provide interventions to help them practice just that.

We think of instructional practice as "doing math." This mantra is key to the success of all students, including students with disabilities. The relative scarcity of high-quality interventions in mathematics that fully develop the Standards for Mathematical Practices (CCSS-M; NGA, 2010) and focus on multiple representations emerges from two related concerns. First, a hyperfocus on procedurally focused activities (frequently timed) is often generated by the nature of overly specific computationally based IEP goals (e.g., those measured by increases in the

> *We find that student success in learning mathematics isn't about something you buy; it's about instructional strategies you try.*

> *Candidly, this book is not intended for someone who wants a script or a "foolproof" guide filled with procedures or quick-fix tricks as the recipe for student success.*

> **The Individualized Education Program (IEP)** is the cornerstone of a quality education for each child with a disability. Public school students who receive special education and related services *must* have an IEP. The IEP guides the delivery of special education supports and services for the student with a disability. The IEP creates an opportunity for teachers, parents, school administrators, related services personnel, and students (when appropriate) to work together to improve educational results for children with disabilities (U.S. Department of Education, Office of Special Education and Rehabilitative Services, 2000. https://www.ed.gov/sites/ed/files/parents/needs/speced/iepguide/iepguide.pdf).

> *We are not looking to fix children (they do not need fixing) but to fix the structural obstacles that hinder inclusion and prohibit student progress.*

correct number of digits in computation-focused assessment responses; Lambert et al., 2023). Second, it is difficult to find high-quality intervention tasks that combine conceptual and procedural understanding. Again, this book aims to help.

As mentioned in the Preface, we firmly believe in the Priming Approach guiding us in this solution. This direction requires us to set the following goals in three different areas:

- working cohesively and systematically from strengths while building rapport and relationships,
- regularly using formative assessment to monitor student progress, and
- using rich tasks focusing on students *Doing Math* to prime important intervention-focused mathematics learning proactively.

We encourage noting the important connections between strengths-based instruction, classroom-based formative assessment, and progress monitoring tools (e.g., unit tests and district benchmarks required in many school settings). Embracing this approach and understanding the connections will help you directly monitor and plan to assess individual learning needs using the most effective teaching. Additionally, it is essential to highlight activities and lessons that provide engaging interventions to support student success. We are not looking to fix children (they do not need fixing) but to fix the structural obstacles that hinder inclusion and prohibit student progress.

The message throughout this book is deliberately incorporating well planned, strengths-based instruction to support students who struggle to learn mathematics as they strive to make substantial learning gains. We know that strengths develop best "in response to other human beings" (Clifton & Nelson, 1992, p. 124). So effective teachers are indeed central to making a difference and delivering this message! Worksheet packets, digital options, or even AI-enhanced digital feedback cannot replace an effective teacher engaging students in *doing* math.

What Is the Research Base?

Is this book based on research? YES. We base our research framework for this work on a variety of sources, including the research-based recommendations outlined within the Institute of Education Sciences (IES) Practice Guide *Assisting Students Struggling with Mathematics: Intervention in the Elementary Grades* (Fuchs et al., 2021), the NCTM's (2014) *Principles to Actions: Guidelines for Teaching Mathematics*, Universal Design for Learning (CAST, 2024), the NCTM/CEC *Joint Position Statement* (2024), CEC and the CEEDAR Center's High Leverage Practices in Special Education (2017), and the NCTM's (Huinker et al., 2020) *Catalyzing Change in Early Childhood and Elementary Mathematics*.

The co-authors and researchers of the IES Practice Guide initially identified 2,635 research studies between 2009 and 2020 for review. The final number that met the standards of strong research aligned with the target population of elementary students with disabilities was 47 studies. In this book, we suggest how to adapt and use elements of the Practice Guide's recommendations.

We aim to guide thinking regarding an intervention "plan" and how it should play out in schools. We uniquely weave together the following six Practice Guide recommendations, along with research from mathematics education throughout the book:

Recommendation 1: Systematic Instruction

Provide Systematic Instruction During Intervention to Develop Student Understanding of Mathematical Ideas. We use this thinking to guide our instruction and approach in all intervention lessons. Systematic instruction uses models with concrete, semiconcrete, and abstract (CSA) representations interwoven throughout. However, systematic instruction does not equate to teaching by telling. It is not "I do, we do, you do." Instead, systematic instruction focuses on learning progressions. It provides purposeful, structured learning opportunities. These opportunities can build on conceptual and procedural understanding, promote inquiry, and draw out students' prior knowledge as a foundation for rich mathematical thinking and discussion. So we have designed the tasks in this book to do just that.

Recommendation 2: Mathematical Language

Teach Clear and Concise Mathematical Language and Support Students' Use of the Language to Help Students Effectively Communicate Their Understanding of Mathematical Concepts. Language is key to helping students become "ready to learn" new content. Words they've already learned can be reviewed with them, and they should be prepared to hear those words again as new content is introduced. We suggest ensuring the content is as familiar as possible so students can hit the ground running. However, we don't advise preteaching mathematics vocabulary. Liljedahl (2020) discussed this idea by suggesting that the names of concepts should come when students have experience with those concepts. So it's best to name the term while the model, image, or idea is being explored (Dixon, 2018; Van de Walle et al., 2023), not before (too early) and not after (too late). This connection of vocabulary to actions and representations in context is crucial for students with language-based disabilities. Indeed, it's how we introduce new concepts in this book.

Recommendation 3: Representations

Use a Well-Chosen Set of Concrete, Semiconcrete, and Abstract Representations to Support Students' Learning of Mathematical Concepts and Procedures. The CSA representations are presented simultaneously. When students grasp their interrelationships, that is a sign of their level of understanding. In this book, we also include applications and situations as representations. These may include the use of children's literature and relevant contexts that we hope will align with students' interests. To avoid buggy procedures (Thompson, 1999)—because the steps of the procedures are mixed up, forgotten, or used for the wrong operation—we shift to multiple representations as a significant focus of the tasks (Van de Walle et al., 2023).

Recommendation 4: Number Lines

Use the Number Line to Facilitate the Learning of Mathematical Concepts and Procedures, Build Understanding of Grade-Level Material, and Prepare Students for Advanced Mathematics. You may be surprised that number lines are being called out separately as a recommendation, as they are a representation. Nevertheless, the research and importance of using number lines were so compelling that this important representational tool was elevated to a "must use" status. Number lines need everyone's full attention regardless of grade and should be used and carefully sequenced across the grades. As another influential IES Practice Guide entitled *Developing Effective Fractions Instruction for Kindergarten Through 8th Grade* (Siegler, 2010) noted, number lines should be used as a central representational tool from the early grades onward.

Recommendation 5: Word Problems

Provide Deliberate Instruction on Word Problems to Deepen Students' Mathematical Understanding and Support Their Capacity to Apply Mathematical Ideas. Word problems are a first step to having students apply the number-focused computational mathematics they are learning in various contexts and situations. Rather than just taking the numbers and "doing something with them," students learn that different problem scenarios drive them to set up equations and solutions. For example, unlike in the past, where a keyword strategy suggested the term *more* was equivalent to saying the solution automatically required addition, students are encouraged to imagine situations and determine what operation is needed (Hardy et al., 2025; Karp et al., 2019).

This moment is also ideal for weaving in the mathematical practices and processes that enhance their learning.

> The **keyword strategy** involves identifying words in a word problem such as *altogether*, *share*, *more*, or *left* and thinking that alone determines which computation to use. Research suggests this approach is not useful, particularly with multistep problems (Powell et al., 2022).

Recommendation 6: Timed Activities

Regularly Include Timed Activities as One Way to Build Students' Fluency in Mathematics. Although named "timed activities" in the practice guide, it is important to note that the aspect of the recommendation we will focus on is "fluency activities."

Timed activities may be a concluding component of the process, but an important trajectory comes first (Table 1.2).

> **Fluency activities** help students gain flexibility, accuracy, and efficiency with appropriate strategies for such skills as basic number facts or computational procedures.

Table 1.2 Three Phases of Fluency

Phase	
Phase 1	**Modeling and/or counting (e.g., counting by ones/skip counting) to find the answer** • Example: Solving 6 + 4 by drawing 6 dots and 4 dots and combining them by counting the dots
Phase 2	**Deriving answers using reasoning strategies based on known facts** • Example: Solving 8 × 7 by thinking 7 × 7 equals 49 and adding one more group of 7 equals 56
Phase 3	**Fluency (efficient production of answers)** • Example: Knowing that 8 + 5 = 13 or 5 × 5 = 25

Source: Adapted from *Elementary and Middle School Mathematics: Teaching Developmentally*, by Van de Walle, Karp, and Bay-Williams, 2023, Pearson. Adapted with permission.

Based on the movement from Phases 1 through 3, fluency approaches based on automaticity, or the efficient production of answers, come only after students have moved through modeling and counting (skip counting in the case of multiplication and division) and then derive answers from known facts. Only then do they move toward the efficient generation of answers. As teachers, we focus on first building students' confidence in their capacity to figure out a solution, as opposed to a focus on recalling memorized answers. Therefore, we do *not* endorse any activities that focus on speed. Instead, we encourage strength-based learning activities that build fluency through efficiency, flexibility, and accuracy (Bay-Williams et al., 2022). Such activities will emphasize effort and persistence, leading students to achieve their "personal best" in achieving automaticity and building their mathematics strategic competence.

To increase our understanding of this subject, let's now attend to the nine major (revised) Guidelines of Universal Design for Learning (UDL 3.0; CAST, 2024).

First, Table 1.3 presents the guidelines that are written with a new focus on equity that includes critical teacher actions:

Universal Design for Learning (UDL) is a framework that supports students' access to learning. Teachers provide flexible materials, methods, and learning environments. UDL promotes multiple means of representation, expression, and engagement.

Table 1.3 Revised UDL Guidelines 3.0

	Teacher actions to support students	Examples
1	Welcome interests and identities	Provide choices and make relevant
2	Sustain effort and persistence	Optimize challenge and foster belonging
3	Enhance emotional growth	Promote individual and collective reflection
4	Engage perception	Offer multiple ways to perceive information
5	Clarify language and symbols	Support decoding of mathematical notation and symbols
6	Build knowledge	Connect prior knowledge to new learning, explore patterns, develop generalizations
7	Generate interaction	Vary response methods, provide access to tools
8	Enhance expression and communication	Build fluencies and encourage discussion
9	Develop strategies	Plan, anticipate challenges, and organize information

Source: Adapted from *UDL Guidelines 3.0*, by CAST, 2024. Adapted with permission.

Second, in 2017, the CEC and CEEDAR Center (2017) identified important areas of effective teaching called high-leverage practices (HLP). We emphasize the examples of those HLPs in special education in this book, including collaboration, assessment, social/emotional/behavioral supports, and instructional practices. Each contributes to improving learning environments and outcomes.

Finally, we turn to the landmark work of the National Council of Teachers of Mathematics (2014) in their policy document, *Principles to Actions*. Their six principles for effective teaching include articulating goals, making connections, fostering engagement, differentiating challenges, structuring lessons, and promoting fluency and transfer. We embark on this work with these research-driven themes and powerful guidance.

When We Talk About Intervention, What Do We Mean?

Let's explore what interventions are and are not (Table 1.4).

Table 1.4 Interventions

Math interventions are:	Math interventions are not:
Proactive and preventative	A repeat of the grade-level lesson either before or after it takes place
Addressing the foundational knowledge that students may not have learned or have forgotten and providing "on-ramps" to the new lesson's mathematics content	Time provided to do homework, study, or make corrections to graded work/assessments
Focusing on students' strengths *and* targeted to their specific needs	Focused on test preparation
Providing students with opportunities to be "doing math" in small groups	Passive worksheets or computer activities
Engaging, challenging, and memorable	Computer-driven sessions focused on speed
Structured to allow communication and interaction	A replacement for the general education class
Co-orchestrated and often co-taught	The class students get pulled from for other things (e.g., instrumental music, breaks)
Developed to build students' confidence, positive attitudes, and math identity	One size fits all
Dependent on dedicated planning time during the school day	Overreliance on extrinsic motivators like "compliance charts"
Focused on reasoning and sense-making	"Math made easy"
An opportunity for students to shore up essential mathematics content and process knowledge and skills	A behavior management plan

Source: Adapted with permission from *Strengthening Math Intervention in the Middle Grades*, by Brodesky et al., 2022, Education Development Center.

Icon Source: thumbs up and thumbs down icon by istock.com/vlan Yarovyi

Because schools are mandated to provide intervention opportunities for students with disabilities, many issues must be considered. These issues may include important decisions related to student identification and intervention implementation. Additional areas to consider (e.g., program size, instructional materials) are discussed in Chapter 4. Consider the following policy and program issues:

- **Identification:** How will students be screened and identified for participation in your school's intervention offerings? Which will include responses to the following:
  - What will such screening involve? What mathematics, length of screening assessment, or aspect of the resulting data will be useful to plan future instruction?
  - How much time should be spent screening students for the intervention offerings?
  - Who will screen students, share results, and determine the next instructional steps?

- **Implementation:** How will the intervention opportunities be implemented, and when will they occur (e.g., daily, three times a week; before school, after school, online)?
  - If the intervention opportunity occurs online, how will students receiving intervention still receive the same amount of class/grade-level instructional time as their classmates?
  - What materials will be used? How will online students have access to those materials?
  - How will you determine that the materials address important mathematics based on the standards?
  - Will both print and online materials and related activities be used?

Although critical and impacting much of what you will engage in while reading this book, state and school district regulations primarily determine student identification and intervention implementation issues. Therefore, they are not the focus of this book.

It is also important to understand that interventions are not a "fix" without full coordination with Tier 1 instruction. Our goal is to provide engaging and memorable intervention sessions in which children run to the table, door, or interventionist excited to do mathematics!

We clear the path to reasoning and sense-making. Here are some ways to make that happen:

- Engage students using *Doing* Math Tasks (in Part 2 of this book, we provide a complete collection of Intervention Tasks and variations).

- Plan deliberate acts of teaching that address instructional needs. They are not "the next page in the textbook."

- Work together, as mathematics intervention opportunities are collaborative. Special education teachers, interventionists, paraprofessionals, teaching assistants, general education teachers,

Our goal is to provide engaging and memorable intervention sessions in which children run to the table, door, or interventionist excited to do mathematics!

- coaches, tutors, and school-based support staff (e.g., occupational therapists or speech therapists) are often partners in intervention planning, instruction, and assessment.

- Follow recommendations from the *IES Practice Guide* (Fuchs, 2021), UDL 2.0 (CAST 2024), *Principles to Actions* (NCTM 2014), *Catalyzing Change* (Huinker et al., 2020), the NCTM/CEC (2024) Joint Position Statement, and other research-informed sources.

- Plan intervention offerings within sessions devoted to this work and potentially at varied additional times (e.g., before or after school).

What Do We Mean by Tutoring?

We define intensive tutoring (sometimes called high impact or high dosage) as supplemental instruction within a school-based setting. However, it is offered before or after school (out-of-school time, also known as OST) instead of during the school day. These programs are often sponsored with Title 1 funding or with recent infusions of funding to improve student achievement, such as the 2024 budget proposal of $8 billion (D'Souza, 2024; The White House, 2024). With small groups of one to four students, such intervention tutoring can be delivered face-to-face or virtually, customarily two to five times a week, using high-quality instructional materials. Its purpose is to enhance the performance of students who previously were considered "at risk of failing" (Ludwig & Guryan, 2023; Robinson & Loeb, 2021). Again, like the model for interventions, we suggest a preventative approach for heading off issues rather than responding to poor performance. This must be done as a collaboration with the person providing the tutoring and the classroom teacher providing the grade-level mathematics content. Any tasks we share in this book can be used in these proactive tutoring sessions just as they are used in the interventions.

One concern is that states sometimes use an "all-call" approach to attract possible volunteers to guide intervention tutoring programs. There is preparation for the volunteers, but there are likely not enough hours to give these potential educators the mathematics pedagogy and content they may need to align and provide coherence with the schools' instructional plan. Again, we hope this book can help with that process as we provide high-quality materials through sets of tasks.

What Are Some Key Elements of the Intervention Instruction?

We will examine more deeply the importance of key elements to intervention-focused mathematics instruction such as instruction that regularly engages students in learning via multiple representations, collaboration across teams, and our preemptive and proactive approach—Priming.

Employing Multiple Representations

As mentioned, the CSA (concrete, semiconcrete, abstract) approach in various forms has been an important instructional consideration in mathematics education for many years (Fuchs et al., 2021; Heddens, 1964; Heddens & Speer, 2009).

Based on Bruner's (1966) theory of enactive, iconic, and symbolic reasoning, this model reflects simultaneous and fluent movement between an instructional focus on concrete representations/models (manipulative materials), semiconcrete (using drawings, math sketches, and graphs), and abstract (incorporating symbols, numerals, equations, mentally solving problems, or using stories with mathematical ideas). The need for multiple representations does not cease as students age (Bismark & Prosser, 2024; Kestel & Forganz, 2024; Thomas et al., 2024). The use of models and visuals must not fade away.

Importantly, this approach should not be rigid, where one only moves to abstraction after the long-term experiences with the other phases. Instead, there should be intertwined and parallel modeling of number symbols throughout this use of multiple interpretations of the situations. In this way, students can directly relate the concrete models and visual representations to the corresponding numerals and equations (Figure 1.1; Van de Walle et al., 2023). Modeling the mental conversations about reasoning and sense-making in a teacher's mind may help students articulate their thinking through what is known as a "think-aloud" (Evans et al., 2024).

CSA stands for the concrete, semiconcrete, abstract approach that presents these models to show the concept from a connected set of representations. The integrated model suggests using various representations in the same time frame to highlight the linkages between them.

A **think-aloud** is where the teacher gives an account of the thinking going on in their head by saying it out loud to the students. The idea is that hearing another's mental actions can support a learner's thinking process.

Figure 1.1 Interwoven CSA - Concrete-Semiconcrete-Abstract Model

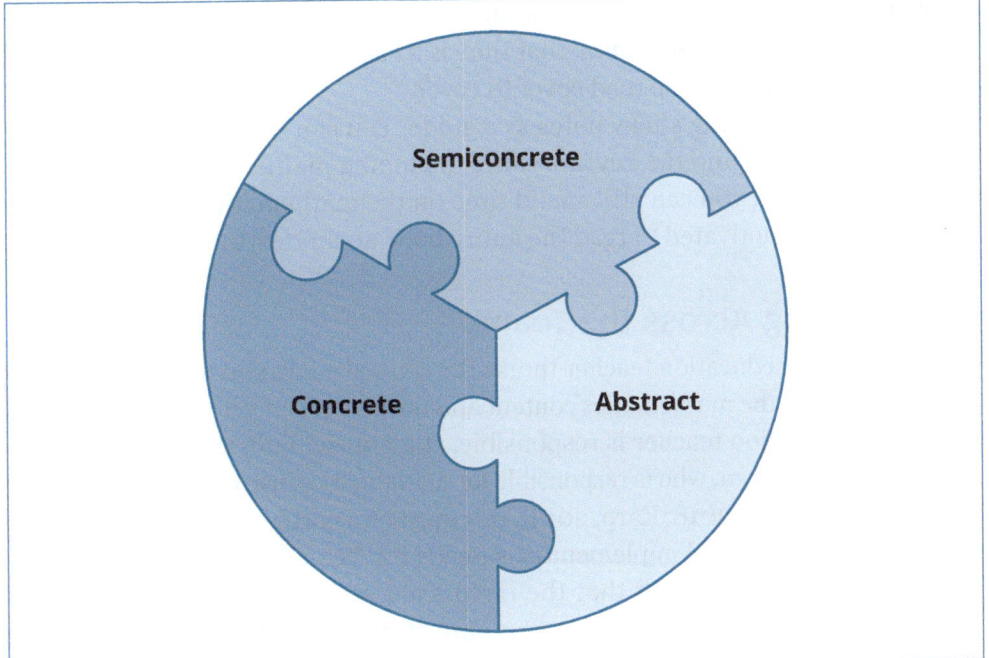

CSA often centers on combining manipulative materials, math sketches, and symbolic representations on the same concept. CSA aims to develop a concept and equip students with thinking strategies and skills needed for more independent learning. The model initially emphasizes conceptual understanding on a continuum to procedural knowledge (Skemp, 1978). In this way, it honors that both are critical components to students' mathematics proficiency (Findell et al., 2001).

This model should be aligned across the general education classroom in Tier 1 and the interventions in the Tier 2 or Tier 3 stages of RtI. Aligning instructional interventions with the CSA approach in general education classrooms provides continuity. No matter what CSA may be called (e.g., CRA—concrete-representational-abstract, CPA—concrete-pictorial-abstract), results of studies indicate that interventions and whole-class mathematics lessons regularly using the CSA instructional representations are successful with children with disabilities (Jitendra et al., 2023).

Another way to build connections between representations of mathematical ideas is by using children's storybooks. Recommended children's literature is included in the tasks found in Part 2 of the book, and there is an overall bibliography in Appendix A on the companion website. When woven into instruction, children's literature has been shown to positively affect achievement, attitudes toward mathematics, and the ability to make connections between representations (Zhang et al., 2022). These opportunities can help students consolidate their previous knowledge. They can use that foundation to develop new ideas (which are precisely aligned with the direction of our intervention approach). Reading part or all of a children's literature book together creates a shared context from which all students can draw. So reading facilitates the opportunity for interactive learning.

In addition, these stories and literature provide examples and realistic situations of how mathematics is a human endeavor. Such connections are important to honor. They connect with the students' lived experiences, thus rehumanizing mathematics (Zavala & Aguirre, 2024) and bringing a personal relationship to the learning process. Some may think that time is a barrier to this integration, but all the books do not need to be read cover to cover.

An example is using sticky notes as a guide. You can share excerpts containing the main points and the key storylines, including mathematical situations or wisdom. That way, you can still spend time on the mathematics discussion, and students will be motivated to read the entire book at another time.

Collaborating Across the Team

When the special education teacher thinks the general education teacher is responsible for teaching the mathematics content and the general education teacher thinks the special education teacher is responsible, students with disabilities sometimes become invisible. Then, who is responsible for mathematics instruction? Yes, it's both of you (Blanton et al., 2018; Karp, 2013); this must be a partnership. Collaboration is critical, and planning and implementing should be a reciprocal activity.

We know from research that the results are powerful when teachers consciously enter a professional learning opportunity, whether within a formal professional learning community (PLC), a partner PLC, a small group, or a self-selected thought partner. "Communal support structures" (Pilotti et al., 2023, p. 15) build shared leadership and collective learning. They lay the groundwork for personal growth. What may begin as a group of individuals who have "siloed practices" can eventually move to "joint responses" and a shared vision—characteristics of a team (Pilotti et al., 2023, p. 14).

We suggest beginning with the interventionist or special education teacher and the general education teacher regularly co-planning at least 4 weeks ahead of instruction planned for the general education classroom. Everything revolves around the instructional planning to prepare for the upcoming mathematics content by refreshing students' prior knowledge or constructing needed foundational knowledge. The general education teacher partners in this work by using the opening move of the new lessons based on the interventions that just occurred. They will ask questions extending the precise prior knowledge selected and practiced during the intervention (i.e., Liljedahl, 2020). We suggest the following aspects to agree on:

- consistently co-planning and co-teaching the mathematics content, with teachers co-owning and co-orchestrating the content delivery

- capitalizing on the strengths of general education teachers, special education teachers, interventionists, and math coaches, compounding the power of jointly made contributions

- working together to avoid repetitive IEP goals (i.e., needs to learn multiplication facts) that travel from year to year. Did you know the U.S. Supreme Court ruled that it is against the law to repeat an IEP goal from year to year (*U.S. Supreme Court Endrew F. et al. vs. Douglas County Colorado School District RE-1*, 2017)? That's not making adequate progress, as mandated by legislation. Think about it! How many students have IEPs that are not aligned with the law?

- avoiding IEP goals that emphasize low expectations by focusing on narrowly defined arithmetic skills, instead focusing on conceptual understanding and the Mathematical Practices

- agreeing as a team to practice "never say anything a kid can say" (Reinhart, 2000, p. 20), instead vowing to ask questions to prompt thinking. This includes when a child is working on a strategy that the teacher won't interrupt, erase what they've done, work the manipulative materials for them, or move the materials, using brief questions that are closed rather than open (Jacobs et al., 2014). We also know that "teaching as telling" does not reach an outcome of long-term learning (Lobado et al., 2005).

- addressing the role of administrators, coaches, and supervisors—what should they look for in this new collaborative arrangement? Do they know that CSA and "*doing* math" are the focus? How can they be supportive in providing time and resources to these teams?

- Aligning paraprofessionals with the intervention may require providing them with experiences related to learning more mathematics content and pedagogy background to deliver instruction.

Using a Preemptive and Proactive Approach

We repeat this here because it is essential to the success of our instructional model. We also believe this paradigm shift can "switch the script" and "change the narrative" for many students who may feel that they are not a "math person."

Rather than consistently using interventions to work backward as a reaction to a student's failure, we suggest working ahead in a proactive mode.

We call this a Rewind as we flip an old intervention model on its head. This Rewind may require a do-over of some things that have been practiced for several decades. This sequence aligns with our suggested strengths-spotting approach (see Chapter 2), in which strengths are identified and intentionally developed (Kobett & Karp, 2020). What students know and can do—their strengths—become the starting point for the intervention provided.

Again, interventions provide a preventative point where students get Primed for what is coming mathematically with a carefully choreographed set of preparatory sessions. So interventions are reoriented away from a reaction to classroom performance where the student isn't showing grade-level growth on a standard. This standard may be set as a way of working ahead of time to anticipate and then provide the perfect groundwork for upcoming lessons. This preemptive approach anticipates the needed foundational understanding of the mathematics topic using the strength-based learning trajectory-based instructional approach (Suh et al., 2022). The approach proactively provides the language, representations, and actions so that they become familiar and known to the student (Bahr et al., 2023). Interventions are reoriented from a reaction to classroom performance where the student isn't showing grade-level growth on a standard to a way of working ahead of time to anticipate and then provide the perfect underpinnings for upcoming lessons.

Delivering a solid background early on generates student confidence. It builds competence in students' mathematical ability so that students can feel and experience the agency to say, "I am a math person." In a study that interviewed secondary students with disabilities, rather than receiving sessions of reteaching content they were learning, they preferred a model of teachers presenting what they had as unfinished learning in advance of the new content (Munk et al., 2010). Students said this approach built their self-confidence and made them feel they could participate actively in class. These are important reflections to consider.

Preloading material during intervention sessions is a teaching turnaround that aligns with a strengths-based instructional theme. Through this proactive approach, students gain knowledge and can contribute to their grade-level mathematics lessons. By now, they will already be equipped with the background knowledge that ensures success in connecting to the mathematics. Let's look at the planning model in Figure 1.2.

By capitalizing on what the teacher knows is coming up instructionally, students' best qualities as learners are fueled by prior knowledge. That "on-ramp position" leads to greater success than prior approaches that chase perceived deficits or weaknesses. Applying strengths-based approaches early on opens the chance for familiarity with the materials and, ultimately, higher performance. We will discuss more strengths-based approaches in Chapter 2. We will go deeper into the planning model in Chapter 4.

> *Rather than consistently using interventions to work backward as a reaction to a student's failure, we suggest working ahead in a proactive mode.*

> *Interventions are reoriented from a reaction to classroom performance where the student isn't showing grade-level growth on a standard to a way of working ahead of time to anticipate and then provide the perfect underpinnings for upcoming lessons.*

CHAPTER 1. Introduction   19

Figure 1.2   Planning Model: General Education and Special Education Teacher Teams

4 weeks before new content lesson
- Review curriculum sequence as a collaborative team.
- Plan out necessary foundational concepts and skills, anticipate common hiccups.
- Align CSA representations and terms to be taught/refreshed and used.

3 weeks before new content lesson
- Select intervention priming lessons/activities from this book and existing curricular materials to prepare students for upcoming lesson(s).
- Use agreed-upon language materials and representations (CSA) for foundational concepts and skills taught in intervention sessions to support and connect to gen ed instruction.

2 weeks before new content lesson
- Priming intervention is presented over multiple sessions.

1 week before new content lesson
- Meet to compare notes to see how students in the intervention are doing.
- Prepare to use exact components from the interventions and vocabulary to launch the new grade-level whole-class lesson.

New Content Lesson
- Present all students in the class with a bridging component drawn directly from the intervention to review foundational ideas and to activate knowledge.
- Encourage and support students who took part in the intervention to provide answers and share ideas to build their confidence and math identity—let them shine!

Repeat Planning Sessions
- The same process continues to provide a preventative approach to the next instructional topic.

online resources   Available for download at **https://companion.corwin.com/courses/ProactiveMathIntervention**

To **leverage** students' prior knowledge, use purposeful questioning and thoughtfully selected tasks. The goal is to provide opportunities to make meaningful connections between something students *know* and something we want them to learn. To do this effectively, teachers must attend to students' ideas *with curiosity*—a desire to truly understand the students' thinking.

What Should We Think About Next?

As noted, we are considering what students can do and refusing to accept limitations. To do so, we need to pair formative assessments with data already collected in summative assessments that can be used as benchmarks. Understanding the learning progression allows teachers and tutors to see where students are in the learning trajectory (Clements & Sarama, 2014), build on what they know, and move them forward. This strength-based approach (Kobett & Karp, 2020) prompts the following questions:

- What are students bringing to the new mathematics content?
- How can they leverage that prior knowledge to be prepared for upcoming mathematics content?
- How can we facilitate that leveraging of prior knowledge?
- How can our actions be preventative rather than after-the-fact responses to repeated errors and confusion?
- How can we proactively build on where students are and simultaneously build their mathematics confidence and mathematics identity?

Reflection Opportunities for Chapter 1

1. How are you involved, in any way, with providing mathematics interventions at the classroom or school level? How is your school district involved?

2. To what extent does your school provide your students with mathematics tutoring or intervention opportunities? Is there evidence that these opportunities are making a difference?

3. Describe essential elements of a mathematics intervention program that you want to see provided for your students. How would YOU be involved in such a program?

4. What does the following statement mean to you: *"A math intervention program identifies instructional needs and starts with student strengths, instructionally, monitoring progress along the way as students become engaged in doing math they are learning"*?

5. How are you currently involved in providing or just having students participate in mathematics tutoring or intervention? What challenges do you see with such programs, and how would you suggest that things change?

6. We hear a lot about "learning loss" and how tutoring and intervention may answer such concerns regarding mathematics achievement. What do you think?

7. What about families? How should your intervention activities and student progress in math intervention opportunities be communicated to family members?

8. Have you analyzed the IEPs in your district? Do they focus on mathematics concepts and procedures? Are any IEPs repeated from a previous year? How can individual IEPs become a source of intervention goal setting and planning?

Characteristics of Effective Strengths-Based Interventions

CHAPTER 2

As I approached my 5th year of teaching, having taught second grade my first 2 years and fourth grade since then, I was asked by my principal to help create and implement a math intervention program for our school. I spent a lot of time reviewing a variety of online and commercially available intervention or tutoring programs. Not interested! I want a program for our kids that genuinely engages our faculty and staff in both its development and implementation. As I met with grade-level teams, one of the issues we discussed was the relative stigma given to both tutoring and intervention as programs that determine deficiencies and address them instructionally. Not us! We think intervention is about starting with and advocating for our students' strengths!

—A fourth-grade teacher

What Is Strengths-Based Teaching and Learning?

Imagine a learning experience in which your strengths, skills, and talents take center stage—that's the heart of strengths-based mathematics teaching and learning. Strengths-based mathematics teaching and learning is an approach that focuses on students' unique abilities, skills, and talents rather than concentrating on their weaknesses or deficits. You can use this approach to create a supportive, positive, and productive learning environment that fosters students' confidence and engagement and a growth mindset.

As suggested by research within the framework of strengths-based teaching and learning, teachers actively identify and acknowledge their students' mathematical strengths and use them as a foundation for further learning (Clifton & Harter, 2003). Students can develop their mathematical abilities through effort, persistence, and learning from mistakes to build a growth mindset (Dweck, 2006). Strengths-based teachers also plan lessons that cater to individual students' strengths and learning needs to more fully engage students and help them succeed in mathematics (Tomlinson, 1999).

Strengths-based learning environments develop a supportive and inclusive atmosphere, encouraging students to take risks. Students can share their thinking to help them feel more comfortable and confident in their learning (Hattie, 2009). Strengths-based teachers provide such opportunities for students to discuss, question, and justify their mathematical thinking. Through this process, teachers can help students develop a deeper understanding of concepts and recognize the strengths of their peers (NCTM, 2014). Therefore, as teachers, we are responsible for empowering our students to learn and grow in their mathematics knowledge.

Similarly, effective strengths-based interventions and in-school tutoring are supported by the same strong emphasis on strengths-based teaching and learning. Like you, students want to be recognized for who they are—special and unique learners with their own strengths, challenges, and interests. They want to feel seen as whole, capable humans in every interaction. Students' interest in learning and their perceptions of their abilities as learners are associated with their beliefs about the degree to which their teacher cares about them and recognizes them as unique (Ryan et al., 2023; Sethi & Scales, 2020; Wang, 2023). Based on that understanding, we suggest examining school intervention policies and practices that create structures that may dehumanize students. Such structures may make recognizing every student's valuable contribution to learning spaces difficult.

When you work with students in interventions or tutoring settings, especially those students with disabilities, strengths-based teaching asks you to lean in—to listen, observe thoughtfully, and ask strategic questions actively. These moments help you uncover and amplify student strengths. You can then use those strengths as a springboard to tackle challenges and grow student confidence in learning mathematics. The essence of this intervention approach lies in valuing the individual strengths of each student rather than comparing them to their peers. Using strengths as comparative measures diminishes the potential to harness and cultivate a student's unique talents. So we encourage acknowledging, nurturing, and using the evident strengths of every student.

What Is the Strengths-Spotting Process?

To truly become doers of mathematics, students should feel encouraged to build a rich toolkit of mathematical understanding, skills, and ways of thinking. Their mathematical strengths shine through their grasp of content, habits of mind, *and* use of mathematical practices—those ways they reason, problem-solve, and engage with mathematics on a deeper level. Strengths spotting hinges on the focus on the eight original mathematical practices (CCSS-M; NGA, 2010), which are represented in some form in every U.S. state, to be developed alongside the state's content standards. Some people may imagine that students who learn mathematics the quickest will be considered the strongest in mathematics. However, we encourage thinking about how students *respond to* and *interact with* mathematics learning via these process components. We suggest that in doing so, whispers of those students' previously undetected strengths may be revealed.

Four Steps to Spotting Strengths

You may use the 4-Step Strengths Process to identify, acknowledge, and use students' unique mathematical strengths. These steps will occur throughout the Priming Approach. Through this process, you can create a learning environment that fosters mathematical growth and supports students to develop positive mathematical identities.

Notice the Strength

Plan to recognize your students' strengths. Observe and record the strengths you see using specific, observable details (see Figure 2.1).

Figure 2.1 Math Strengths Checklist

- ☐ Finds patterns
- ☐ Makes connections
- ☐ Writes explanations to record thinking
- ☐ Uses mathematics representations
- ☐ Writes explanations to record thinking
- ☐ Communicates with peers
- ☐ Attends to precision
- ☐ Connects to prior knowledge
- ☐ Explains to justify strategies
- ☐ Communicates mathematics ideas verbally
- ☐ Uses math sketches
- ☐ Reasons about mathematics
- ☐ Uses representations and tools
- ☐ Makes generalizations
- ☐ Collaborates with peers
- ☐ Makes sense of problems
- ☐ Models with mathematics—writes equations to match a situation

online resources Available for download at **https://companion.corwin.com/courses/ProactiveMathIntervention**

Some teachers use checklists like this one. Others use formative assessment tools for observations and interviews, while others record evidence in journals.

When recognizing students' strengths, you may state the following:

- ▸ I noticed that you have a strength in _____.
- ▸ When you did _____, you showed a strength in_____.

Name the Strength

When you notice a student's strength, please share it with them immediately! Let them know exactly what they did and why it stood out. Then, name that strength so that they can recognize it in themselves. This simple conversation builds their confidence and helps them see their potential as confident and capable doers of mathematics. You may state the following:

- ▸ When you _____, this demonstrates your strength in _____.
- ▸ You show your _____ strength when you _____.

Translate the Strength

Help students see how their strengths make a difference! Explain how the strength they demonstrated is helping them grow their mathematical understanding. Connecting their strength to their learning demonstrates how to use it to tackle challenges. This kind of conversation helps students recognize the power of their abilities and builds their confidence as they apply those strengths to new and complex mathematics concepts. You may state the following:

- Your strength in _____ helps you _____ because _____.
- I wonder if we can use your strength in _____ to help you solve this problem?

Leverage the Strength

At this stage, it's all about helping students see how their strengths can be powerful tools to overcome challenges. Although it's common to focus on areas for improvement or past mistakes, remind yourself and your students that their mathematical strengths aren't accidental—they're real skills they can rely on! Students can tackle tough problems with greater confidence and resilience by strategically leveraging these strengths. You may ask the following:

- What challenged you about this problem?
- What would you like to improve?
- How can you use a strength you have to tackle this problem?
- You have a strength in _____. How can you use that strength to solve this problem?

You may have noticed that these examples don't include traditional praise! Although praise can be useful, it differs greatly from strengths spotting because students walk away from the strengths-spotting exchange with a better and more specific understanding of their mathematics learning. They need to know what you're celebrating them for! It is important to note that recognizing, championing, and nurturing students' mathematical strengths goes beyond simply praising them for their achievements (Dweck, 2008; Koenig & Buckley, 2018; Shumow & Schmidt, 2016). The process involves acknowledging and celebrating the unique ways students approach and understand mathematical concepts.

To effectively recognize and nurture students' mathematical strengths, we suggest viewing each student as an individual with distinct abilities and gifts. Thus, you move away from using the traditional model of comparing students to one another and, instead, focus on each student's qualities and strengths. You can build your strengths-spotting repertoire to recognize students' mathematical strengths without relying solely on praise. A few ways you can do this are discussed in the following subsections.

Emphasize Growth and Effort

Instead of focusing on correct answers, highlight the importance of growth, effort, and persistence in problem-solving. Students can then understand that their hard work and dedication are valuable, even when they struggle or face challenges. For example, you may notice a student persevering through a mathematics task even though they don't know exactly what to do first. You can recognize this strength by saying, "I see you persevering and trying so hard to solve this problem. This strength in persevering will help you learn and grow" (see Figure 2.2).

Figure 2.2 Teacher Noticing a Student's Strength

Source: Student images by istock.com/zuperia; teacher images by istock.com/lemono

Value Diverse Problem-Solving Approaches

Acknowledge and celebrate different ways of thinking and problem-solving in mathematics. By valuing diverse approaches, students can learn that they have unique mathematical perspectives and abilities (Star & Verschaffel, 2016). Students build confidence in their abilities by demonstrating what they value and how they appreciate their strengths. Asking open-ended questions and expressing genuine interest in a student's thought process will encourage them to share their ideas and promote a deeper understanding of mathematics (Boaler & Humphreys, 2005). For example, you may say the following:

- I never thought of solving the problem this way. Can you tell me more about your idea? Seeing different ways that students think about the mathematics they are doing helps *me* understand math better.
- You are showing a strength in creativity in solving this problem. Can you walk me through your thought process step-by-step?
- It's fascinating to see how you approached this problem from a different perspective. What led you to try this method?
- Your approach to this problem is thought-provoking. How do you think it compares to other methods we've explored?
- Your idea has given me a new perspective on this problem. Thank you for sharing!
- Your unique approach to this problem highlights the importance of diverse thinking in math. Can you think of any other situations where this method might be useful?

Encourage Self-Reflection and Goal-Setting

Help students identify their strengths and areas for improvement by incorporating self-reflection and goal-setting into your teaching practice (Howard, 2016; Marrongelle & Foltz, 2014; Schoenfeld, 2011). When students see their progress and set goals, they feel more in control of their learning and discover how to use their strengths to grow even further. Here are some prompts:

- Let's take a moment to think about your experience with today's math lesson. What do you feel were your strengths? What were some challenges you faced?

- Let's set some personal goals for our next mathematics topic. What concepts would you like to improve upon? How can I help you reach these goals?

- Think about your extraordinary mathematics progress so far this year. What have been your most exciting accomplishments?

- Think about when you felt proud of your work in a mathematics learning experience. What made that experience stand out for you? How can you apply what you learned from that experience to what we are learning now?

- As we begin _____, let's take a moment to consider your current understanding of the _____. What do you already know? What questions do you have? How can we work together to build your mathematics knowledge?

Provide Opportunities for Collaborative Mathematics Learning

Create opportunities for students to collaborate, discuss mathematics, and learn from one another while they work on tasks (Henningsen & Stein, 1997; Sun & Tan, 2019). When you facilitate collaborative learning experiences and prompt students to engage in mathematical discussions, you create an environment in which students learn from and appreciate the strengths of their peers. This approach helps students develop a deeper understanding of mathematics and promotes essential social and communication skills. You can achieve this goal by using the following prompts:

- Work with a partner or in a small group to solve this math problem. Discuss your ideas and strategies, and be prepared to share your approach with everyone.

- Choose a math problem from today's lesson and visually represent the problem and its solution. Then, exchange your work with a partner and discuss each other's representations.

- As you work on this problem, feel free to talk to your partner and discuss your thinking. Remember, we can learn much from each other's strengths, perspectives, and ideas.

Offer Strengths-Based Feedback That Promotes Growth

It can be challenging to break the "praise only" cycle! Rather than simply praising students for their accomplishments (e.g., "good job" or "nice work"), provide formative feedback that helps them grow as mathematical thinkers. This process can involve asking probing questions, providing guidance, and assisting students to develop a deeper understanding of the concepts they are learning. For example, you may say the following:

- I noticed that you used an interesting strategy to solve this problem. Can you explain your thinking? I think your approach could help others in our group.

- You've made some great progress on this concept, and I can see you've been working hard. Let's discuss the next steps in your learning and how you can continue building on your strengths.

- I can see that you've really grasped the concept of _____. Have you thought about how you might apply this understanding to other areas of mathematics, like _____ or _____?

- Your perseverance in solving this challenging problem is impressive. How did you manage to stay focused and engaged throughout the process? What strategies can you apply to future problems?

- Your ability to explain your thinking and justify your answers is a valuable skill in mathematics. Developing clear and concise explanations to help others understand your reasoning is important.

Example of the Process Using a Student's Representation

A representation created by a student is a window into their mind. For a teacher, a student's representation is an invitation to learn about how that student is reasoning about a problem they are working to solve. As teachers, we value how students think. So we should consider how to make it possible for all students to represent that thinking by sharing their ideas. For example, suppose a student draws a math sketch to represent a word problem. In that case, we can see how the student breaks down the problem into visual components and identifies mathematical relationships. Many students may create an equation to represent a problem. In this instance, we can gain significant insight into a student's understanding of mathematical symbols and their ability to translate a story into mathematical expressions or equations.

To increase your understanding of this process, let's consider some student work (see Figure 2.3) and the student–teacher conversation about the work and transcript of the student–teacher conversation. This work may contain what we are calling common hiccups, emerging conceptions, or differing understandings (Lewis, 2018).

A representation created by a student is a window into their mind.

Figure 2.3 Student Work on Ordering a Set of Fractions From Greatest to Least

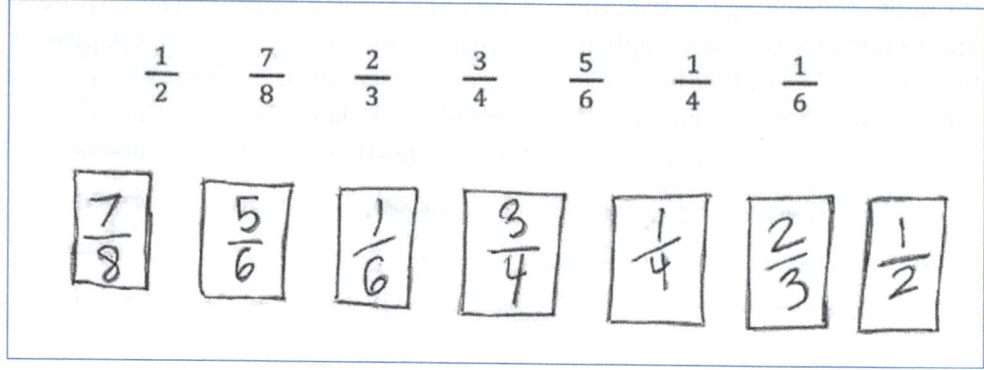

Student:	I did greatest to least. I looked at the numbers and I was looking at the numbers that were big and at the end, I was looking at the numbers that are small.
Teacher:	Why is this [points to $\frac{7}{8}$] the greatest?
Student:	It has the most fraction tiles. Then, five-sixths has the next most.
Teacher:	Why does this [points to $\frac{1}{6}$] come next?
Student:	Because 6 is bigger than all of the other numbers.
Teacher:	Why is this [points to $\frac{1}{2}$] the least?
Student:	Because 1 and 2 are small numbers.

At first glance, it's natural to focus on what the student does not yet know. However, what if we instead highlight what the student *does* know and use that as a foundation for growth?

Strength

The student demonstrates a clear plan and can articulate their reasoning for placing the fractions. They try to apply what they know about whole-number reasoning to build fraction understanding. They have applied their understanding of whole numbers in a new context. Later, the student references fraction tiles and visualizes that "there are a lot of fraction tiles" in $\frac{7}{8}$, indicating they know that manipulatives are a good way to show understanding.

Challenge

This student's interpretation is part of the natural developmental mathematical progression of understanding the size of fractions. When first working with fractions, many students apply whole-number reasoning to fractions. The student's reasoning reflects the belief that fractions with "bigger" numbers are automatically larger than other fractions without yet considering the critical relationship between the numerator and denominator.

Leveraging the Strength

To build an understanding of fractions, we can first acknowledge the student's understanding of whole numbers and say, "Fractions are numbers, too, so it makes sense that you want to compare them in the same way. Let's explore how we need to consider how fractions can represent parts and wholes." Students are helped with more time and opportunities to explore fractions using manipulatives, which can support their conceptual understanding. Now, let's take a sneak peek at some Priming Tasks you can use to build fraction understanding.

Close to a Whole? Close to Zero? Encourage the student to construct fractions using manipulatives such as fraction tiles and compare their sizes. For example, have them explore how close fractions like $\frac{7}{8}$ and $\frac{5}{6}$ are to 0 or 1 whole.

Context-Based Problems. Introduce real-world scenarios that involve comparing fractions. This type of problem encourages reasoning within meaningful contexts and helps students make sense of fractions in everyday situations. For example:

Remy ran $\frac{1}{8}$ of a mile, and Terry ran $\frac{2}{3}$ of a mile. Who ran farther? How do you know?

As you can see, learners develop mathematical ideas when they feel supported in successfully applying their strengths. Whenever we, as teachers, refuse to draw limitations on what a student can do, we play a critical role in improving their performance. We assume there may be cognitive differences but never deficits (Lewis, 2014). Identifying that they need more experience builds on evidence of their math power, which is fundamental to working with interventions. Your embodiment of those beliefs plays out in your questions, connections, conversations, and creation of the learning environment. All children can learn; thus, we encourage you to create personal learning experiences to maximize their strengths and abilities, as discussed in the following section.

> *We know all children can learn. We just have to create personal learning experiences to make the most of their strengths and abilities.*

How Do We Build Rapport and Relationships?

Students with disabilities have historically faced some barriers in accessing grade-level mathematical concepts (NCTM/CEC, 2024). Now, even more than before, their time away from their school setting during the pandemic challenged them twice—academically and through lost relationships with their peers, teachers, and the school's support team. Building rapport, trust, relationships, and a sense of belonging has become more essential. Interventionists and their general education partners must create intervention plans that foster a welcoming environment and affirm what students can do academically.

"It is the reason that good teachers can change lives, helping students find unsuspected gifts and inner purpose" (Aspen Commission, 2019, p. 6). This goal of responding to the cognitive, social, and emotional needs of students with disabilities brings the partnership between general education and special education to the forefront. The expertise of both is needed to support and respond to student behaviors (e.g., recognizing potential signs of anxiety, frustration, or giving up)

Sense of belonging is the student's perception of the level of their school's and their teachers' support, connection, respect, and concern for their future.

while expanding foundational knowledge and positive dispositions centered on mathematical reasoning and sense-making.

"Students' sense of belonging impacts their affect, motivation, well-being, and academic success" (Young et al., 2023, p. 1515). To provide evidence for the academic performance component of this statement, Cobbold (2024, as cited in Cassidy, 2024) reported a "close relationship" between students' sense of belonging and their performance on the international PISA mathematics test. Cobbold (2024, as cited in Cassidy, 2024) stated, "A sense of not belonging at school can hinder learning and lead to disaffection and active disengagement from learning" (p. 20).

To ward off "not belonging," as teachers, we can mix in and closely consider enhancing student well-being by showing care and concern. In intervention settings, we can't help but recognize that students who have been told they are not doing well in math are now assigned some form of double math. That may not feel like a good thing. There are good intentions to support students in these plans, but how do the students perceive it?

Relationships and support play a central role when thinking about what truly makes a difference in students' well-being in math instruction. In a study in Australia (Hill et al., 2022), researchers analyzed data from approximately 500 students to find what affects students' well-being during mathematics instruction. The study revealed that positive relationships with their teacher and peers were most often mentioned (aligns with the work of Clarkson et al., 2010). Other important areas included experiencing a sense of engagement and gaining increased mathematical understanding. Students described a "good" teacher as engaging, clarifying mathematical explanations, understanding their learning needs, and providing individual assistance when needed.

Another approach to creating this relationship-strong environment is when teachers exhibit "warm demander" pedagogy (Bonner, 2014; Delpit, 2013; Ware, 2006a, 2006b). When students experience warm demander pedagogy, as discussed in what follows, they are held to certain standards. The learning culture and atmosphere foster a warm, caring setting filled with encouragement, support, and opportunities for growth. Unsurprisingly, this environment is where mistakes are recast as normal instances of the learning process and confidence-building rather than as instigators of anxiety.

> **Well-being** includes students' attitudes toward mathematics, whether they exhibit or experience math anxiety, their level of engagement with the content, perseverance, and student-to-teacher and peer-to-peer relationships.

What Does It Mean to Be a Warm Demander?

Teaching children—including and especially those with disabilities—requires a balance of warmth and high expectations, which can be described as being a "warm demander." This approach involves building strong, trusting relationships with students while maintaining high standards and supporting their growth and achievement (Bondy & Ross, 2008; Ware, 2006a). We suggest incorporating several components into this pedagogical approach. These include establishing trust and generating warmth:

- *Establishing trust* is crucial for creating a safe and supportive learning environment for students (Bryk & Schneider, 2002). Teachers can build trust by being consistent, predictable, and genuinely invested in their students' well-being and success. When students feel valued and understood, they are more likely to engage in learning and take risks.

- *Generating warmth* refers to teachers' empathy, kindness, and respect toward their students. This process can involve expressing sincere interest in students' lives, celebrating their successes, and providing emotional support during challenging times. By showing this outer expression of warmth, teachers can create a sense of belonging, foster positive relationships with their students (Cornelius-White, 2007), and support learning (Pianta & Stuhlman, 2004; Wentzel, 1997).

When teachers hold high expectations for their students, they communicate a belief in their abilities and potential, encouraging them to strive for new learning. Teachers can support students in meeting these expectations by providing "just right" accommodations, scaffolding, and individualized learning opportunities (Hord & Xin, 2001).

For you, the key to being a warm demander is finding the right balance between exuding acceptance and warmth and holding high expectations for your students. You should strive to be supportive and understanding, encouraging students to take responsibility for their learning and reach for new understanding, even when it feels hard and scary. When found, this sweet spot helps create a learning environment that is both nurturing and challenging. This approach acknowledges and values students' unique and brilliant strengths. You also provide the structure and support they need to overcome learning obstacles they may face. Your students will flourish in a space in which they find it is safe to be wrong (Matthews et al., 2021).

How Can We Integrate Culturally Responsive Mathematics Teaching as We Engage in Intervention Activities?

Zavala and Aguirre's (2024) framework for culturally responsive math teaching offers educators a way to integrate teaching practices that honor students' knowledge and identities while maintaining rigor and support and promoting power and participation during mathematics learning. When working with students, learning about their cultural and community knowledge and experiences affirms positive mathematical identities and honors their thinking and ideas (Aguirre et al., 2013; Carpenter et al., 2014; Civil, 2007). Through this process, they can feel a sense of belonging and visibility. Asking questions about family practices and lived experiences that connect to mathematics situations and contexts can help students make sense of mathematics, deepening their understanding of skills and concepts.

You can also use this approach to access cultural and community knowledge accurately. For example, asking students to share a cultural practice that involves a "making task" (https://eqstemm.org/tasks/making-tasks/), like making cascarones or mooncakes for a cultural celebration, connects to mathematics involving multiplicative reasoning. You will encourage them to scale up a recipe that can make proportional reasoning come to life! Asking students about family road trips can also help students understand how speed is a ratio and how they can use mathematics to estimate their travel time. Finally, a variety of games from different cultures offer mathematics connections, such as Mancala from Africa and Asia or Toma Todo from Mexico.

In summary, there are many ways to build intervention instruction, celebrations of cultural practices, and diverse ways of knowing. These methods honor the wealth of understanding students bring to mathematics classrooms, small-group interventions, or in-school tutoring sessions. The proactive intervention model will set students up for success by building their knowledge and identities. As you recognize and celebrate the students' growing strengths, they can highlight their brilliance. This practice disrupts stereotypes and status hierarchies that may exist. It can foster positive social relationships and interactions among students with diverse strengths.

How Can We Leverage Family Involvement in Supporting Math Proactive Intervention?

Family involvement is a powerful catalyst for strengthening math interventions. When families actively participate in their child's mathematics education, it can significantly amplify the effectiveness of these interventions. Engaging with parents and caregivers allows for a supportive learning environment beyond the school while fostering a collaborative atmosphere in which children can thrive in their mathematical journey. In addition, this positions parents and caregivers as intellectual resources (Civil & Andrade, 2003), empowering them to be advocates for their children's education.

It's important to value how families engage with mathematics through cooking, managing finances, or other daily activities. These everyday experiences are rich learning opportunities and can be used to support formal mathematics education. Research has shown that when parents and children talk about math together, it boosts the students' numeracy skills (Hurtado, 2018; Ishimaru et al., 2015). These interactions create a math-friendly environment at home, making mathematics a natural part of everyday life. Based on this information, we encourage using these interactions through the creation of home connections (see Figure 2.4 and Appendix B on the companion website).

Figure 2.4   Samples of Home Connection Activities

Home Connection: TOPIC—Linear Measurement With Fractions, Precision

DIY Project Fun: Being Precise

Involve your child when you are doing a DIY project by asking them to help you measure with a tape measure. Fractional units will be useful in being more precise with measurement of a garden or hanging curtains. Ask your child to count by fractional units.

Source: DIY measuring stick image by istock.com/stevecoleimages

Home Connection: TOPIC—Estimation and Multiplication

Seats for Everyone?

When you are at a movie theater, field, stadium or concert hall, or even a restaurant, ask your child, what do you think is the seating capacity? These modeling questions will require estimation skills and multiplication that can support mathematical reasoning.

Source: Seats image by istock.com/Image Source

How Can We Help Students Avoid or Overcome Math Anxiety?

We mentioned that students can find situations in which the posing of mathematics problems makes them feel scared or uneasy, and we know even adults can feel that tension. Many students suggested that math anxiety stemmed from missed learning opportunities (Brewster & Miller, 2020). It seems the students can achieve the greatest levels of learning when they are in a space in which they are both challenged and feel comfortable, connected to their surroundings and their teacher(s). This comfort mirrors the warm demander pedagogy, which balances high expectations with a supportive and caring environment (Panicucci, 2007).

Cognitive researchers have found emotion to be essential to reasoning (Bechara et al., 2000). Children learn from people who show them goodwill and demonstrate caring actions; in the counterexample, when children are in a state of fear, that impacts their ability to learn (Aspen Commission, 2019). Based on this knowledge, we encourage open communication and dialogue between you and your students. You can ensure that your students feel comfortable expressing their ideas, feelings, and needs before and throughout the lesson.

Reflection Opportunities for Chapter 2

1. When you look at the Strengths Checklist (Figure 2.1), which strengths have you noticed in your students receiving intervention? What strengths should be added to the list?

2. How does translating students' strengths support their problem-solving skills and ability to overcome learning challenges?

(Continued)

(Continued)

3. How can teachers redirect students' thinking to emphasize the role of their strengths in tackling problems?

4. What are some strategies for celebrating diverse problem-solving approaches and unique mathematical abilities with students and their families?

5. What might help you balance warmth and high expectations in your teaching practice, particularly when working with students with disabilities?

6. What strategies have you used or could you use to establish trust with students and create a safe and supportive learning environment?

7. What have you noticed about the role emotions play in your students' mathematical performance? For example, what might you say when you see a student "shut down"? What might you do?

8. How can you continue to adapt your instruction to support the varying strengths of learners while aligning with UDL principles?

Formative Assessment

Targeting and Monitoring Interventions

CHAPTER 3

As a teacher with more than 2 decades of experience, I have learned to value the importance of classroom-based formative assessment. As I think about our intervention and tutoring programs—both before and after school—I am convinced that much of what I do in using formative assessment with my students can easily be adapted, as well as connected with our intervention and tutoring programs, and should support my efforts with my collaborating teacher in special education. I also think that my involvement with classroom-based formative assessment has allowed me to monitor student progress regularly. And as I consider the feedback I provide, have students provide me, and engage students in giving feedback to each other, my students have become partners in the assessment–learning process.

—A fourth-grade teacher

We know that planning, instruction, and assessment are everyday considerations and responsibilities for the classroom teacher, special educator, or interventionist. Intervention-focused formative assessment directly impacts teacher planning and instruction. It also provides useful student-centered and individualized insight when monitoring student progress. We recognize the influence of summative assessment results on instructional priorities and as an element of placement decision-making for classroom-based or schoolwide intervention efforts. However, the feedback you provide your students, the feedback they offer you, and the feedback your students provide to each other are also integral components of any intervention initiative.

As we think about interventions, assessments identify the instructional starting line for those students engaged in the intervention and serve to monitor their progress regularly. Summative assessment results, progress monitoring, and daily classroom-based formative assessments intersect to identify the planning focus for a student's intervention. The actual instructional activity occurs next and is monitored using formative assessment. Such assessments provide indicators of student progress and identify particular student strengths and challenges. Thinking about and providing feedback opportunities are critical components of all intervention initiatives. Formative assessment is essential in providing students with timely and meaningful feedback and the "just-right instruction" to advance students' understanding. To better understand these ideas ourselves, let's briefly consider formative and summative assessments and feedback.

What Is Formative Assessment, and Why Is It Important?

Based on a review of hundreds of studies, of which 250 were directly relevant to formative assessment, Black and Wiliam (1998) defined formative assessment "as encompassing all those activities undertaken by teachers and/or by their students, which provide information to be used as feedback to modify the teaching and learning activities in which they are engaged" (p. 7). The Council of Chief State School Officers (2018) defined formative assessment as "a planned, ongoing process used by all students and teachers during learning and teaching to elicit and use evidence of student learning to improve student understanding of intended disciplinary learning outcomes and support students to become self-directed learners" (p. 2). Fennell et al. (2024) suggested that the focus of formative assessment "is on the everyday use of formative assessments during instruction to monitor, probe, and provide feedback designed to impact student learning and your planning and teaching" (p. 6).

What's the Role of Summative Assessments?

Summative assessments are typically implemented after a learning experience. Such assessments may include unit assessments, assessments required by the school district, or high-stakes and high-profile end-of-year state assessments. Summative assessments are often used to compare student progress, which may be class-to-class, grade level–to–grade level, student-to-student, or the extent to which the summative assessment data address predetermined standards (state or school district) or expectations.

As teachers, we often use summative assessment scores and class profiles to define individual or group levels of student success (e.g., advanced, proficient, developing, basic, not yet met, or a level as determined by a student's IEP). Of course, summative assessment results or performance on particular items of a summative assessment (e.g., a unit test) can also be used formatively. This may occur when a grade-level team analyzes student work on such assessments. They may help determine instructional goals and related classroom activities, perhaps at the Tier 1 and Tier 2 levels within a multitiered system of support (MTSS).

We intend to identify and provide specific suggestions for adapting classroom-based formative assessment techniques for individual students, small groups, whole-class instruction, or related support efforts for use in interventions. We will also suggest how summative assessment results may help to identify initial student needs and related planning for Priming Interventions and strategic tutoring. As described in Chapter 2, and as you'll see in the intervention activities in this book, spotting strengths while implementing formative assessments is key. You will want to continually use various methods of formative assessment and ongoing analysis of summative assessments to provide the proper diagnostics that drive intervention opportunities for your students. So yes, assessment happens at the beginning of instructional decision-making for students with disabilities in mathematics. Nevertheless, it is also frequent, consistent, and intentionally connected to proactive planning and targeted teaching.

And What About Feedback?

Feedback is an essential element of the assessment process. Teachers provide feedback to students, students give feedback to teachers, and students provide feedback to each other (Figure 3.1). Feedback is connected to intervention planning, instruction, and the ongoing assessment of the intervention plan.

Figure 3.1 Planning – Priming – Teaching – Assessing (and Feedback) Cycle

Feedback can also play an important role in how a student sees themselves as a "math person." Cohen and Lotan's (2014) work on complex instruction helps us understand how assigning competence to students' mathematical contributions affects how we position them in the learning community. This statement is so important, as it advises us to consider not just providing feedback but the impact and potential "next steps" of the input provided.

The type of feedback you provide during an intervention activity, as well as how and when you provide it, has the potential to either motivate or inhibit student progress. We suggest recognizing that your feedback launches your planning for the next intervention activity that day or perhaps the next time you meet with your students, indicating that you assess while teaching. So as you consider the feedback that you provide to your students, the feedback that students provide to you, and the feedback that students provide to each other, think about how such feedback supports and influences a student's learning. Although feedback may include acknowledging student effort with comments like "great job" or "I like your graph," such comments only indirectly relate to a student's progress. In addition, feedback should encourage students beyond representing a "ticket" to do more! Also, remember that Fennell et al. (2024) stated, "Feedback is often thought of as teachers making statements about students, not about the impact and influence of their teaching … thus diluting the impact and benefit of truly connecting assessment and feedback" (p. 16).

Your feedback launches your planning for the next intervention activity that day or perhaps the next time you meet with your students, indicating that you assess while teaching.

How Can We Focus on Formative Assessment in Intervention Settings?

Summative assessment and feedback are both critical legs in the three-legged assessment stool. However, for the sake of this book, let's explore three formative assessment techniques (adapted from Fennell et al., 2024). You can use the following techniques in an intervention setting for maximum impact on planning and enacting your instruction: observations, interviews, and Show Me.

Observation Tools

Teachers observe students all day long. Thinking about what you would like to observe and expect to observe and documenting such observations allows you to see how your students approach mathematics tasks, strategize solutions, and work with other students. Of course, observation is an everyday assessment element during the planning and implementation of your interventions. Still, the following questions are designed to help you plan for using this technique as students engage in the intervention activities you provide:

- What will you expect to observe?
- How will you know "it" if you "see" (observe) it?
- What particular student strengths or challenges might you observe?
- How will you record and provide feedback on what you observe (Fennell et al., 2024, p. 23)?

To help you better observe your students, the following subsections present two beneficial tools for use in intervention settings: the Small-Group Implementation and Recording Tool and the Individual Student Mathematics Strengths Log.

The Small-Group Implementation and Recording Tool

You may find this tool helpful when you observe small groups in interventions or work with an entire class (Figure 3.2). You may decide to use this tool during your intervention activity. You may note comments as you observe or wait until the end of the activity to record comments.

Figure 3.2 Observations: Sample of a Completed Small-Group Implementation and Recording Tool

Intent of Observation	Brief Description/ Comments	Observed?
Mathematics content:	Students working in a small group of five, adding fractions with three addends	Yes, for the most part. I need to provide for more immediate access to the tools the students need.

CHAPTER 3. Formative Assessment

Intent of Observation	Brief Description/ Comments	Observed?
Mathematics Practices/ Processes	make sense of problems reasoning using tools	Yes, for both reasoning and making sense of problems, but 3 of the students seemed unsure about how the number line could be used when they saw Finn's model.
CSA	Interestingly, Finn and Lani both focused on using math sketches. They also used equations.	General comment. I am going to really focus on the use of tools, including math sketches and number lines for representing fraction operations.
Student engagement	Entire group engaged. Finn showed strengths in describing his model to the others. Lani also persevered by creating more than one model and selecting the best one.	Very pleased with the progress and engagement of all 5 learners. They enjoyed sharing their work and thinking with others.

General Comment:

Comments below are specific to Finn and Lani.

It was interesting that Finn (image A)* added two of the fractions first and then added the other as two different equations (not shown). Lani (image B)* made one math sketch that was crossed out (not shown here) and then wrote she had a better model. She used drawings of the crops to effectively show the three addends. Her equation points to her knowledge of what adds to a whole as she circles ten tenths but doesn't provide the answer to the problem.

Feedback:

General for the whole group:

Discuss with all 5 students in this intervention group how to partition a whole into a particular number of equal-sized parts. Include a session where both effective explanations of their thinking (e.g., I divided the whole into the number of parts of the same size that I needed) and ineffective explanations (e.g., I did the math) are shown, discussed, and improved.

*Attached student work samples.

 A blank fillable version of this observation template is available for download at **https://companion.corwin.com/courses/ProactiveMathIntervention**.

*Finn's Work – A

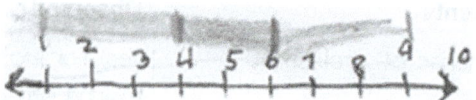

*Lani's Work – B

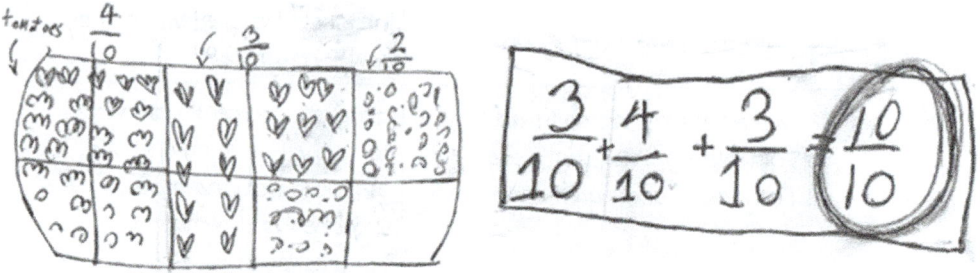

Source of the Tool: Reprinted with permission from Fennell et al., *Annual Perspectives in Mathematics Education: Assessment to Enhance Teaching and Learning*, copyright © 2015, by the National Council of Teachers of Mathematics. All rights reserved.

Individual Student Mathematics Strengths Log

The Individual Student Mathematics Strengths Log (Figure 3.3) will be most helpful when documenting individual students' progress. You can provide documentation of a student's responses to a mathematics activity. At the same time, you can assess the student's mathematical disposition, CSA use, memory, attention (to tasks within the mathematics activity), socioemotional engagement, and organizational skills.

Figure 3.3 Observations: Sample of a Completed Student Mathematics Strengths Log

Learner Profile for ___Tory_ (see student work below*) _____		
*Here is Tory's work for the problem "I have 38 crayons I would like to put in boxes of 8. How many boxes are needed?"		
Mathematics concepts/skills	List the student's strengths with specific content, concepts, and skills.	Tory understands the meaning of division and is working on how to interpret remainders.
Mathematics disposition	What types of content, tasks, and activities does the student respond to with positivity, interest, and engagement?	Tory responds positively to tasks that have a context of interest. Tory likes soccer, cooking, art, and pets. He seems genuinely interested in engaging in the mathematics he is learning.

CSA	How does the student make connections among varied representations?	Tory connects representations by making a concrete or semiconcrete model to represent the problem then translates that into an equation (abstract) CSA.
Memory	What kinds of things does the student remember?	Tory can remember previous models he has made to solve problems and can connect new models to existing knowledge.
Attention	What strengths does the student demonstrate? Does the student attend to particular types of activities?	Tory's strengths are acting out problems, making concrete models with manipulatives, drawing math sketches, and talking aloud through the solution process.
Socioemotional engagement	How does the student: • Work with others? • Productively struggle? • Persist?	Tory works well when collaborating with peers, especially when working on multistep problems. When a challenge is presented, Tory benefits from encouragement to persevere. Given this support, Tory experiences success.
Organizational skills	How does the student organize/record their mathematical thinking?	Tory orally explains his work to show his mathematical thinking, often focusing on how the model connects to the context of the problem and how that was translated into an equation.

Source of the Tool: Adapted from *The Formative 5—In Action*, by F. Fennell, B. Kobett, & J. Wray, 2024, Corwin, p. 46. Adapted with permission.

 A blank fillable version of this student strengths log is available for download at **https://companion.corwin.com/courses/ProactiveMathIntervention**.

Teachers like to use this tool in conjunction with an interview. Some teachers also like to use the tool to document student strengths and performance over time (e.g., use periodically or within particular mathematical content topics). Providing background information about mathematics concepts and skills in terms of learning progressions can be helpful for many tutors or adults working with students through intervention. They can situate where the learner is within this continuum to assess their strengths and define "the edges of their understanding" (Berger, 2004, p. 336). This process will pinpoint the knowledge educators want to build to advance their mathematical understanding. Interviews and observations of students during instruction can inform professionals (tutors and interventionists) who work with these students about their socioemotional needs, cognitive strengths, and executive functioning skills. These can be leveraged to tap into students' strengths and build on their mathematics skills and understandings.

Interview Tools

Interviews essentially extend an observation. For example, let's say you observe a student within any element of their math intervention activity and need to know more. What did they do? How did they determine that solution? Why did they work through a response in a particular way? Asking such questions will allow you to dig deeper into student thinking and understanding of a mathematics concept or procedure.

Interviews are a valuable formative assessment technique within any intervention program because they become conversations that allow you to get more than just a snapshot of student thinking. You will find that you will begin to frequently use observations and interviews as your everyday or "go-to" formative assessment techniques to identify strengths and needs. These will also help monitor student progress.

We suggest asking the following questions to help guide your use of this formative assessment technique:

- What evidence would make you decide to work with a student one-on-one or with a small group? How does your collaboration with your special education or general education teaching partner play a role in this decision-making?
- How frequently will you interview students within an intervention program? Daily? Multiple times each day?
- What interview questions might you ask? How might the questions be different for particular students?
- What responses will you anticipate from students? (Consider asking about understandings *and* possible challenges.)
- What follow-up interview questions might you ask, and how would such questions be connected to the feedback you might provide to the student or group?
- How will you provide a record of student interviews that will become cumulative over time, thus providing evidence of progress? And how will this information be shared with others (e.g., other teachers, family members, IEP team members) (Fennell et al., 2024, p. 23)?

As with observations, we would like to present two invaluable interview tools to use in intervention settings: the Interview Planning Tool and the Individual Student Prompt.

Interview Planning Tool

This tool is helpful as a planning guide for an interview (Figure 3.4) and a record of a student's interview responses. The tool can also be used in conjunction with required assessment tools. The elements of this tool will help you decide how you can learn more about a student, including important monitoring of their progress on:

- understanding your mathematical goal(s) for a day or after several intervention sessions
- conceptual understanding and/or procedural fluency
- use of particular strategies during the interview (e.g., halving, doubling)
- prerequisite knowledge or perceived challenges
- positive mathematical dispositions

Figure 3.4 Interview Planning Tool

Student Name: | | | | **Grade Level:**
Teacher(s):
Interview Goal(s):

Components	Purpose	Action(s)	Materials	*Doing Math* Questions	Feedback to Student
Introduction	Build rapport	How will you open the interview and explain the purpose of the interview?	Manipulatives		
Mathematics Assessment Concept 1:	Determine students' strengths and challenges	How will you present the task and collect student responses?	Manipulatives, tools, graphs, number lines, etc.	Task:	
Mathematics Assessment Concept 2:	Determine students' strengths and challenges	How will you present the task and collect student responses?	Manipulatives, tools, graphs, number lines, etc.	Task:	

online resources Download printable at https://companion.corwin.com/courses/ProactiveMathIntervention

Individual Student Prompt

This widely used interview prompt (Figure 3.5) is helpful and easy to use. Of course, you can adapt it to meet your needs. The prompt provides an opportunity for student responses to how and why questions you may typically ask in an interview as you extend an observation or seek responses to student solution strategies. Note that such student responses are essentially student-to-teacher feedback related to the prompt's interview questions.

Figure 3.5 Interview: Sample of a Completed Individual Interview Prompt Recording Sheet

INTERVIEW PROMPT		
Word problem task: Two friends were making pies; one from apples, the other pie from peaches. Paige used $\frac{4}{8}$ of the apples and Rudy used $\frac{3}{4}$ of the peaches. If they each used 8 pieces of fruit, who used more of their fruit?		
Name: Ben	**Date:** 2/13	**Math Topic:** Comparing fractions using a set model
QUESTION	STUDENT RESPONSES (teacher recorded)	
How did you solve that?	I made 8 circles for apples and 8 circles for peaches to show each person's total fruit. And then I colored the circles to show what was used.	
Why did you solve the problem that way?	Making the model made most sense to me because then I don't have to keep things in my head. It helps when I can see the math.	
What else can you tell me about what you did?	Then I figured out what $\frac{4}{8}$ of the apples was and $\frac{3}{4}$ of the peaches. I know that $\frac{6}{8}$ is the same as $\frac{3}{4}$ and $\frac{4}{8}$ is the same as $\frac{1}{2}$.	
How did you represent your thinking?	$\frac{6}{8}$ is greater than $\frac{4}{8}$. So Rudy used more fruit.	
How could you connect your representation to another representation?	I could make a model with manipulatives, real fruit, or maybe using a double number line. Either way, I show Rudy using more of his fruit.	

Note: Attach completed work samples

*Ben's work sample:

Source: Recording Sheet Adapted from Figure 2.6 Individual Student Interview Prompt found on p. 76 of *The Formative Five in Action*. Corwin. Adapted with permission.

 A blank fillable version of this recording sheet is available for download at **https://companion.corwin.com/courses/ProactiveMathIntervention**.

Show Me Tool

To an extent, the Show Me assessment technique is a student demonstration of what they know. Like the interview, it extends an observation. Show Me occurs when one of your students, a pair of students, a small group, or perhaps an entire class is asked to demonstrate how something works, how a problem is solved, how a particular representation (e.g., base 10 materials or number line) was used, or how three representations demonstrating CSA relate within a student response. Teachers using the Show Me technique regularly comment that their students' responses validated information initially developed from an observation or interview. This validation provided the "tipping point" for redirecting student responses and helping to determine their next steps instructionally.

We suggest considering the following questions as you anticipate your use of the Show Me technique:

- How is your Show Me prompt different from an observation and interview?
- Consider an intervention activity you are planning to bridge to an upcoming grade-level lesson.
  - What might you use as a Show Me prompt for foundational knowledge related to that lesson?
  - How would that help you determine the selection of an intervention to prepare students for the lesson?
  - How would that influence your implementation of the intervention activity and connect it strategically to the upcoming lesson?
- What might you want your student(s) to demonstrate and say as they describe their Show Me responses? How might concrete, semiconcrete, and abstract representations (CSA) or Universal Design for Learning (UDL) support the structure and implementation of your Show Me prompt?
- Recognizing that a student response to a Show Me prompt is student-to-teacher feedback, when would you provide teacher-to-student feedback to a Show Me response?

Again, let's examine two tools for this Show Me technique—one for a small group and one for individuals.

Small-Group Record Tool

The Small-Group Record tool provides an actual record of responses from a Show Me prompt. It can be easily adapted for a single child or small group of students (Figure 3.6). The Show Me responses provide indicators to help you plan for students' next intervention opportunity. An actual photo of student responses is an efficient way to document progress.

Figure 3.6 Show Me: Sample of a Completed Show Me Recording Sheet

SHOW ME: Mathematics Content: Grade 4

Show Me Prompt: Use math sketches, the number line, fraction circle templates, or fraction tiles to order the following fractions from greatest to least.

$$\frac{1}{2} \quad \frac{7}{8} \quad \frac{2}{3} \quad \frac{3}{4} \quad \frac{5}{6} \quad \frac{1}{4} \quad \frac{1}{6}$$

Activity Focus/Standard:	Anticipated Show Me Responses:
Comparing and ordering fractions with different numerators and denominators. I gave some students who are challenged to write fraction cards with the fractions written on them so they could move them around. I also had fraction tiles and circular templates for shading. Unless shown on the recording sheet, the students responded without manipulative materials.	I want to see and hear their responses related to ordering fractions. In particular, how they approach their representation of the comparison and ordering of the seven fractions. I think something I should watch for is students who may base their decision-making only thinking about the size of the denominators. Other students may consider the overall size of each fractional representation in math sketches, number lines, or with fraction tiles or circles, essentially considering both the numerators and denominators of each fraction.

Student: Jasmine

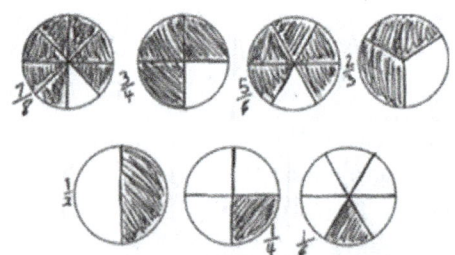

"I knew that $\frac{7}{8}$ was bigger than the others because it takes up more space. Then I compared which were bigger than $\frac{1}{2}$. Then the rest only take up $\frac{1}{4}$ and $\frac{1}{6}$ of the whole. $\frac{1}{4}$ is bigger as it takes up more space than the $\frac{1}{6}$."

Student: Gabe

"I see seven, eight, five, six, one, six, three, four, one, four, two, three, one, two. So out of these two numbers [points to $\frac{7}{8}$], these are the greatest—seven and eight. Seven and eight are big. Then, five six is the biggest number after. Then, one and six because six is a bigger number. Then one-fourth because four is a bigger number than three-four. Two-thirds is not bigger than three-fourths, but it is bigger than I don't know how to say it [points to $\frac{1}{2}$.] This is the one I am stuck on."

Student: DeShawn

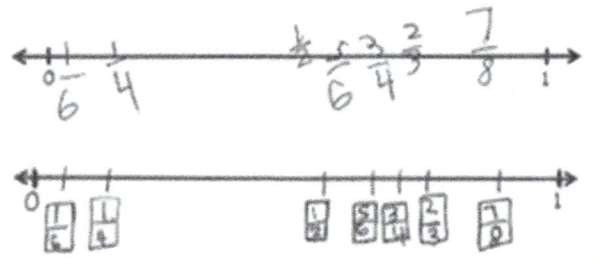

Student: Layla

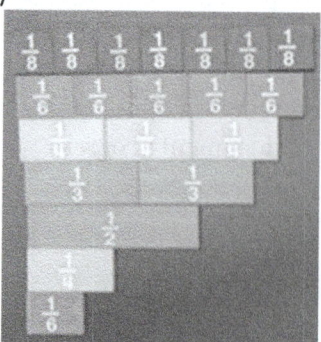

[the first number line is the student's initial writing; the second with the support of an occupational therapist]

"I put $\frac{1}{2}$ at the halfway point, halfway between 0 and 1. Then, I put $\frac{5}{6}$ and $\frac{2}{3}$ after the $\frac{1}{2}$ because they are closer to 1. I put $\frac{3}{4}$ between $\frac{5}{6}$ and $\frac{2}{3}$ because I think it is between these two numbers because it has a big numerator and a smaller denominator than $\frac{5}{6}$. $\frac{7}{8}$ is the biggest number because it is a big numerator and a big denominator, so that is why it is at the very end near 1. $\frac{1}{6}$ is at the very bottom because it has a big denominator and a tiny numerator. $\frac{1}{4}$ is doing the same thing as $\frac{1}{6}$ because it has a 1 as its numerator and its denominator is pretty big compared to it. But, I think it is bigger than $\frac{1}{6}$ because fourths seem bigger than sixths, but smaller than halves."

$\frac{7}{8} \quad \frac{5}{6} \quad \frac{3}{4} \quad \frac{2}{3} \quad \frac{1}{2} \quad \frac{1}{4} \quad \frac{1}{6}$

"I used the fraction tiles to help me put them in order from greatest to least. $\frac{1}{6}$ is the smallest of the fraction tiles because of its length. The length told me how big the fractions were. $\frac{7}{8}$ has the longest length. [Teacher: She came back after class to tell me she kept thinking about the problem and wanted to explain more]. "$\frac{1}{2}$ is first because in fractions the smallest fraction is the biggest. [Teacher: I asked her what she meant by that]. "If you think about fraction tiles, the 8ths would be the smallest and the halves would be the greatest. That is why I put them in this order. $\frac{3}{4}$ is greater than $\frac{1}{4}$ because there's more fraction tiles on the table if you build it. $\frac{5}{6}$ is greater than $\frac{1}{6}$ for the same reason."

Source of the Tool: Adapted from *The Formative 5 - In Action*, by F. Fennell, B. Kobett, & J. Wray, 2024, Corwin, p. 98. Adapted with permission.

A blank fillable version of this recording sheet is available for download at **https://companion.corwin.com/courses/ProactiveMathIntervention**.

You can use the recording sheet to document and interpret a student's mathematical understanding as well as to debrief with colleagues as appropriate. For example, in Figure 3.6, **Jasmine** wanted to use materials and had available fraction templates (all cut into individual square cards). We know that many children are challenged to divide wholes into equal-sized parts, and this helped her get a jump-start on comparing the fractions. She selected those she needed, shaded them, wrote the fraction, and then moved them around to compare the size. Then she relied on the actual matching of the fractions to see the relationships between them and guide her decision-making. Jasmine's process indicates a strength in knowing what representation would work best for this task.

Gabe sees the numerators and the denominators of the fractions as two different whole numbers. He does not yet understand that a fraction is a number. He focuses on the individual whole numbers in the fractions rather than comparing the fractions as parts of a whole. He refers to the numbers in the denominator, saying, "Six is a bigger number," implying that he believes a larger denominator represents a larger fraction. He orders the fractions according to their denominators, which suggests he has some intuitive awareness of the relationship between numerators and denominators. However, his reasoning afterward becomes inconsistent. He relies more on the size of the denominators than on the understanding that a larger denominator means smaller pieces of a whole. Gabe needs more support when comparing fractions such as $\frac{3}{4}$, $\frac{2}{3}$, and $\frac{1}{2}$, particularly when comparing $\frac{2}{3}$ and $\frac{1}{2}$. His hesitation in explaining which fraction is greater and the admission of being "stuck" on how to compare them indicates that he isn't sure about fractions that are farther from benchmarks like 1 and 0. Gabe did not apply the concept that the size of the pieces matters as much as the number of pieces in the fraction. His intuition about fractions, like $\frac{7}{8}$ being greater than the others, is an emerging strength and suggests new learning, which may be leveraged if we introduced using benchmarks (like 0, $\frac{1}{2}$, and 1) and manipulative representations such as fraction tiles or circular templates to compare the fractions.

DeShawn's explanation demonstrates a strong understanding of fraction comparison and the relative size of fractions. He has strengths in using a number line effectively to compare fractions based on their proximity to the benchmarks of $\frac{1}{2}$ and 1 whole. He has a good grasp of the relative magnitude of fractions, with a hiccup with the positioning of $\frac{5}{6}, \frac{3}{4}$, and $\frac{2}{3}$ that can be supported by using concrete materials. DeShawn also uses clear comparisons, noting that $\frac{1}{2}$ is greater than $\frac{1}{4}$, and $\frac{1}{4}$ is greater than $\frac{1}{6}$, demonstrating a firm understanding that smaller denominators (when the numerator is 1) represent larger parts of the whole.

Layla uses fraction tiles as a concrete tool to help visualize and compare fractional amounts. This action indicates that she finds a physical representation helpful for making sense of abstract concepts like fraction size, which is a significant strength. The strategy of using fraction tiles to show physically the length comparisons can be leveraged to support a transition from concrete lengths to the semiconcrete number line. She also identifies halves as the largest fraction pieces and eighths as the smallest, demonstrating an intuitive understanding of the relative size of fractions. She recognizes which fractions with like denominators are larger or smaller, suggesting she has a conceptual understanding of comparing fractions. However, it is noteworthy and a strength that she continues to reflect on the problem later in the day and wishes to give the teacher added explanation to describe her thinking.

What Is the Importance of Task Selection in Assessment?

Tasks are an integral component of this book. We developed these tasks to encourage regular consideration of problem-based activities. You may use these tasks to engage students in interventions in *doing* the mathematics they are about to be learning. Importantly, you can also use the tasks to monitor progress and determine student strengths. The tasks are your next instructional steps when assessment tools reveal what students know.

"Effective formative assessment involves using tasks that elicit evidence of students' learning, then using that evidence to inform subsequent instruction" (NCTM, 2024, p. 95). We developed tasks to open an intervention-focused lesson, perhaps engaging mathematics learned from prior activities. The task may be used as the key component of the day's work on refreshing prior knowledge for an upcoming new mathematics topic or teaching something that was missed due to absences. We developed these tasks to focus on engaging students in *doing* math!

A **task** connects concepts with procedures, multiple representations, and various mathematical ideas. There are multiple paths to solving tasks. Thus, students use prior knowledge and experiences to work toward a solution (Smith & Stein, 1998).

How Can We Connect Assessment and Instruction?

In this section, we explore key considerations for connecting intervention opportunities with a student's daily grade-level mathematics class. We aim to create a cohesive and supportive learning environment that benefits all students by addressing these connections. Some possible discussion prompts include the following:

- What is the focus of the student's (or group) intervention opportunities, and how has this been determined?

    This discussion is important to the planning that will probably occur at the district and school levels. There are lots of things to consider—think about it! How will you choose students for intervention programs you may provide? What is the mathematical focus of the interventions? What resources may be needed?

- How frequently do the classroom teacher and interventionist meet to examine, discuss, and update individual students and/or class progress in mathematics, considering formative progress within the intervention program and during mathematics class?

    Time is a precious commodity in any school. Still, regular meetings involving classroom teachers and interventionists are critical to monitoring progress, determining Priming initiatives, and sharing anecdotal comments on instructional efforts. These meetings also prompt necessary program modifications.

- How frequently is the monitoring of progress noted above communicated to others (e.g., parent/family; those monitoring IEPs, special education, or general education teacher) to support effective collaboration?

    This question is essential and often overlooked because of the focus on day-to-day instruction and related intervention opportunities. Still, as the classroom teacher and interventionist establish planning and meeting routines, a plan to communicate and involve and truly engage "others" (e.g., students' parents/family members, special education coordinators, supervisors or service providers such as speech pathologists or occupational therapists) must be developed. In addition to providing regular updates to principals and supervisors, this discussion may include inviting them to your meetings. Regular updates to parents/family members are stand-alone communication links. Some programs may entail the general education classroom teacher and interventionist developing and disseminating a newsletter to parents and family members. The letter can include at-home math activities (also see Home Connections in Appendix B on the companion website) and updates on their instructional programming. Other intervention programs include parent/family math nights for communicating student progress and instilling a positive pride-focused opportunity relative to their children's progress and the importance of mathematics learning.

▸ How do formative assessments fit with progress monitoring, more formal assessments, and IEPs?

As noted in this chapter and Chapter 1, you are encouraged to daily consider the links between your planning (including co-planning with those providing math intervention opportunities), teaching, assessment, and feedback. How and when you use the observation, interview, and Show Me techniques and your documentation of student progress (consider the tools presented in this chapter) will, no doubt, also include how your formative assessments connect with student IEPs. Your school's or school district's summative assessments are elements of discussion within the regular meetings you have between classroom teachers, interventionists, special educators, administrators, and others. Your formative assessments will provide data about student progress within the intervention opportunities you provide.

Reflection Opportunities for Chapter 3

1. How do you use formative assessment in planning and teaching? How would such planning be adapted to include planning for and providing intervention opportunities for your students?

2. How does your feedback to students, their feedback to you, and providing opportunities for your students to give feedback to each other influence your planning and instructional next steps?

3. How should summative assessments required by your school district or state be used as you consider placement decisions regarding mathematics intervention opportunities and tracking student progress on the mathematics being learned during such intervention opportunities?

4. Consider a recent lesson in which you observed your students engaged in the mathematics they were learning.

    a. What did you observe that indicated that the students seemed to be doing well, understanding, and succeeding with the lesson's activities?

    b. What did you observe that suggested an interview with the student or students?

5. How do you see the interview technique being used as you monitor student progress (individual and small group) within a mathematics intervention or tutoring program?

6. Use one of the interview tools presented in this chapter (Figures 3.4, 3.5) to plan for or implement a math intervention activity.

    a. Which tool did you use?

    b. Would you adapt the tool for future use? If so, how?

7. Fraction equivalence is a foundational understanding that is crucial for continuing work with fractions. Suggest a Show Me activity using manipulative materials or the number line to engage students in creating or extending their knowledge of equivalent fractions.

8. How might you use each of the following formative assessment techniques to monitor student progress within mathematics intervention opportunities?

    a. Observation:

    b. Interviews:

    c. Show Me:

9. What and how frequently should you communicate to families regarding student progress in your school's mathematics intervention program?

Your Turn

A Proactive Preventative Mathematics Intervention/ Tutoring Model in Action

CHAPTER 4

*It feels like we have worked forever in thinking about our school's mathematics intervention program. So many things to consider. But one of the things I have learned is what my grade-level team and I have started to call the three C's: Collaboration, Connections, and Coherence. Here's what I'm talkin' about! As we began our planning, just finding the time for our grade-level teams to meet and then agree on issues related to placement, assessment, and truly buying into our strengths-based approach and everyday use of formative assessment was critical. As was working with our special education teachers and the paraprofessionals assisting in our math intervention and tutoring program—**collaboration** with a capital C! It was also all of the intervention program's **connections**—our pre-teaching Priming connecting both our math intervention and tutoring lessons and regular class math lessons, the bridging prompts linking intervention task activities to regular class math lessons, even the time of the day and location of our math intervention efforts, and so much more including our team-by-team discussions to ensure the **coherence** math-wise between intervention and tutoring opportunities and regular class lessons and within and across math content topics and standards. Our intent is to reflect on the three C's—collaboration, connections, and coherence regularly as we meet in our grade-level and school teams.*

—A fourth-grade teacher

This chapter focuses on providing the meaningful connections necessary to assist you in launching math intervention sessions or tutoring opportunities for Priming your students. We hope that the brief but focused connections back to the book's foundational chapters (Chapters 1–3), discussion of the intervention model (Figure 4.1), and the many suggestions and tools provided in this chapter will help support you when designing, implementing, and monitoring your mathematics intervention program. Let's get started by first introducing elements of mathematics interventions.

Considering Elements of Your Mathematics Intervention Program

We encourage asking a vital question early on when considering mathematics intervention opportunities: "What are the nonnegotiable mathematics concepts and related understandings at each grade level?" Essentially, what is the math? Analysis of summative assessments, including results from your school, school district, or state assessments and recent NAEP scores, will suggest content topic areas of importance and those that may respond to specific needs.

Of course, the objectives of mathematics interventions are not just about student success on tests. We know that in school and the rest of life, mathematical proficiency includes much more than success with the most popular and talked-about content strand—number and operations. As you think about math intervention opportunities, you've most likely focused on developing a well-established, important understanding and proficiency within and across other content domains (e.g., geometry and measurement). In many cases, these domains may have included elements of number and operations (e.g., perimeter, area, and volume) and rely on other kinds of mathematically related thinking (e.g., spatial thinking, algebraic thinking, and data literacy).

We suggest aligning math intervention preplanning to mathematics content emphasis and focusing on the following:

- learning trajectories or progressions, such as those identified by research (Clements & Sarama, 2011/2019; Confrey et al., 2014; Daro et al., 2011);
- best-practice initiatives, such as the CCSS-M Progressions documents from Achieve the Core (2013) (www.achievethecore.org);
- recently released state standards for mathematics (e.g., Virginia Standards of Learning for Mathematics | Virginia Department of Education, 2023); or
- other widely available mathematics progressions support documents.

You can use such preplanning to identify critical standards at particular grade levels.

How Is Our Approach Different?

As previously mentioned, some key differences in our Priming Approach to intervention and tutoring may seem new from others in the past. Some new elements include the timing and nature of the instructional planning.

What Does Planning for Priming Look Like?

There is an important distinction between typically discussed tutoring relative to a day's math topic or help with homework as math intervention approaches and what we promote in this book. So if you say, "Well, that's what we are or have been doing," here's what we advocate instead. As we alluded to in the Preface and Chapter 1, we suggest a process called Priming. Through Priming, you can use intervention or in-school tutoring sessions to address foundational skills that need strengthening for students to become successful in grade-level content that will be presented in the near future.

Priming requires planning approximately 4 weeks before the content coming up in the curriculum (see Figure 4.1). Such preparation magnifies the power of instructional support as your students are prepared to hear about new grade-level ideas with newly learned or recently refreshed "backfill" of the needed foundational knowledge in place. We plan to prepare your students for what's to come. Let them get in front of the new mathematics content. This focus on foundational ideas should be systematic, cohesive, and framed to integrate students' prior knowledge of mathematics concepts and skills.

Let's begin by taking a "deep dive" into our Planning Model for General Education and Special Education teams, initially presented in Chapter 1 (Figure 1.2).

Note. The planning model intends for grade-level classroom teacher teams, interventionists, and their special education colleagues to come together and do the necessary planning for instructional connections related to general classrooms, mathematics interventions, and in-school tutoring. Let's consider each of the following:

> *We plan to prepare your students for what's to come. Let them get in front of the new mathematics content.*

- **Four weeks before new lesson (mathematics) content:** This first step in planning entails reviewing the curricular sequence of a particular mathematics topic (e.g., fraction equivalence or perimeter). It also involves fully considering the concepts and related skills of that topic, including the "common hiccups" that teachers with classroom experience or targeted research-informed resources can identify. This step also provides an early opportunity to gather, consider, and align the CSA representations and related mathematical terms that may be used instructionally.

> *Our team never thought we needed to think about what we might do intervention-wise so early! We were wrong! Our grade-level team discussions about the common "hiccups" we have all experienced when teaching particular topics and concepts were SO helpful and frankly helped us to prepare for our use of observations and interviews to monitor progress [see tools 3.2 and 3.3]. And the time we spent discussing CSA and what representations we may use led to Show Me prompts we filed electronically—to be used once our intervention program began.*
>
> —A third-grade teacher

- **Three weeks before new lesson content:** This is the time to bring elements of prior planning together. It includes the Priming Tasks and Activities from this book (see Part 2) and curriculum materials the interventionist or the grade-level teacher intends to use to prepare students for upcoming units and lessons. It's also time to discuss the use of agreed-on mathematics language, materials, and representations (CSA) with your grade-level team to build the foundational concepts and skills

PART 1. INTRODUCTION TO PRIMING

Figure 4.1 Planning Model for General Education and Special Education Teaching Teams

4 weeks before new content lesson
- Review curriculum sequence as a collaborative team.
- Plan out necessary foundational concepts and skills, anticipate common hiccups.
- Align CSA representations and terms to be taught/refreshed and used.

3 weeks before new content lesson
- Select intervention priming lessons/activities from this book and existing curricular materials to prepare students for upcoming lesson(s).
- Use agreed-upon language materials and representations (CSA) for foundational concepts and skills taught in intervention sessions to support and connect to gen ed instruction.

2 weeks before new content lesson
- Priming intervention is presented over multiple sessions.

1 week before new content lesson
- Meet to compare notes to see how students in the intervention are doing.
- Prepare to use exact components from the interventions and vocabulary to launch the new grade-level whole-class lesson.

New Content Lesson
- Present all students in the class with a bridging component drawn directly from the intervention to review foundational ideas and to activate knowledge.
- Encourage and support students who took part in the intervention to provide answers and share ideas to build their confidence and math identity—let them shine!

Repeat Planning Sessions
- The same process continues to provide a preventative approach to the next instructional topic.

online resources Download printables at **https://companion.corwin.com/courses/ProactiveMathIntervention**

taught during math intervention opportunities. We recommend planning to discuss how they support and connect to mathematics within general education lessons.

> *We met across several grade levels, including general education and special education colleagues, for this several-weeks-out planning time. This allowed us time to dig into the Priming lessons. We tried them out, sometimes revising them a bit for clarity and once again were able to talk about how we might assess student progress with these Priming lessons and think hard about the connections between the Priming lessons and our class lessons. We also had great discussions about the mathematics language, which we will both use and develop within our mathematics intervention time and general education lesson. Importantly, we began the necessary collaboration between all of us—responsible for the mathematics learning of our students at our respective grade levels—general classroom teachers, special education teachers, math interventionists, and paraprofessionals. I can't think of a time when I have felt so connected and actually understood how my colleagues can and will work together for the betterment of OUR math learners.*
>
> —A mathematics coach, an interventionist, and two teachers

- **Two weeks before the new lesson content:** This is when the intervention in the form of the Priming lessons is taught. The special education teacher, interventionist, paraprofessional, teaching assistant, general education teacher, coach, or tutor (whoever is responsible for the intervention Priming lessons) engages students in *doing* math through the carefully selected tasks the team has planned.

> *Teaching the Priming lessons felt so much better than just using intervention time to help with homework or redo the same lesson that happened in math class! Students were so engaged. They even asked when they could come back! We chose tasks that would help fill in some gaps in their learning, and they are really making progress. I'm feeling really confident about how they will do with the new content in math class next week.*
>
> —A third-grade teacher

- **One week before the new lesson content:** Now's the time to meet grade-level colleagues and interventionists to review formative assessments from the intervention Priming lessons and get a sense of the actual progress of your students in the math intervention. Then, we recommend preparing to use the actual components of the math intervention activities and related math vocabulary to launch the new grade-level whole-class lesson. These will be notated as "bridging activities" in the tasks in Part 2 of the book.

> *In fairness, the 4-week- and 3-week-out planning got us ready for "launch time." We have considered the observation [3.3] and interview tools [3.4] to use, actually created Show Me prompts, and have a plan for our intervention activities. But this is so different instructionally. In the past, we've used our intervention times to preteach the new content lesson before it happened in the classroom. We thought that would help students feel empowered and engaged during the math class, but it didn't really work. Now, I really love how we use formative data to plan intervention activities that help fill in gaps in students' prior learning—the concepts and skills they need to be prepared for the new content. I feel like we are really equipping our students with the tools they need, and because it's not just a preview of the same lesson, the students don't get bored.*
>
> —A fourth-grade teacher

- **New content lesson:** A new class lesson begins with a *bridging* prompt or activity (see Part 2 tasks) drawn from recent math intervention activities designed to review foundational ideas and activate knowledge as the general educator's class begins a lesson on a new topic. We recommend encouraging students involved in the math intervention activities to provide responses and share ideas with the full class to build their confidence and enhance their math identities. Such strength-supported connections are invaluable.

> *Our team had never really thought about the importance of our lesson's bridging prompt. Talk about strengths-based teaching! The bridging question instantly engaged my students in the math intervention group in the day's lesson. Its impact is as much about generating and increasing student confidence as it is about connecting math intervention success to the day's class lesson. Our weekly math planning team lessons now always focus on thinking about, organizing, and electronically "filing" bridging activities and questions.*
>
> —An interventionist

- **Repeat Planning Sessions:** The long-term process from 4 weeks before the new topic lesson implementation continuously repeats. This repetition helps encourage the preventative approach to mathematics instruction.

> *We've learned! Our math team—all of us—general class mathematics teacher, special education teacher, math interventionist, and paraprofessionals—meet regularly, certainly to plan, but also to review, refine, and add activities and importantly provide weekly progress updates to our school administrators and our students' family members.*
>
> —An elementary mathematics specialist

The Planning for Priming: Intervention Planning Tool (Figure 4.2) was developed to assist you in implementing the planning model (Figure 4.1). Note the backward design format, which should help the team of grade-level general educators and special education colleagues address each stage of the planning process. They may begin with analyzing upcoming math topics, referring to the learning progressions, and co-planning proactive Priming interventions (4 weeks before grade-level implementation). They then will move to select and co-plan lessons and related resources to support interventions for Priming math topics (3 weeks before grade-level implementation), teaching the math intervention Priming lessons (2 weeks before the actual grade-level implementation), analyzing formative assessments of the students in the math intervention, and then co-planning bridging prompts designed to support connections to the new content (1 week before grade-level implementation). These steps set the stage for the delivery of the new content topic in your grade-level mathematics classroom.

Figure 4.2 The Planning for Priming: Intervention Planning Tool

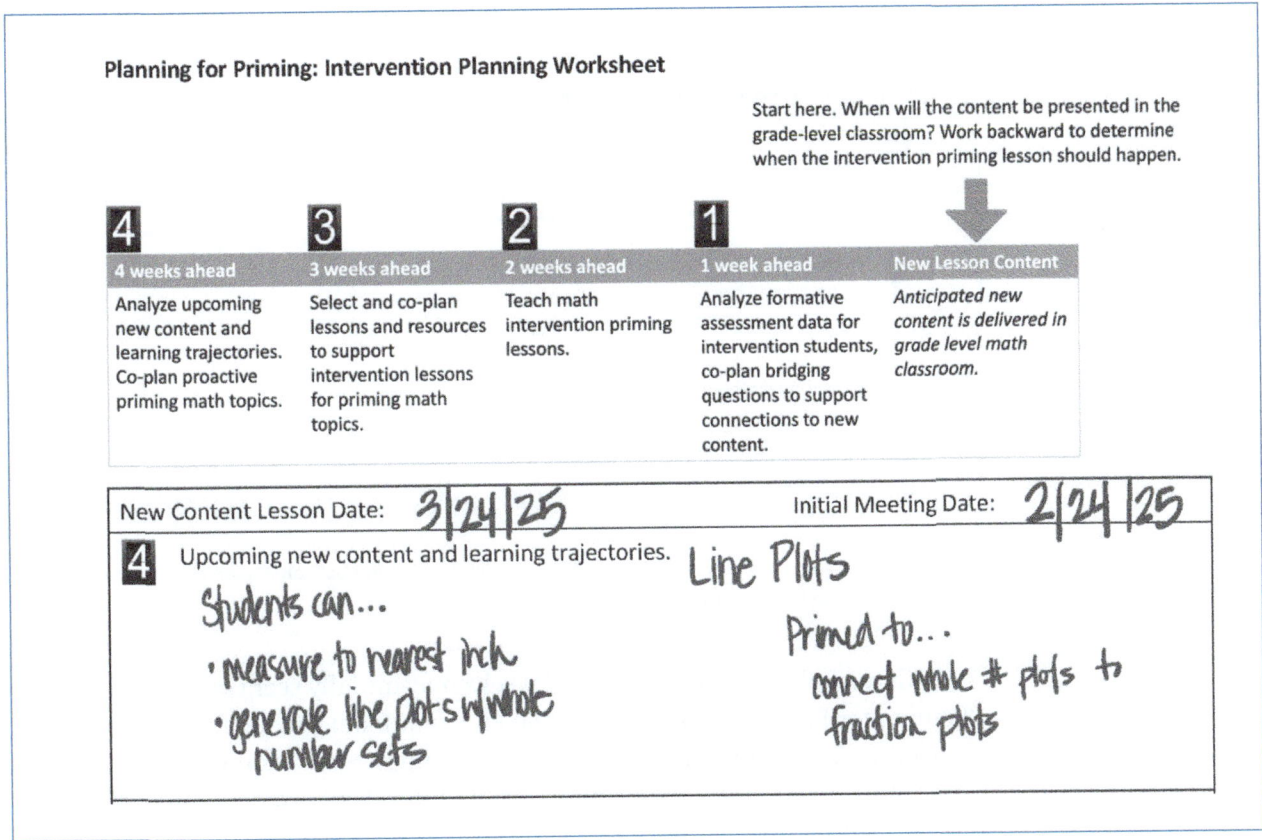

(Continued)

(Continued)

Planning for Priming: Intervention Planning Worksheet

3 Intervention priming lessons selected:
Task 43: Tip & Plot   Task 20: Line 'Em Up
                                  (fractions on a # line)

Resources needed:
- fraction manipulatives
- clothesline & clothespins
- Line 'Em Up cards
- collections of things to measure (should measure to 1/4, 1/2, 3/4 units)
- adding machine tape
- sticky notes

Who's responsible:
- Tamara
- Sara
- Mateo
- all of us!
- Sara
- Tamara

2 Date(s) these Intervention priming lessons will be taught:
Week of 3/10 : Line 'Em Up (Mon, Tues, Wed)
                Tip & Plot (Thu, Fri)

1 Formative assessment notes on these intervention priming lessons:

Bridging prompts selected to support connections to next week's new content:

A blank version of this tool is available for download at https://companion.corwin.com/courses/ProactiveMathIntervention.

Although planning across several weeks may initially seem overwhelming, the Planning for Priming Tool will help you and your team consider the steps leading to Priming. Then, you can consider the actual implementation within grade-level classrooms. In addition, the Priming for Planning Tool provides a record of your progress within the planning stages.

What Does a Planning Conversation Sound Like?

You can read the following planning discussion as an example of what a planning conversation might sound like. The discussion took place 4 weeks in advance and, again, 2 weeks before the new topic of perimeter was introduced. This topic was covered over six lessons in the grade-level curriculum. Dawn is the special educator leading the interventions, and Chonda is the classroom general education teacher (third grade).

Chonda: At the end of the month, we will be doing perimeter. This is a big and important idea for third grade, and it seems like some of my students may be missing some prior knowledge.

Dawn: Let's talk about some Priming activities we might do in intervention before those lessons.

Chonda: Yes. We also need to decide how I can introduce the lessons in class in a way that connects to those Priming intervention activities you are presenting. If I know what they've been Primed for and what prior knowledge they've practiced, I'll feel confident asking the students with disabilities in the class to lead more during the lesson launch. Then, as we've said, they become the stars.

Dawn: I am not sure of what the longitudinal progression is of the skills, and so I'm going to become dependent upon you here. I'm not sure where to start with perimeter. What prior knowledge will students rely on? What might we need to shore up to get them Primed?

Chonda: Well, many students seem to struggle with the difference between area and perimeter. If we focus on perimeter for the intervention, I think it will be important *not* to use 1-inch square tiles to measure or anything that has an area. We need to be sure students see perimeter as a length measure.

Dawn: I also think it's important that we should be linking perimeter to addition.

Chonda: Yes, great point. We also have to review length and the components of measuring length like using the ruler properly and paying attention to no gaps and no overlaps.

Dawn: We need a length tool to measure length, so it's going to have to be a ruler. We will need to measure the length of a couple of things and then add the lengths together. Then, at some point, I'm going to need to think about the best way of bringing up the idea of something close to a rectangle.

Chonda: In both perimeter and area is where I see geometry and measurement start to really overlap. I found a task (see Task 39) that addresses our goals. It starts with students measuring one line segment, then two lines that are connected at a right angle. In other words, we won't give them the whole rectangle at first. When we meet next week, we can talk through the intervention Priming lesson and bridging activities and questions. Keep in mind which students I can pair to start the launch of the actual lesson. I'll need your help to select two kids who are on the cusp, and their demonstration for the whole class would be another good way to cement the ideas for them.

Note: See Task 39 for a complete description and related bridge activities and questions for this perimeter intervention task.

How Do We Connect Strengths-Based Teaching and Formative Assessment?

In Chapter 2, we shared the importance of cultivating a strengths-based mathematics teaching and learning approach. This approach encourages cultivating a positive and supportive learning environment that recognizes students' unique abilities and skills. Focusing on students' strengths, you can boost students' confidence and engagement before, during, and after the Priming lessons. As you plan your Priming lessons, consider recording your students' strengths using the tool below (Figure 4.3) to inform your instructional decision-making. For example, you may adapt the tool to include the specific content of the task or lesson and a variety of representations, such as manipulatives, number lines, etc.

Figure 4.3 Monitoring Student Strengths Tool

Strengths	Student 1	Student 2	Student 3	Student 4	Student 5
Content 1					
Content 2					
Content 3					
Problem-Solving					
Strategic Thinking					
Perseverance					
Communication					
Representations 1					
Representations 2					
Representations 3					

online resources — This tool is available for download at https://companion.corwin.com/courses/ProactiveMathIntervention.

Chapter 3 addressed the role and importance of formative assessment and presented assessment techniques for you to consider as you regularly engage in planning, implementing, and continuously assessing your mathematics intervention or in-school tutoring initiatives. The use of observations, interviews, and Show Me assessment tools will become an everyday consideration for you as you monitor student progress. Importantly, we want to note that these classroom-validated assessment tools can be adapted to meet your needs: Observation Tools: Figures 3.2, 3.3; Interview Tools: Figures 3.4, 3.5; and a Show Me Tool: Figure 3.6. Finally, you can use the Progress Monitoring Tool (Figure 4.4) to consolidate your mathematics students' observations, interviews, and Show Me responses related to all Priming Tasks (Part 2). We recommend updating this helpful tool regularly based on variations of Priming Tasks, with comments summarized and considered in planning your launch of the grade-level math lessons for the whole class.

Figure 4.4 Progress Monitoring Tool—Priming Task

Activity Variation	Students	What Did You Observe?	Interview Questions & Student Responses	Show Me Prompts & Student Responses
Learning Target				
Summary Comments				

This tool is available for download at **https://companion.corwin.com/courses/ProactiveMathIntervention**.

As you plan for and implement the Priming Tasks provided, including formative assessment as part of your planning and each task's implementation will be important. To assist you in that process, we suggest using the observation, interviews, or Show Me techniques for particular elements of the *Engage in Math Discourse (Make the Mathematics Visible)* portion of the planning guide for each of the 43 tasks provided in this book. We signal these formative assessments through icons with a magnifying glass for observations, two students talking for interviews, and a hand working with manipulatives for Show Me opportunities.

These icons should not prevent you from adding questions and approaches to what is shared. We propose considering the following example showing the formative assessment icons displayed within an excerpted portion of an intervention task (Task 16).

CHAPTER 4. Your Turn

Multiplication Chains

TASK 16

Variation 1

Learning Target: *Students will create related multiplications by adding or subtracting areas.*

Engage in Math Discourse (Make the Mathematics Visible):

As students work, encourage them to notice important relationships and push on early and partial understandings.

Observation: Monitor the room to ensure that students have changed models in ways that maintain a rectangular area. For example, watch for models that may look like this:

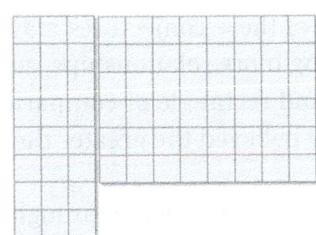

If you see models like this, ask students to use the factors in their expressions to describe their models in terms of "rows of square units." Ask, "How many rows are there in your model? How many square units in each row?" When students notice that all the rows don't have the same number of square units, help them think about how to adjust their model to show the area represented by their expressions.

Show Me: Show me where in your model you see the (8 × 2) part of this equation.

How are these two products related? How much greater/less is this product? How does that connect to the part you added/took away?

After the activity, bring the intervention group together for a discussion. Suggested prompts include:

- **Interview:** What did you notice about your chain of facts?
- **Show Me:** *Choose one of the student models and ask students how they could draw open-area models to represent the changes. Challenge each group to return to their equation chain and sketch a series of open-area models to show the changes.*

Regularly using the formative assessment techniques discussed here and in Chapter 3 will help as you monitor student progress, provide feedback that addresses students' mathematical strengths and challenges, and consider your next steps in planning for instruction. Consistently using formative assessment and focusing on strength-based teaching and learning will allow you to look beyond procedural proficiency to connecting conceptual understanding and the Mathematical Practices. You can then identify and focus on students' strengths; determine and diagnose the source of occasional or frequent errors, including partial conceptual understandings; and describe and show how students can represent and explain what they know. Your instructional intent with math intervention and tutoring is to identify students' prior understandings while considering tentative and sometimes fragile or partial conceptualizations and common hiccups.

All of us have misinterpreted something, regularly made a particular computational error, or just didn't get it at first! Many of these occasions would fit into our description of a "common hiccup." Maybe it's not addressing the place value of a number (e.g., saying 27 for the number 2.7) or determining the area of a figure rather than the perimeter. Within an intervention lesson or related activity, we encourage you, as the instructor, to identify "common hiccups" in the students' discussion or represented in student work. We suggest avoiding treating instructional barriers or common hiccups as errors or mistakes. Instead, you can use such hiccups to generate a discussion that pulls from what students know and helps them make logical conjectures from past information that doesn't always make sense. As noted, some of these conjectures may not be "mistakes" but show the movement toward finding more relationships, such as seeing the subtle relationship between a square and rectangle or comparing decimal and whole numbers where students initially believed the greater the number of digits a quantity has, the greater the value.

You may occasionally see hiccups emerge through examples and nonexamples on a student's learning path to building generalizations (e.g., van Garderen et al., 2020). The same idea about probing common hiccups is explored when you see if students can hold firmly to an idea when a "fooler" or distractor is introduced. If students could consider and reason about examples and nonexamples, that could ensure what Hattie et al. (2016) noted as an important distinction between surface and deep understanding. Therefore, we suggest considering the following approaches to reaching that goal. Students have achieved a deep understanding if they can

- state their understanding of the problem situation and what is being asked,
- verbalize thinking,
- recognize an error (their own or someone else's),
- describe how someone else solved a completed problem, and
- "teach" the teacher or another peer a mathematical idea.

As discussed in Chapter 3, feedback will be an important consideration in planning, teaching, and assessment. We suggest using feedback positively. Feedback can help you truly define students' next learning steps and shape your next planning steps (e.g., Fennell et al., 2024, p. 12). To obtain this goal, you are encouraged to regularly think about and provide your students with feedback, consider and present opportunities for them to provide you with feedback (perhaps using an interview or Show Me prompt), and provide opportunities for your students to

---

A **common hiccup** is neither a misconception nor a mistake. Instead, it is what research and practice point to as a common barrier to students' accurate thinking about a concept or skill. We don't want to ignore these hiccups, and in fact, we want to fend them off.

provide feedback to each other. Such formative assessment and related feedback demonstrate that you assess as you teach within the math intervention lessons and the general education classroom. When students provide feedback to one another, they also rehearse ideas. That rehearsal increases the likelihood that they will share their ideas with you or the rest of the group. In essence, the deliberate linkage between strengths-based teaching and learning and the everyday use of formative assessment promotes assessment as a means of student advocacy.

How Can We Communicate Student Progress?

Communicating the intent, influence, and impact of your mathematics intervention effort will be an important responsibility. This process will include how you determine and document student progress. Communication will also involve sharing these results with family members and collaborating teachers. The results may be particularly shared among general education and special education colleagues who regularly interact with your math students in math intervention, school principals, and others (e.g., math coaches, paraprofessionals).

Figure 4.5 provides an example of a record-keeping tool for student growth during interventions. You may use or adapt this Intervention Update tool as you regularly anticipate, look for, and then document student progress. You can use the tool to enter your comments and student-related work samples. These samples are necessary not only to consider your next steps instructionally but also to provide varied indicators of progress to support your conferences. These conferences may include students' family members, administrators, and intervention colleagues. You can also use such record-keeping and progress-monitoring tools to assist in the required documentation for your RtI program or IEPs.

Figure 4.5 Intervention Updates

Communicating Progress—Math Intervention Look Fors

Student Names	Math Focus	Observation Comments	Interview Comments	Show Me Comments	Progress Summary	Next Steps

This printable is available for download at **https://companion.corwin.com/courses/ProactiveMathIntervention**.

What Are Other Considerations We Need to Think About?

As discussed in the following sections, other critical elements are involved in planning an effective intervention program.

Staffing

Who will be providing math intervention or tutoring in your school? The general education classroom teacher? Special education teacher? Title I teachers? Math specialist/coach? Interventionist? Paraprofessionals? Some or all of the above? Collaboration will be most important at the planning level, during math intervention lesson implementation, and related to student and program progress. Collaboration will also be important as progress is determined and shared with others, including your school's administrative staff, family members of students in intervention, and others.

Time and Location

Will your mathematics intervention program exist within individual grade-level classrooms in conjunction with the daily math lesson? Will mathematics intervention opportunities be offered outside of the daily math lesson time, perhaps in another location, classroom, or area of the school? What will students miss, if anything, when attending the intervention? Research suggests that given sufficient learning opportunities and time to learn, students with initial challenges can become students with strong understanding. However, they may need up to five times as long as their peers to master content (Usiskin, 2007). This finding suggests that the most effective interventions layer on support, providing added learning opportunities for students (Larson et al., 2012).

If we remove students from grade-level math instruction to receive interventions, we risk widening the same learning gaps we have tried so hard to narrow. If a pull-out structure must be used, it's important to find an appropriate time when the intervention will not interfere with other mathematics learning opportunities. Our point is that the planning for the timing and location of the math intervention opportunities is essential to student success.

Program Size

It will be important to consider various factors when determining appropriate limits for the number of students in your intervention program. How many staff will be involved (see the first point about staffing)? When will the intervention lessons happen? What data will be used to identify students for the intervention, and when will those data be collected? Each of these factors will impact the size of your intervention program. Suppose only one person is available to teach the intervention. How will you ensure that person can plan and maintain communication and collaboration with the classroom teacher(s)? What if it is the classroom teacher who also provides the intervention program? When and where will that take place? If the intervention program is offered as a pull-out program, how will you ensure that participation doesn't exclude students from opportunities to learn grade-level mathematics content? Keeping in mind that the goal is to move students out of intervention, not move more students in, what data will you collect to analyze student progress? At what point will students "graduate" out? Balancing these critical factors will help to ensure that your intervention program is well situated to Prime students for progress.

Instructional Materials

What materials (e.g., concrete, semiconcrete, and abstract representations) will you create, adapt, and use within your math intervention lessons? How can they be easily accessed for use with math intervention learning opportunities? To what extent will these materials be precisely the same as materials used within general education class lessons? How will they support the math tasks and related instructional activities presented in Part 2 of this book? How will online activities and/or programs be used within your math intervention opportunities? These questions and the issue of materials are far from being trivial. Any online search will suggest that math intervention is for sale. We do not suggest avoiding materials designated explicitly for RtI or MTSS programs. However, we suggest remaining wary of the marketing and related advertising promises such programs may provide. As previously noted, successful interventions are not something you buy but instructional strategies you try! Ongoing collaboration with all educators in delivering your school's math intervention will be an important element of the materials-related decisions behind your cohesive math intervention program.

Successful interventions are not something you buy but instructional strategies you try!

Instruction and Assessment

Finally, we recommend considering this question each time you and your colleagues come together to use the intervention planning model (Figure 4.1) discussed earlier in this chapter: How will you connect student strengths to formative assessment techniques when determining student progress with your responsive instructional planning and implementation?

Where Do You Begin Priming Intervention Lessons?

We hope you feel more knowledgeable about what we consider to be the most important facets of a successful intervention program: a proactive approach to learning content through Priming, a strengths-based view of students' skills and abilities, regular and productive use of formative assessment and feedback, the need for team collaboration and communication, and the right kinds of "doing math" tasks for the job. However, you may question what you should actually aim to do now that you have this information. In answer to this question, we have prepared some helpful steps to follow.

Here we go! Begin your math intervention session with "*doing* math" (see the 43 Tasks, which include more than 100 activities, in Part 2). Get your students engaged and hooked while they are ready for action. If you want students to practice basic facts for 5 minutes daily, save that activity for the end of the intervention session. Otherwise, you start with a passive launching pad, which does not work best to encourage active thinking (Liljedahl, 2020). Introduce the "doing math" task verbally and have a visual on a screen or paper to help support students' attention and working memory. You may want to record, list, or visually represent information such as a table of particular elements of the problem or directions for the activity so that reading-related issues will not get in the way. Note that most of the suggested tasks are within a series of variations lasting more than 1 day. The other exciting component is that these tasks can be repeated multiple times, especially the games. The tasks have been developed to focus on the following:

- having foundational mathematical concepts and practices that represent essential "must have" knowledge components of the grade-level standards
- using CSA representations that tap into the current depth of understanding and build from there
- developing mathematics vocabulary that has been taught but may need to be refreshed (not taught in isolation but as needed during instruction)
- having a sequence of activities that ramp up to the new mathematics content
- asking probing questions that prompt thinking, such as "Is there another way to do this?"
- engaging students in the assessment process, such as having students talk through or use "show me" to identify their solution path(s), followed by asking why such paths were used
- having home connections, which provide quick tips for families to continue to engage their child's math skills and deepen math concepts through games and easy-to-understand and apply at-home activities including some used in the intervention sessions

What Can We Learn From Listening to the Voices of Others Trying Priming in Their Schools?

One of the authors interviewed individual teachers as well as groups of teachers who have implemented the proactive Priming practices detailed in this book. The following excerpts come from a selected sampling of their wisdom of practice:

- Some teachers were skeptical after the PD on the Priming approach. They'd say instead, "But I've always done it this way." We heard crickets for a while but then found that teachers were using the ideas from the PD in their instruction and changing their small group practices. We thought it might be tricky to get "buy-in," but then they felt this was a move in the right direction. I found that the people I would call the strong math teachers—and you would likely call the early adopters—really helped.
- I told the kids that they were getting "sneak peeks" in their intervention sessions that would be helpful in their standards-based lessons. Then they started to feel like an expert with a leg up as I told them I'd like them to take the lead on the first day (I teach both the interventions and the

general classroom lessons). They latched on to that idea. Then during the session on the new math topic when the familiar material from the intervention was shown, they'd say "Oh, I know what to do here." The students' confidence grew in leaps and bounds.

- In the past, we were planning by pulling skills from here and there as our topic lessons. We'd just try to keep ahead of it. But this summer, we did all the work to build in Priming by giving kids an early preview at the skills they needed to respond well to the lesson. We created documents with lessons listed from the curriculum for all the fifth-grade teachers—special education teachers and general education teachers—to use daily including Priming suggestions by cycle.

- Previously, the kids were sitting in the room when the math lesson was being taught, but they were watching the world go by. Quiet didn't mean they understood. They were trying to hide when we were doing math. Now they are saying, "This looks like something I've seen before."

- Some thought that students might be confused by working on two math topics at once as that happens sometimes with Priming. Like we are doing fractions right now but kids are being Primed on volume. But it was a nonissue. Kind of like me reading a book orally to the kids while also having them read different books in their reading groups.

- The strengths-based shift was different. We were ready to do what we always did, which was to identify kids from their tests, say they needed interventions, send home letters to parents when all of a sudden, we hit the brakes. By focusing on the unit assessment after Priming, some students graduated out, some others came in for the next topic. Intervention went from a life sentence with a label of being called a Title 1 kid to no one talking about that. We recognized that most students didn't need all the skills—just this particular chunk. There is now a better use of our time and the students' time, and we completely lost track of who are the "strong kids" and who we expect to struggle.

- Our superintendent came to the roll-out for the Priming Approach. He was asking people behind the scenes what they thought. He went over to one teacher—a person who could be skeptical and was counted on for her honesty. At first, we worried. What would she say? But then she said, "This just seems to make sense." We were thrilled.

- When the team came back from the session on the Priming Approach that was presented by one of the authors of this book, they were all pumped as they shared their learning. One thing that resonated was that they were all working so hard to make what we were doing in math the same as what was being done in literacy. They felt validated by hearing that it's not the same—it can't be the same—as teaching math isn't teaching reading. They thought, "We were right." So, we changed our progress monitoring to focus on the Tier 1 unit assessments as we have no state requirement of giving multiple progress-monitoring instruments.

- We know that changing practice is hard. We had three camps during the PD—the Yes, I can do this group, the OK this is what the "boss" is telling me to do and I'm compliant, and the third, I liked what I was doing. But after they started looking at the improvement in the student assessments and the student work they were getting, they changed their thinking. All students didn't master the content right away, but there was evidence that students who had a profile of not historically benefiting from the interventions were showing real progress. It's motivating. Lots of people are very excited.

- Teachers need structures to support a different system like Priming. They are simply too busy to fully engage in these shifts, especially in the beginning, without leaders supporting them differently. So we are planning for that.

And Now, It's Your Turn!

We hope you are ready to engage fully in our math tasks in Part 2 of the book. The focus on strengths-based teaching and learning and formative assessment can help you plan, develop, and implement math intervention/tutoring opportunities for your students. So go for it!

Reflection Opportunities for Chapter 4

1. What are examples of nonnegotiable mathematics concepts at your grade level(s)? Share your examples with grade-level or school colleagues. How do you build consensus with your colleagues?

2. What are the challenges, perhaps dangers, of pulling instructional ideas off the web each week?

3. What is the difference between preteaching upcoming lessons in intervention sessions and Priming students for the new content using interventions?

4. How will you use formative assessment to monitor student progress? How will you communicate such progress to those in any way connected to the mathematics learning of your students in math intervention?

5. How would you describe the difference between student errors or mistakes and a misconception? Provide some examples, including common hiccups you regularly encounter, and discuss these with your grade-level or school-based team.

6. How would you have changed Dawn and Chonda's dialogue about Priming for the perimeter lesson? What might you have added? Where did you want to hear more?

7. How will you find the time to plan intervention opportunities, and who will you involve in the planning?

8. Why is Priming students for the upcoming new mathematics content a good way to develop their sense of self as a person who can do math?

9. What are your plans for providing feedback to your math intervention/tutoring students? And how will you provide opportunities for them to provide feedback to you and each other?

PART 2
PUTTING PRIMING INTO ACTION

Mathematics intervention starts by identifying individual student strengths, and then addresses, daily, the planning to both support and supplement student learning.

—An elementary math coach

Part 2 of this book contains a collection of 43 tasks by mathematics topic with each having variations that, together, total more than 100 Priming Activities. Teachers who have used this approach and these tasks found them to be most helpful for developing the kinds of engaging Priming interventions we described in earlier chapters. In the coming pages, you will find a variety of instructional formats for these activities including games, hands-on experiences, and bite-sized lessons. We know that no two intervention programs are alike, and, as discussed in Chapters 1 and 4, your instructional periods may vary in time and location – meaning some intervention or tutoring sessions may be longer (45 minutes or more), while others may feel all too short (less than 30 minutes), with some taught outside the general education classroom with another teacher and others taught within the classroom with the classroom teacher. Our hope is that the variety of tasks in this collection of Priming Interventions will provide you with that "just right" adaptable resource no matter what the situation at your school. Some of the activities in this section are brief and may be completed in a single intervention session. You may choose to implement other activities across multiple intervention sessions. For example, some tasks and their multiple variations can be the intervention curriculum for a week or more. Remember they can be repeated too!

However delivered, take your time. Unlike covering content across the curriculum to meet every standard, you are building a strong and solid foundation for the future. Have students use the manipulative materials, make a "math sketch," and write all the corresponding expressions and equations. Ask the questions we suggest. Discuss students' work, their solutions, and their thinking in the intervention group. Use the guidance in these tasks to build carefully, slowly, and purposefully. You will note that some task variations blend seamlessly into each other in growing sophistication; others will add a distinctive but connected shift to a higher grade level within a topic of study. Select the variation that focuses on the prior knowledge you are working to shore up for your student group.

Use the Interventions at a Glance feature in Appendix C of the companion website to find areas of need for your students and the intervention tasks or variations that are a match. You'll find that most of the activities here offer "gamification" so students can play them over and over again. And they will want to! Also, doing the same activity again builds students' confidence and prepares them to take the lead as the new topic on grade-level work is introduced to the whole class.

We try in each case to give you the resources you need to carry out a task or activity. Each task includes a list of suggested materials, including concrete manipulatives, digital resources, and custom printable resources to support your instruction. We recognize that printing budgets are tight, so printable resources are designed to be used and reused! Consider using sheet protectors and dry-erase markers for student activity pages or laminating card decks for games. Our printable materials may require cutting, back and front printing, and following a few easy directions. You may download printables at https://companion.corwin.com/courses/ProactiveMathIntervention.

After the suggested materials for each task, we also include suggestions for related children's literature that might be used to support (or introduce) the mathematical context (Appendix A on the companion website). Suggested books shouldn't be read cover to cover. Pick and choose the parts that center on the mathematical situations and draw students into the intervention activity through a problem-solving approach. We use sticky notes to mark those portions to read. Use excerpts judiciously to create scenarios that engage the students and build a shared context. Students may be inspired to read the whole book themselves at a later date! A bibliography of these books is available on our resources website (Appendix A). If you don't have the book, many are available as read-alouds on the web.

As you explore the instruction provided, you will also notice a wide variety of concrete, print, and digital resource suggestions shared in the Priming Tasks (Appendix D on the companion website). As first discussed in Chapter 1, the development and connection between and among concrete, semiconcrete, and abstract representations (CSA) is critical to Priming Interventions and students' long-term learning. Please be sure to use concrete materials and instruct students on how to properly handle them. Time spent now with supporting students in building mental models of concepts through CSA will pay off later because you won't need to expend time in continuous remediation and backtracking. Students must not see these physical models as "toys" (avoid using the word *play* when working with them), but instead, they should be branded as thinking tools. This is no different from the other representations we encourage.

Also note that suggestions have been provided for your use of the observation, interview, and Show Me formative assessment techniques from Chapter 3, using the following icons to identify suggested use of these techniques:

PART 2. PUTTING PRIMING INTO ACTION

The tools provided in Chapters 2, 3, and 4 will be helpful to you as you plan for and implement the Priming Intervention Tasks, assess and monitor student progress, and consider student feedback. This includes providing feedback to them, engaging them in providing feedback to you, and providing opportunities for your students to provide feedback to each other.

So again, welcome to 43 different tasks that are inclusive of several domains of study such as number and operations, algebraic thinking, measurement, geometry, and data. Accounting for the variations within the tasks, you have a series of unique learning experiences for use during intervention sessions to build students' strengths and thereby their success. Our aim is to support you in your work to ensure that each and every student is Primed for success in the grade-level classroom. Explore and enjoy.

Intervention Tasks

CHAPTER 5

TASK 1: Mystery Number Riddles	80
TASK 2: Come on Down!	88
TASK 3: Trading Places	95
TASK 4: Adding it Up	105
TASK 5: What's the Difference?	114
TASK 6: Targeted Sum	128
TASK 7: Targeted Differences	134
TASK 8: Valuable Digits	140
TASK 9: Picture This	149
TASK 10: Balancing Act	158
TASK 11: Escape Room	166
TASK 12: Things that Come in Groups	170
TASK 13: Garden Spaces	180
TASK 14: Build it Bingo	187
TASK 15: Rearrange It	197
TASK 16: Speed Estimator	209
TASK 17: Multiplication Chains	220
TASK 18: Magic Pot Patterns	234
TASK 19: Fraction Frenzy	241
TASK 20: Cake Time	250
TASK 21: Line 'Em Up	260
TASK 22: To Be or Not to Be: Equivalent Fractions	270
TASK 23: Fraction Equivalence through Quilts and Fringes	276
TASK 24: Sport Stats	284
TASK 25: Fraction Zap	290
TASK 26: Fraction Sums	295
TASK 27: Fraction Multiplication: Fundraiser	304
TASK 28: Fraction Multiplication: Farmland	308
TASK 29: Fraction Races	313
TASK 30: Shape Sort	320
TASK 31: Alike and Different	324
TASK 32: Shape Shifting	331
TASK 33: Shapes: Build It!	337
TASK 34: What's My Rule?	343
TASK 35: Counting Coins	352
TASK 36: Smart About Money	359
TASK 37: Measure Up!	365
TASK 38: Comparing with Measurement	369
TASK 39: On the Edge	373
TASK 40: Field Day	377
TASK 41: Broken Clock	382
TASK 42: Tip and Plot	388
TASK 43: Picture Perfect	396

TASK 1

Mystery Number Riddles

General Objective:

Students will work on place value, renaming, rounding, and comparing three-digit numbers.

This activity might be used with:

Students who are able to (prerequisite knowledge):	Students who are Primed to (getting ready to learn):
• understand 2- and 3-digit place value. • recognize that 100 can be thought of as a bundle of ten 10s—called a "hundred." • mentally find 10 more or 10 less than the number, without having to count; explain the reasoning used.	• read, write, and identify the place and value of each digit in a whole number, with and without models. • count within 1,000; skip count by 5s, 10s, and 100s.

Materials:

- Place value blocks
- Index cards
- PRINTABLE: Mystery Number Riddle Slides
- PRINTABLE: Hundreds Chart
- PRINTABLE: Place Value Mat

 Printables for this task are available for download at **https://companion.corwin.com/courses/ProactiveMathIntervention**.

Recommended Children's Literature:

- *Too Many Frogs!* – Hasset, 2011
- *Zero, Zilch, Nada: Counting to None* – Ulmer, 2010

Task Overview:

In this activity, students will solve mystery number riddles by using place value concepts and strategies such as adding 10, subtracting 10, and skip counting by tens. Students will work in pairs, using tools like place value blocks and hundreds charts to visualize and explain their answers. After several rounds, students will create their own mystery riddles, helping reinforce their understanding of place value and number relationships.

Select (or create) a variation that focuses on the prior knowledge you are working to shore up for your student group.

Variation 1

Learning Target: Students will solve mystery number riddles using place value clues and by adding 10, subtracting 10, and skip counting by tens.

Variation Directions:

Provide each pair of students with a set of riddle cards along with place value blocks and a hundreds chart. In each round, a student will choose a riddle and solve it using place value blocks or a hundreds chart. They will explain their solution to their partner then switch roles. Students will continue taking turns as time permits.

Figure 5.1.1 Mystery Number Riddle variation 1 example

Partner A selects the mystery number riddle:	Partner A explains, "I can use the hundreds chart! I will start at 75 and count 10 more. I notice that I moved down one row from 75. The mystery number is 85!"
Mystery Number Riddles (ten more, ten less) I am 10 more than 75. Who am I ?	 | 2 | 63 | 64 | 65 | 66 | 67 | | 2 | 73 | 74 | ⓞ75 | 76 | 77 | | 2 | 83 | 84 | ⓞ85 | 86 | 87 | | 2 | 93 | 94 | 95 | 96 | 97 |
Partner B selects the mystery number riddle:	Partner B explains, "I can use the hundreds chart! I will start at 42 count back 10 less. I notice that's one row up. The mystery number is 32!"
Mystery Number Riddles (ten more, ten less) I am 10 less than 42. Who am I ?	| 21 | 22 | 23 | 24 | 25 | | 31 | ⓞ32 | 33 | 34 | 35 | | 41 | ⓞ42 | 43 | 44 | 45 | | 51 | 52 | 53 | 54 | 55 |

(Continued)

(Continued)

Partner A selects the mystery number riddle:	Partner A explains, "I need to end at 45 after you give me 10. I'll try 35. Look, if I start at 35 and you give me 10 more, it will be 45. The mystery number is 35!"
Mystery Number Riddles (ten more, ten less) If you give ten more to me, I am 45. Who am I?	<table><tr><td>23</td><td>24</td><td>25</td><td>26</td><td>27</td></tr><tr><td>33</td><td>34</td><td>35</td><td>36</td><td>37</td></tr><tr><td>43</td><td>44</td><td>45</td><td>46</td><td>47</td></tr><tr><td>53</td><td>54</td><td>55</td><td>56</td><td>57</td></tr></table>
Partner B selects the mystery number riddle: If you take ten from me, I am 37. Who am I?	Partner B explains, "I can start at 47 and when you take 10 away, it will be 37. The mystery number is 37!" <table><tr><td>25</td><td>26</td><td>27</td><td>28</td><td>29</td></tr><tr><td>35</td><td>36</td><td>37</td><td>38</td><td>39</td></tr><tr><td>45</td><td>46</td><td>47</td><td>48</td><td>49</td></tr><tr><td>55</td><td>56</td><td>57</td><td>58</td><td>59</td></tr></table>

After playing several rounds, ask students to create their own riddles for mystery numbers on an index card. Add these student-generated riddles to the collection with the answer on the back.

Engage in Math Discourse (Make the Mathematics Visible):

Show Me: As students work, encourage them to notice important relationships and push on early and partial understandings.

- Monitor students to see how they are making sense of 10 more and 10 less riddles. Ask students to show what the riddle means using place value blocks and/or on the hundreds chart.
- Show me how you are making sense of the riddles using the tools. Which tool do you prefer? Why?

Interview: After the activity, bring the intervention group together for a discussion. Suggested prompts:

- What pattern did you notice moving between 10 more and 10 less on the hundreds chart?
- How can the hundreds chart or base ten materials help you with the riddle? How would that help you solve it?

CHAPTER 5. Intervention Tasks   83

Bridging Prompt (Prompt for Classroom Teacher to Use During New Lesson Content):

- Share a mystery number riddle with the class.
- Ask: How can the hundreds chart or base ten materials help you with the riddle? How would that help you solve it?
- Select students who participated in the intervention group to share their thinking with the class.

Variation 2

Learning Target: Students will rename numbers by place value.

Variation Directions:

Begin this variation by playing several rounds with a set of Mystery Number Cards Set #2—Renaming numbers.

Figure 5.1.2 Mystery Number Riddle variation 2 example

Partner A selects the mystery number riddle:

Mystery Number Riddles (rename me)

I have 30 ones and 12 tens. Who am I?

Partner A explains:
- 30 ones equals 3 tens
- 12 tens equals 120
- 12 tens plus 3 tens equals 15 tens or 150

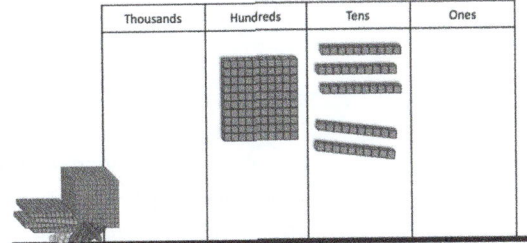

Partner B selects the mystery number riddle:

Mystery Number Riddles (rename me)

I have 24 tens and 8 ones. Who am I?

Partner B explains:
- 24 tens equals 240
- 8 more ones equals 248

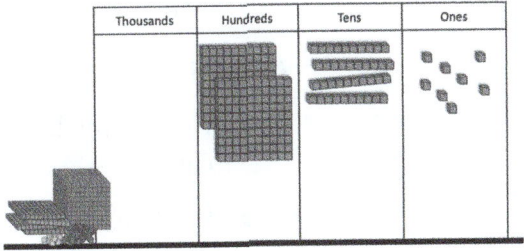

(Continued)

(Continued)

Partner A selects the mystery number riddle:	Partner A explains:
Mystery Number Riddles I have 2 thousands, 4 tens, and 3 ones. Who am I? 	• 4 tens equals 40 • 2 thousands plus 4 tens plus 3 ones equals 2,043 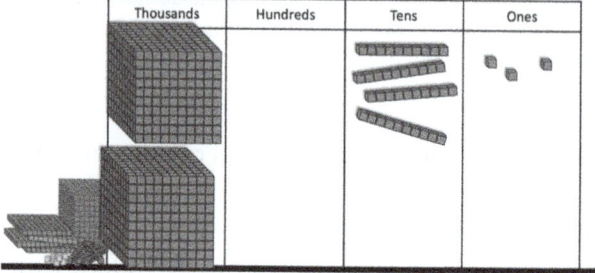
Partner B selects the mystery number riddle:	Partner B explains:
Mystery Number Riddles I have 4 thousands, 4 hundreds, 4 tens, and 3 ones. Who am I? 	• 4 tens equals 40 • 4 thousands plus 4 hundreds plus 4 tens plus 3 ones equals 4,443 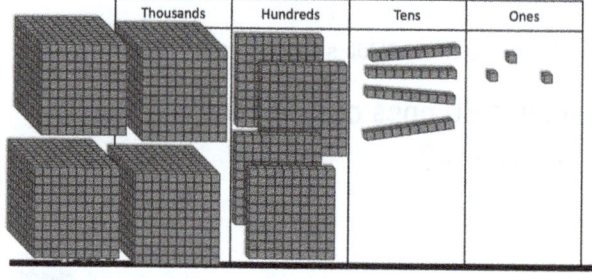

After playing several rounds, ask students to create riddles for mystery numbers on an index card. Add these student-generated riddles to the collection with the answer on the back.

Engage in Math Discourse (Make the Mathematics Visible):

As students work, encourage them to notice important relationships and push on early and partial understandings.

Show Me:

▸ How do you know the number based on the ones, tens, and thousands clues?

▸ Can you break the riddle down step by step by using the materials on the place value chart?

Interview:

- What is the role of each digit in the number (ones, tens, hundreds, etc.)? How do they work together to make the full number?
- What if you don't have any tens in a number in the hundreds? How do you write that?

Interview: After the activity, bring the intervention group together for a discussion. Suggested prompts:

- How does the place value mat help you create a new riddle?
- What strategies did you use to break down the number into its place value parts? How did that help you solve the riddles?
- How does renaming numbers help you better understand the value of each digit in a number?

> ***Bridging Prompt (Prompt for Classroom Teacher to Use During New Lesson Content):***
>
> - Share a riddle created by the students who participated in the intervention group. Have the authors of the riddle come up and pose it to the class. Encourage the class to use base ten materials and the place value mat to figure out or check their answer.
> - Have the class work as pairs to create new riddles. When they are finished, they can trade with another pair to work on their riddle.
> - Ask: What pattern do you see across the place value columns? How many ones equal 1 ten? How many tens equal 1 hundred, and how many hundreds equal 1 thousand?

Variation 3

Learning Target: Students will solve mystery number riddles by rounding numbers.

Variation Directions:

Begin this variation by playing several rounds with a set of Mystery Number Cards Set #3—Rounding numbers. Students should use the base ten materials and their place value charts. Before you begin, discuss the convention for rounding involving situations when you are considering a 5. Although it is halfway and it seems you can go up or down, the convention in mathematics is that you would round to the next number in the sequence.

Example:

Partner A selects the mystery number riddle:	Partner A explains how 488 could be the mystery number because it can be rounded to 500 and all the digits are even.
I am thinking of a number that, when rounded to the nearest hundred, is 500. The digits in the number are all even. What could my number be?	There could be many other answers here: 488, 486, 484, 482, 480, 468, 466, 464, 462, 460
Partner B selects the mystery number riddle: I am thinking of a number that, when rounded to the nearest hundred, is 400. The digits in the number are all odd. What could my number be?	Partner B explains how 399 could be the mystery number using the place value chart. There could be many other answers here: 399, 397, 395, 393, 391, 379, 377, 375, 373, 371, 359, 357, 355, 353, 351
Partner A selects the mystery number riddle: I have some number of hundreds, 7 tens, and 4 ones and can be rounded to 400. Who am I?	Partner A explains how 374 is the mystery number because it can be rounded to 400 and has some number of hundreds, 7 tens, and 4 ones.
Partner B selects the mystery number riddle: I have 5 tens and 8 ones and can be rounded to 500. Who am I?	Partner B explains how 458 is the mystery number because it can be rounded to 500 and has 5 tens and 8 ones.

After playing several rounds, ask students to create riddles for mystery numbers on an index card. Add these student-generated riddles to the collection with the answer on the back.

Engage in Math Discourse (Make the Mathematics Visible):

Interview and **Show Me:** As students work, encourage them to notice important relationships and push on early and partial understandings.

- What do you need to consider when rounding up to the next hundred?
- How do you know the number you selected rounds to the nearest hundred as 500 (or 400)?
- What does the place value of the tens and ones digits tell you about how to round to the hundreds place?
- Can you explain why sometimes, more than one number can fit the riddle? What other numbers would work for the same clue? Do all clues have more than one answer?
- How do you use the hundreds chart or base ten materials on a place value chart to find the rounded number? Can you show your thinking using the tools?
- How does the rounding strategy (round up or round down) help you find the correct number? What is the process for deciding how you round a number?

Interview: After the activity, bring the intervention group together for a discussion. Suggested prompts:

- What patterns did you notice when rounding numbers to the nearest hundred?

- How did the even or odd digits affect the number you selected? Can you explain the connection between the digits and the rounded number?

- How did you approach creating your own mystery number riddle? What clues did you find most helpful?

- When rounding to the nearest hundred, what role does the tens place play in determining whether the number rounds up or down? What role does the ones place play?

- What strategies or tools helped you solve the riddles more easily? How did using a place value chart or place value blocks support your understanding of rounding?

- Looking at the collection of riddles, can you find a number that could fit multiple riddles? How is that possible?

Bridging Prompt (Prompt for Classroom Teacher to Use During New Lesson Content):

- Share a riddle created by the students who participated in the intervention group. Have the authors of the riddle come up and pose it to the class. Encourage the class to use base ten materials and the place value mat to figure out or check their answer.

- Have the class work as pairs to create new riddles. When they are finished, they can trade with another pair to work on their riddle.

- How do you use the hundreds chart or place value chart to find the rounded number?
Can you show your thinking with the tools?

TASK 2

Come On Down!

General Objective:

Students will use semiconcrete representations to make sense of concepts related to place value and decimal notation.

This activity might be used with:

Students who are able to (prerequisite knowledge):	Students who are Primed to (getting ready to learn):
• describe the value of the base ten materials including selecting the right materials when given a number. • compose and decompose numbers by place value.	• write numbers in expanded form. • recognize that the value of each digit is 10 times greater for each place moved to the left and one-tenth as great for each place moved to the right. • compose and decompose numbers by place value. • round numbers to any place. • use decimal notation to represent decimal fractions.

Materials:

- Number cards with place value names at the bottom to match numbers to be called out
- Base ten materials
- PRINTABLE: Place Value Column Labels (to use on the floor)
- PRINTABLE: Teacher Support for Come On Down!

 Printables for this task are available for download at **https://companion.corwin.com/courses/ProactiveMathIntervention**.

Recommended Children's Literature:

- *House With 100 Stories* – Iwai, 2023
- *Fergus and Zeke and the 100th Day of School* – Messner, 2021
- *A Thousand Star Hotel* – The Okee Dokee Brothers, 2017

- *Earth Day Hooray?* – Murphy, 2004 (place value situations with larger numbers)
- *Phantom Tollbooth* – Juster, 1961 (when Milo visits Digitopolis and meets 0.5 of a child) (decimals)
- *Piece = Part = Portion: Fraction = Decimal = Percent* – Gifford and Thaler, 2008 (decimals)

Task Overview:

(This task adapted from a lesson by the late Sandy Cohen, Baldwin, New York)

This task is geared to understanding that in our numeration system, place matters! The position of a digit in a particular place defines its value.

Select (or create) a variation that focuses on the prior knowledge you are working to shore up for your student group.

Variation 1

Learning Target: Given a number read out loud, the students will make concrete images of numbers by standing up and forming the number in the instructional space.

Variation Directions:

Start with a number that is at the correct level of difficulty for your students. Each student in the group should be given one or two cards with a numeral and a place value name on it such as a large 2 with hundreds underneath or 3 tens. Call out a number, and the components of the number must come to the front of the room and stand in the appropriate place to make the number named (see Figure 5.2.1). Make sure you read the number without using the word "and." For example, say one thousand, three hundred, sixty-five instead of saying one thousand, three hundred, *and* sixty-five. The only time you use the word "and" when reading a number is to indicate there is a decimal.

Figure 5.2.1 Students at the front of the room holding cards that were a part of the number called out (1,365)

Source: istock.com/Nahhan

As students are coming forward to be in the number that was called out, other students are creating a model of the number with base ten materials on a place value mat in front of them or with magnetic pieces on a white board or using a projection with virtual base ten materials. To differentiate for those students with cards who may need support, start with placing the names of the place value columns on the floor so students know where to stand (also matches the place value name on the card for confirmation). Choose numbers that match the support students need to be Primed for upcoming lessons on place value or computation. Consider having ones, tens, hundreds, and thousands. Students come forward (mistakes will be made), and when positioned, the number is read and written on the whiteboard by one of the students. Then another number is called. Start removing the place value column labels on the floor as students become more familiar, first taking away the ones and moving to higher place values from there.

Other adaptations: Don't avoid using thousands even if it is not your grade level, just don't assess students on it; also consider presenting an addition or subtraction problem, and the correct answer must come up.

Engage in Math Discourse (Make the Mathematics Visible):

Observation: As students work, encourage them to notice important relationships and push on early and partial understandings.

- In this activity, it is easy to predict that there will be times when a student comes up, thinking they are in the number that was called out, but they are not. Mention from the start that this might happen to reduce any concern. But that scenario is a perfect opportunity for a conversation about what number they do have and what number is needed.

- If a number doesn't come up to the front, take that opportunity to discuss with the group what value is missing and in what place. Actually, that is the time to repeat the number orally and have a student write it on the board or document camera. Then, students can more easily recognize what is needed to complete the number.

Interview: After the activity, bring the intervention group together for a discussion. Suggested prompts:

- In this intervention, we want them to generalize that the value of the number is dependent on where it is positioned. We also want them to recognize that moving to the left in each place value the number represents 10 times more. Ask students what would happen if we switched two numbers. Have them exchange the place value blocks that show what they had and what they have after the shift. These are important ideas to unpack.

> ***Bridging Prompt (Prompt for Classroom Teacher to Use During New Lesson Content):***
>
> - Call out a number (with number cards held by students who participated in the intervention group) and have them come to the front of the room to form the number. Ask another student from the intervention group to place the floor labels down where they belong. Have a student write the number on the board once it is in place and read it out loud. Then, call out another number so the whole class can participate.

Variation 2

Learning Target: Students will be able to round to the nearest ten or hundred.

Variation Directions:

Playing the game described in Variation 1, after a number successfully comes to the front, ask students to round to a given place value. For example, with the number shown in Figure 5.2.1, if rounded to the nearest ten, the students with 5 ones will return to their seat, and the student holding the 6 in the tens place will also sit down. Instead, 7 tens will come up partnered with 0 ones. People should move accordingly. If it hasn't already happened, ask them what happens to the person in the ones place.

Having a large empty number line available where benchmark numbers can be written can help decide whether they are moving to benchmark A or benchmark B when they round. Then repeat in words, "We are rounding to the tens place by changing 5 tens to 6 tens."

Engage in Math Discourse (Make the Mathematics Visible):

Observation: As students work, encourage them to notice important relationships and push on early and partial understandings.

- At first, students may wonder why you round to the next number when you have a five. Ask them to explain why people might find it hard to know what to do because it is "right in the middle."

- Remind students that there are some mathematics conventions or "rules" that we all agree to, and rounding up when there is a 5 is one of those conventions.

Interview: After the activity, bring the intervention group together for a discussion. Suggested prompts:

- What numbers in a place value column would cause you to round "down" to the next-lower number? What numbers would cause you to round "up" to the next-higher number?

> When do we want to round a number? [estimation] When do we not want to round a number? [need a precise answer]

Bridging Prompt (Prompt for Classroom Teacher to Use During New Lesson Content):

- Have several number cards from this task posted at the front of the room with "foolers" such as having both 3 tens and 3 hundreds as choices. Call out a number, and call on students who participated in the intervention group to come up and create the number at the front of the room while the rest of the group is working as pairs or in groups of three to create the number with base ten materials. Compare the results. Ask students how they know, for example, two tens rather than two ones is part of the number.

Variation 3

Learning Target: Students will be able to interpret place value situations when given the digits out of order.

Variation Directions:

Begin with students creating numbers (at the appropriate level for their needs) with base ten materials. Read the number to them while showing the written version, such as 124. Have them explain how they knew which materials to use to represent the number.

After several of these problems, students will be given the following visually as the teacher reads the problem:

Write the number that goes with:

4 ones, 2 hundreds, 6 tens

Let students jot down a response. Some may respond with the correct answer of 264, but others may answer 426. It is a common hiccup for students to lose track of the place values in this abstract variation and just work from right to left or left to right as written. Without saying which is correct, have two students who wrote down different options come to the document camera. Or, if working in a small group, have everyone take out the base ten materials that go with the numbers given (not the ones that match their answer) using a place value mat. Don't ask them to share their answers yet. What number do you have? How do you know? Some students will self-correct their previous answers, but then ask them, "Why was this problem a little tricky?"

Give several more mixed-up numbers "out of order" for students to model and write. Ask teams or pairs of students to write a similar "tricky" three-digit problem for another group, who will solve it using materials and the place value mat if needed. Pass clockwise. When done, students share their response back to the authors of the original problem for solution checking.

Engage in Math Discourse (Make the Mathematics Visible):

Observation: As students work, encourage them to notice important relationships and push on early and partial understandings.

- Ask students to write a mixed-up place value problem for the number 613. Observe what numbers they choose and how they organize them. Talk about what is most important to think about in figuring out how to write the correct version from a mixed-up number.

Interview: After the activity, bring the intervention group together for a discussion. Suggested prompts:

- Ask students how they should write the number that goes with 9 ones and 5 hundreds. See if they can generalize to the lack of a number of tens to be zero tens. Talk about the importance of zero in writing numbers and how it signals that in this case, there is no amount of tens.

	***Bridging Prompt (Prompt for Classroom Teacher to Use During New Lesson Content):*** • Ask students to write the number that is represented by: • 2 tens, 4 ones, 1 hundred • Bring up two students who participated in the intervention group who can create the base ten model on the board and two others who can write the number 124. Ask them to explain why this could be tricky at first.

Variation 4

Learning Target: Given a decimal number read out loud, the students will make concrete and abstract images of numbers by standing up and forming the number in the instructional space.

Variation Directions:

Now, we expand the intervention session to decimals and again give students one or two cards with a numeral and a place value name on it such as a large 1 with tenths underneath or 4 hundredths. Call out a number, and the components of the number must come to the front of the room and stand in the appropriate place to make the number named. Make sure when you read the number you do not use the word "point." It should be read as 1 and 66 hundredths (you now can say the word "and" because it represents that a decimal point is in the number). To differentiate for those students with cards, start with placing the names of the place value columns on the floor so students know where to stand (also matches the place value name on the card for confirmation). You can strategically select a student to hold the decimal point in place, or the teacher can hold it. Other students are creating a model of the number with base ten materials on a place value mat in front of them or with magnetic pieces on a white board or using a projection with virtual base ten materials.

Figure 5.2.2 Students at the front of the room holding cards that were a part of the number called out (1.66)

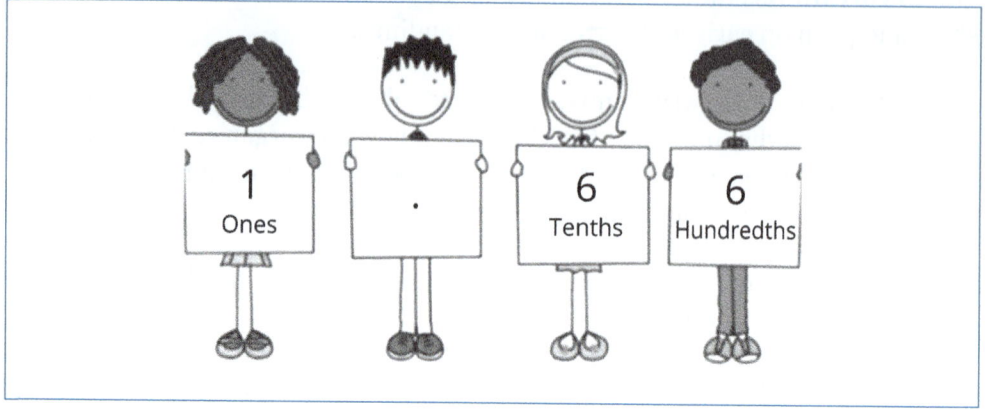

Source: istock.com/Nahhan

Engage in Math Discourse (Make the Mathematics Visible):

Observation: As students work, encourage them to notice important relationships and push on early and partial understandings.

- When we present numbers in various operations and call a 3 in the tens place "three" instead of 30, as in subtracting 3 from 5 in the tens place and not saying 3 tens subtracted from 5 tens, or 30 subtracted from 50, students can lose the value of the actual number. Observe if this makes it hard for them to interpret whether their answer is reasonable.

- Watch to see if students are able to generalize that the value of the number is dependent on where it is positioned.

Interview: After the activity, bring the intervention group together for a discussion. Suggested prompts:

- What happens when we have 0.5 and we need to round it to the nearest one (whole number)?

- What is the relationship between numbers in the tenths and hundredths? What about from the ones to tenths? What do you notice? Why do you think we say we have base ten materials?

Bridging Prompt (Prompt for Classroom Teacher to Use During New Lesson Content):

- Have several number cards from this task posted at the front of the room with "foolers" such as having both 3 tenths and 3 tens or 3 hundredths as choices. Call out a number, and call on students who participated in the intervention group to come up and create the number at the front of the room while the rest of the group is working as pairs or in groups of three to create the number with base ten materials. Compare the results. Ask students how they know, for example, two tenths rather than two tens is part of the number.

CHAPTER 5. Intervention Tasks   95

Trading Places

TASK 3

General Objective:

Students will develop understanding of place value relationships and equivalent representations of whole numbers.

This activity might be used with:

Students who are able to (prerequisite knowledge):	Students who are Primed to (getting ready to learn):
• count within 10 • organize a collection into groups for counting (e.g., groups of 5 or 10)	• regroup (both grouping and ungrouping) between place values (e.g., group 10 ones to equal 1 ten, ungroup 1 ten to equal 10 ones, etc.) • read, write, and represent whole numbers in different ways • generate equivalent representations of whole numbers

Materials:

- Jumbo-sized place value trading chart made on disposable plastic tablecloth or butcher paper (see PRINTABLE directions)
- Place value blocks
- Marker boards
- PRINTABLE: Place Value Mat
- PRINTABLE: Directions for Making a Jumbo Place Value Chart
- PRINTABLE: Place Value Cards
- PRINTABLE: Equivalent or Not? Activity Pages

 Printables for this task are available for download at **https://companion.corwin.com/courses/ ProactiveMathIntervention**.

Recommended Children's Literature:

- *The King's Commissioners* – Friedman, 1994

Task Overview:

In this activity, students will generate physical models of whole numbers on a jumbo-sized place value chart (see printable directions) and equate those models to abstract numerical notations, including standard and expanded notation. They will use the place value trading chart to practice regrouping between place values to make equivalent, nonstandard forms of whole numbers in order to build capacity for addition and subtraction procedures. Regrouping will include grouping (for example, grouping 10 units into 1 ten) and ungrouping (for example, ungrouping 1 ten into 10 units).

Begin each variation with the place value equivalencies left off or covered on the place value regrouping/trading chart.

Figure 5.3.1 Jumbo Place Value Chart with place value equivalencies covered

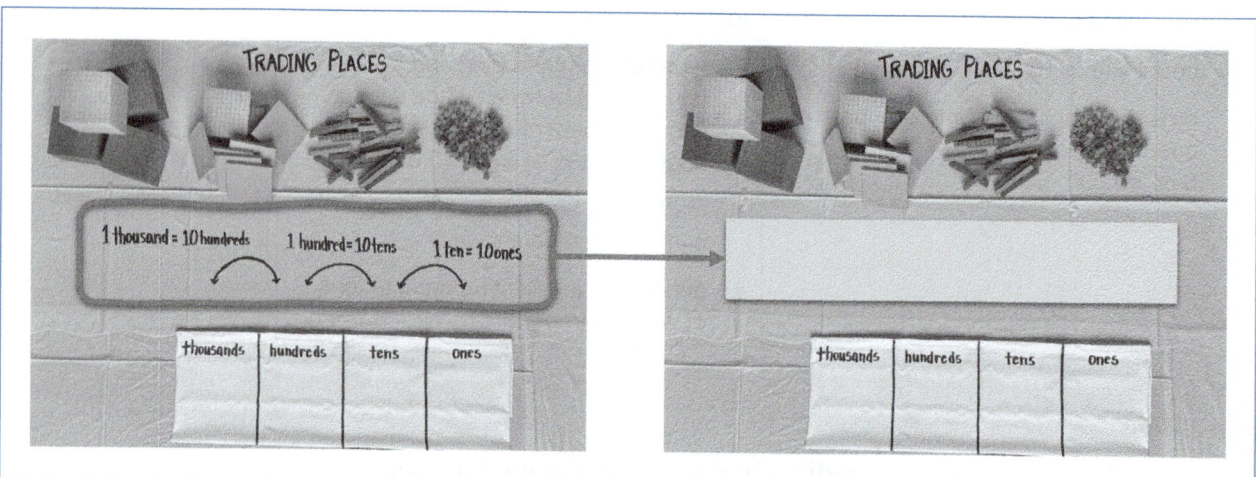

Invite students to explore the base ten manipulatives and ask them to tell you what relationships they notice. As students share (and show) each place value equivalency (e.g., 1 ten = 10 ones), reveal or write the equivalence on the place value trading chart. Next, show students the place value cards and elicit from them which cards correspond to which place value blocks. Once these correlations have been made, students are ready to begin the variation.

Note: Each variation of this task can be adapted to include anywhere from two to four place values. For students Primed to learn about reading and writing two-digit numbers, limit variations to the ones and tens place values. Variations can include the hundreds and/or thousands places for students who are Primed to learn about greater numbers.

Select (or create) a variation that focuses on the prior knowledge you are working to shore up for your student group.

Variation 1

Learning Target: Students will determine the value of a random collection of place value blocks and annotate that value in standard and expanded form.

Variation Directions:

Begin this variation with the place value equivalencies left off or covered on the place value trading chart. Invite students to explore the base ten manipulatives and ask them to tell you what relationships they notice. As students share (and show) each place value equivalency (e.g., 1 ten = 10 ones), reveal or write the equivalence on the place value trading chart. Next, show students the place value cards and elicit from them which cards correspond to which place value blocks.

Place a handful of place value blocks on the place value chart (if desired, invite students to grab handfuls themselves). Provide "extra" place value blocks at the top of the place value trading chart. Challenge students to identify the number represented and make as many trades as they need to until they have the fewest number of blocks possible in each place value. After regrouping (making trades), they will find the appropriate place value card to match to each place and show the number.

Figure 5.3.2 Trading Places Variation 1 example

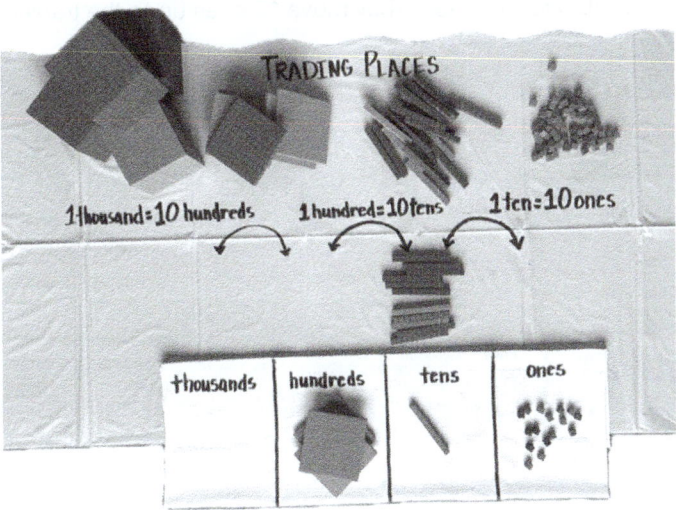

Students are given a collection of hundreds, tens, and ones on their place value chart. Extra place value blocks are available for regrouping at the top of the place value trading chart.

Students decide to trade 10 tens for 1 hundred. They move 10 tens up to the trading space.

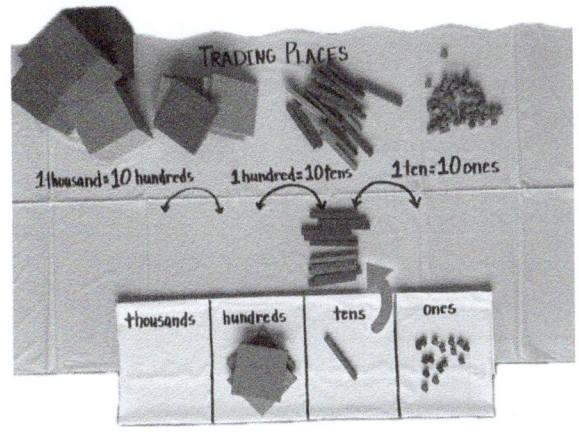
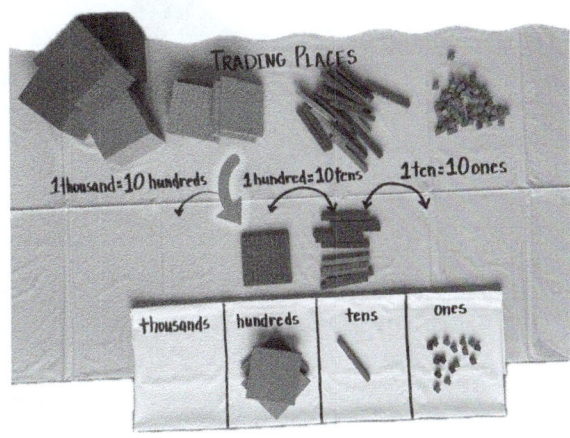

(Continued)

(Continued)

Students get a new hundred to trade for the 10 tens. They move the new hundred onto the place value chart and put the 10 tens in the extra pile at the top of the trading mat.

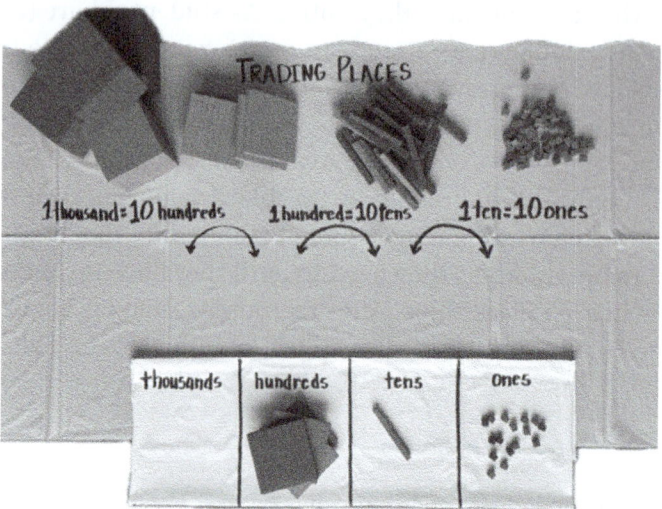

Next, students decide to trade 10 ones for 1 ten. They move 10 ones up to the trading space.

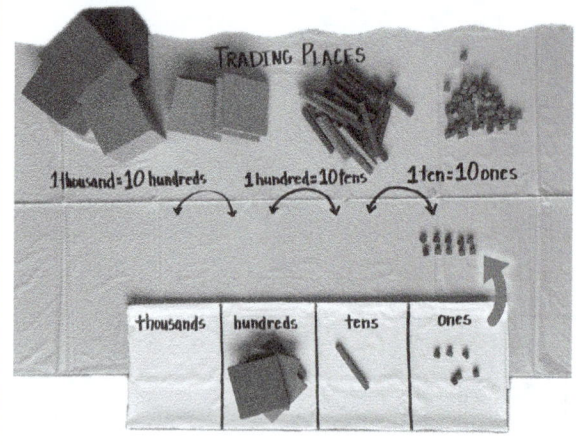
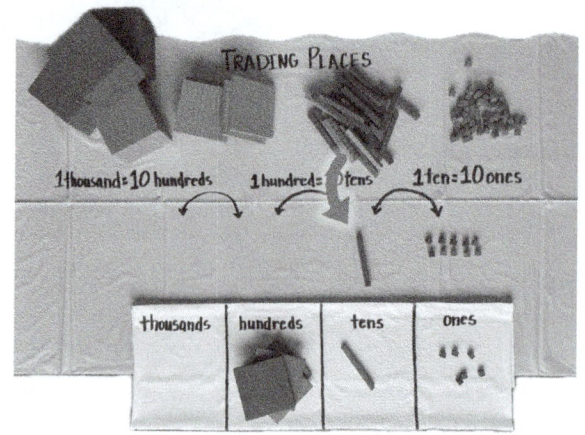

Students get a new ten to trade for the 10 ones. They move the new ten onto the place value chart and put the 10 ones in the extra pile at the top of the trading mat.

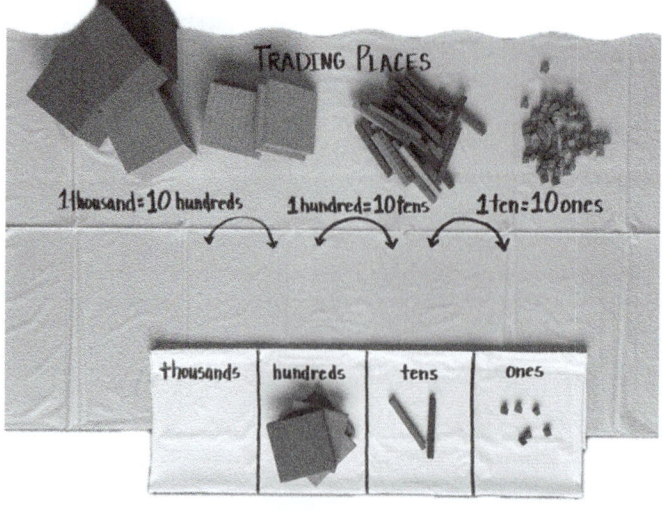

Students decide they cannot make any more trades. They find place value cards to match the number of hundreds, tens, and ones in their number. The total is 426.

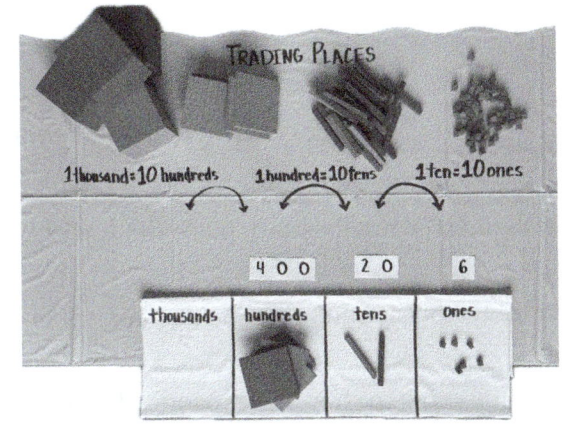 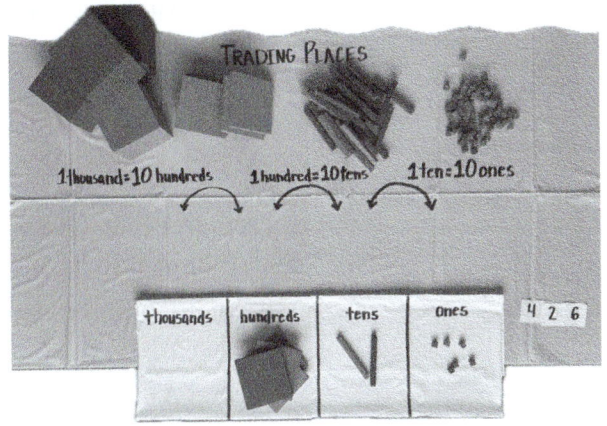

Engage in Math Discourse (Make the Mathematics Visible):

Observation: As students work, encourage them to notice important relationships and push on early and partial understandings.

- How many (ones) will you need to group to make one (ten)?
- Have you changed the total on the place value chart? How do you know?
- What would happen if we ungrouped one (ten)? How would that change what we see on the place value chart? Would the total change?

Interview: After the activity, bring the students together for a discussion. Suggested prompts:

- How did you decide where to start with your grouping?
- How did you know you had enough to make a trade?
- How did you know when you were done? What did you notice about all the places?

Bridging Prompt (Prompt for Classroom Teacher to Use During New Lesson Content)
• Show students an amount of base ten materials on the place value chart where trades are possible. Ask some students who participated in the intervention group, "How can I use groups of ten to make trades between place values?" Ask them to use *place value blocks to support the conversation.*

Source: istock.com/LysenkoAlexander

Variation 2

Learning Target: Students will use ungrouping to generate nonstandard, equivalent forms of whole numbers.

Variation Directions:

Begin this variation by organizing students into pairs. Partners will take turns as follows:

1. Partner A chooses a number, builds the number with place value cards, and shows it to Partner B.
2. Partner B builds the number on the place value chart using place value blocks.
3. Partner A expands the place value cards to show the expanded form of the number.
4. Partner B picks **one** place value and regroups **one** of that place value (for example, Partner B might ungroup 1 ten to get 10 ones).
5. Partner A trades place value cards to show how the composition of the number has changed.
6. Partner B writes an equation showing three equivalent forms of the number (standard form = expanded form = nonstandard expanded form).

Switch roles and repeat.

Figure 5.3.3 Trading Places Variation 2 example

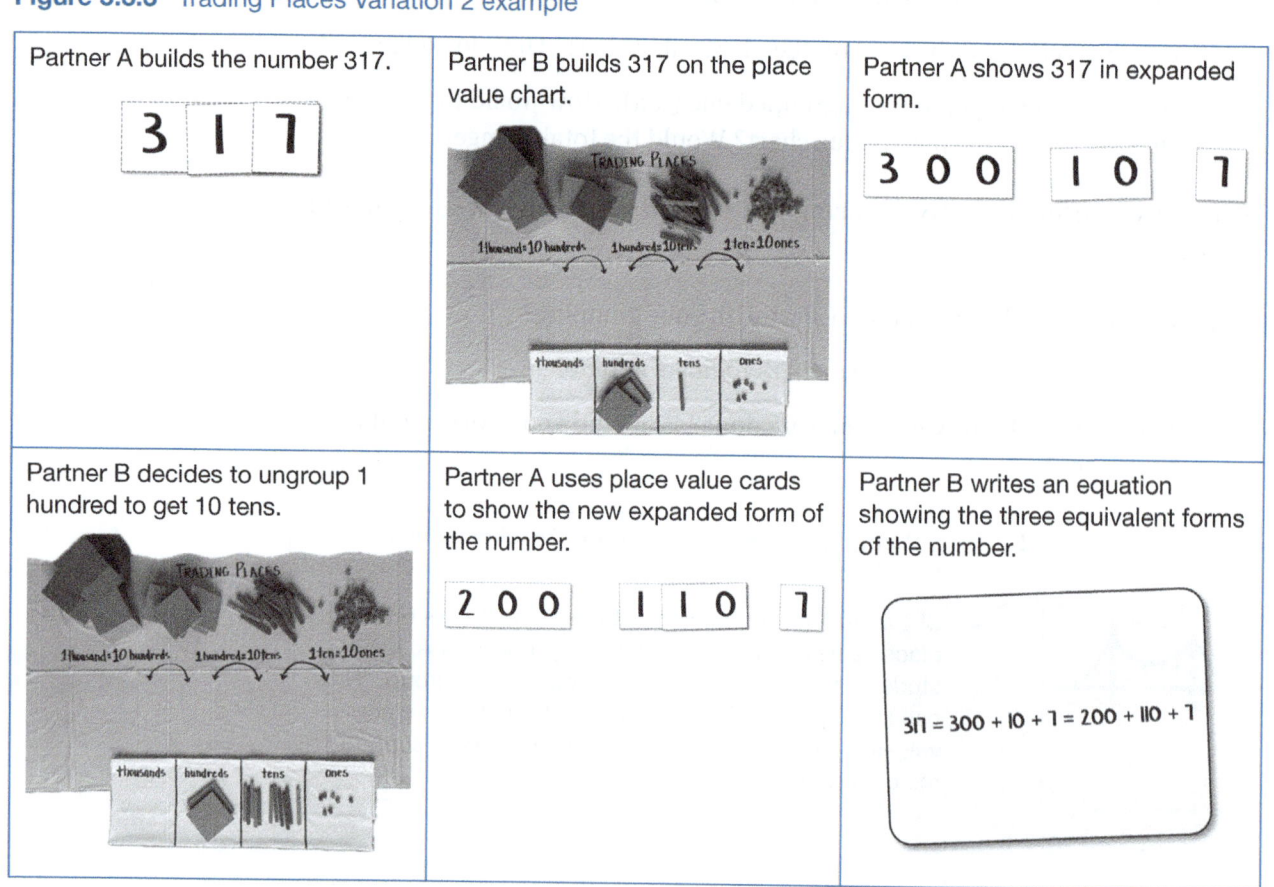

Engage in Math Discourse (Make the Mathematics Visible):

Observation: As students work, encourage them to notice important relationships and push on early and partial understandings.

- How many (tens) did you get when you ungrouped one (hundred)? What would have changed in your equation if you had ungrouped a (ten) instead?
- Have you changed the total on the place value chart? How do you know?
- What would happen if we ungrouped one (hundred) *and* one (ten)? How would that change what we see on the place value chart? Would the total change?

Interview: After the activity, bring the students together for a discussion. Suggested prompts:

- How did you decide which place value to ungroup?
- What would you do if there were no blocks in a place value?
- How did you decide what to do with the place value cards to make your new equation? What did you notice?

Bridging Prompt (Prompt for Classroom Teacher to Use During New Lesson Content):

- Show students base ten materials (thousand/hundred/ten). Ask students who participated in the intervention group, "If we ungroup 1 (thousand/hundred/ten), what will we have?"

Variation 3

Learning Target: Students will use ungrouping to generate nonstandard, equivalent forms of whole numbers.

Variation Directions:

For this variation, follow the same directions as Variation 2, except at step 4, students ungroup in **two** place values.

1. Partner A chooses a number, builds it with place value cards, and shows it to Partner B.
2. Partner B builds the number on the place value chart using place value blocks.
3. Partner A expands the place value cards to show the expanded form of the number.
4. Partner B picks **two** place values and ungroups **one** of each place value (for example, Partner B might ungroup 1 ten to get 10 ones **and** one

thousand to get 10 hundreds).

5. Partner A trades place value cards to show how the composition of the number has changed.

6. Partner B writes an equation showing three equivalent forms of the number (standard form = expanded form = nonstandard expanded form).

Switch roles and repeat.

Engage in Math Discourse (Make the Mathematics Visible):

Observation: As students work, encourage them to notice important relationships and push on early and partial understandings.

- How many (tens) did you get when you ungrouped one (hundred)? What would have changed in your equation if you had ungrouped a (ten) instead?

- Have you changed the total on the place value chart? How do you know?

- What would happen if we ungrouped one (hundred) into **ones**? How would that change what we see on the place value chart? Would the total change?

Interview: After the activity, bring the students together for a discussion. Suggested prompts:

- How did you decide which places to ungroup?

- What would you do if there were no (thousands/hundreds/tens) in the place you wanted to ungroup?
  - *Be on the lookout for groups who may have chosen a number with a zero in a place value and then ungrouped across that place value (e.g., they had a zero in the tens place, but they ungrouped 1 hundred then ungrouped 1 of the new tens). Be sure to discuss what happens there.*

- How did you decide what to do with the place value cards to make your new equation? What did you notice?

CHAPTER 5. Intervention Tasks

Bridging Prompt (Prompt for Classroom Teacher to Use During New Lesson Content):

- Show students one of the numbers from this activity with place value blocks on the place value chart. What happens when you regroup in two places? Engage students who participated in the intervention group in this discussion.

Variation 4

Learning Target: Students will use grouping and ungrouping to identify equivalent standard and nonstandard forms of whole numbers.

Variation Directions:

Begin this variation by organizing students into pairs or small groups. Each group should have a Trading Places jumbo place value chart and collection of place value blocks. Give each group a Trading Places: Equivalent or Not? activity page. Students should build the standard form of the target number, write the expanded form, then use grouping and ungrouping to identify which expressions on the activity page are equivalent to the target number.

Figure 5.3.4 Trading Places Variation 4 example

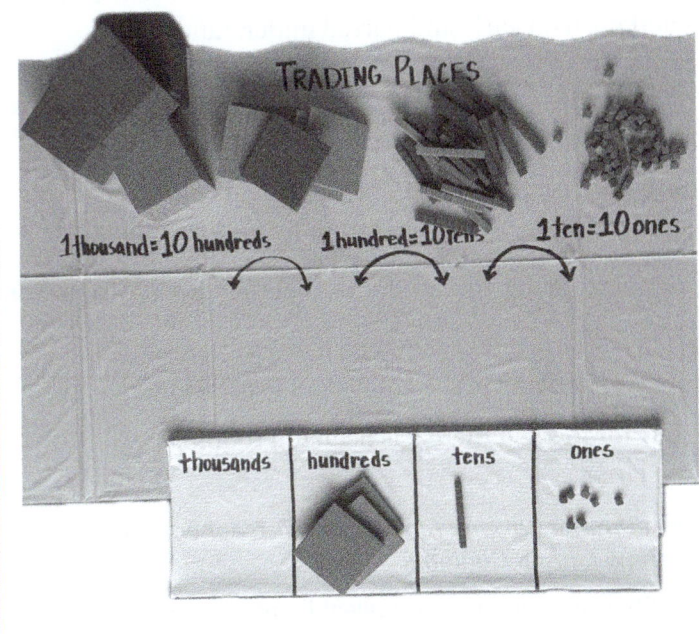

Group 1 gets the activity page for 439. They build the number on their place value chart and write the expanded form in the empty box on their activity page.

(Continued)

(Continued)

Next, they check to see what happens if they trade 1 ten for 10 ones. They discover an equivalent expression and circle it.

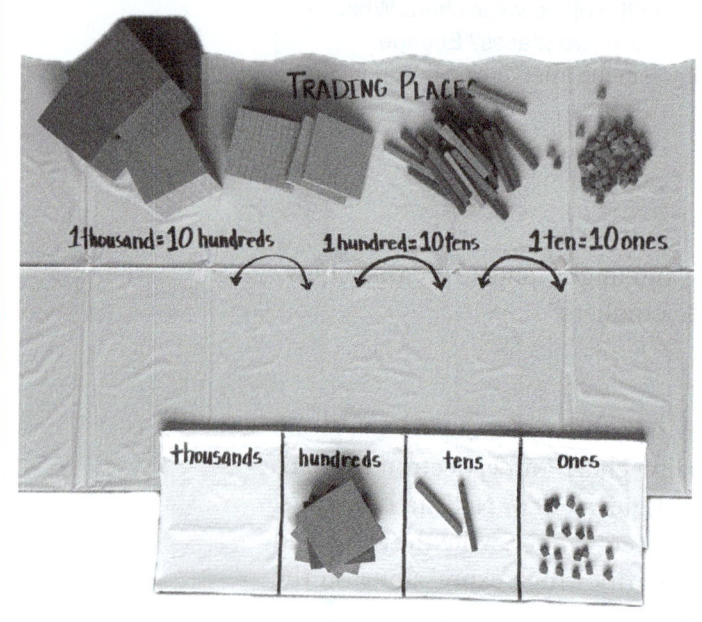

The group continues to work until they are confident they have found all the equivalent expressions.

Engage in Math Discourse (Make the Mathematics Visible):

Observation and Show Me: As students work, encourage them to notice important relationships and push on early and partial understandings.

- How did you decide that this expression is equivalent to (439)?
- What was ungrouped in this expression? How do you know?
- How do you know that this expression is not equivalent? How can you prove it using your place value blocks and grouping or ungrouping?

Interview: After the activity, bring the students together for a discussion. Suggested prompts:

- What patterns were you noticing in the equivalent expressions?
- What might you look for to see if there has been ungrouping?

Bridging Prompt (Prompt for Classroom Teacher to Use During New Lesson Content):

- Show the class two equivalent forms of a number (an expanded form and a nonstandard expanded form). Ask students who participated in the intervention group to use place value blocks to prove that the two numbers are equivalent.

Adding It Up

TASK 4

General Objective:
Students will develop understanding of regrouping in addition problems.

This activity might be used with:

Students who can (prerequisite knowledge):	Students who are Primed to (getting ready to learn):
• describe the relationship between adjacent place values • use models to solve additions within 10 • build a two- or three-digit number using place value blocks	• add two-, three-, or four-digit numbers • regroup in addition

Materials:
- Jumbo-sized place value trading chart made on disposable plastic tablecloth or butcher paper (see PRINTABLE directions)
- Place value blocks
- PRINTABLE: Place Value Mat
- PRINTABLE: Directions for Making a Jumbo Place Value Chart
- PRINTABLE: Place Value Cards
- PRINTABLE: Adding It Up! Addition Mats (suggest laminating so students can use dry-erase markers)

 Printables for this task are available for download at **https://companion.corwin.com/courses/ProactiveMathIntervention**.

Recommended Children's Literature:
- *Great Math Tattle Battle*—Bowen, 2006

Task Overview:
In this activity, students will generate physical models of whole-number addends on a jumbo-sized place value chart and equate those models to abstract numerical notations, including standard and expanded notation. They will practice adding to/combining to find a sum and learn how to use groups of ten to find the standard form of the sum.

Begin each variation with the place value equivalencies left off or covered on the place value trading chart.

Figure 5.4.1 Jumbo Place Value Chart with place value equivalencies covered

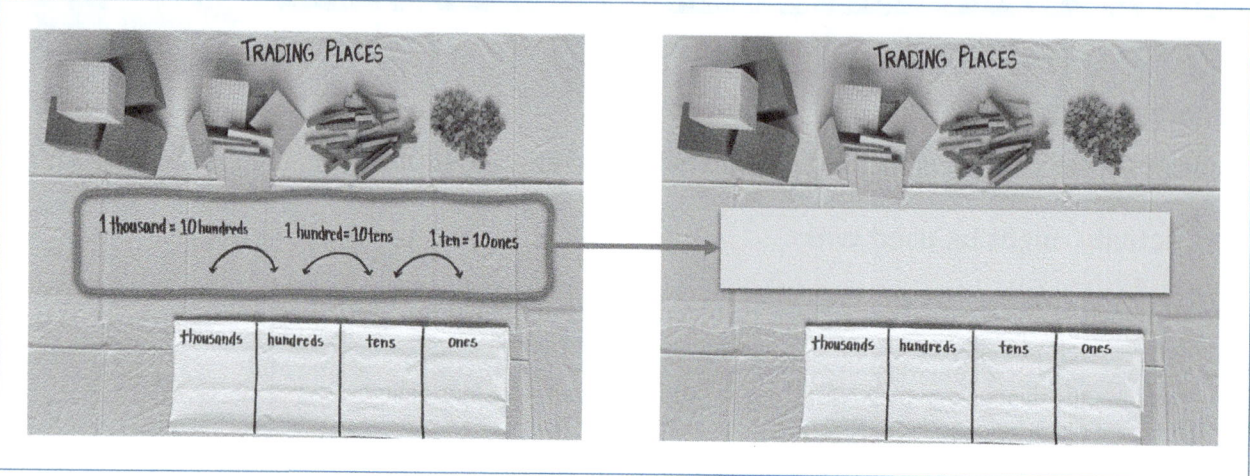

Invite students to explore the base ten manipulatives, and ask them to tell you what relationships they notice. As students share (and show) each place value equivalency (e.g., 1 ten = 10 ones), reveal or write the equivalence on the place value trading chart. Next, show students the place value cards and elicit from them which cards correspond to which place value blocks. Once these connections have been made, students are ready to begin the variation.

Note: Each variation of this task can be adapted to include anywhere from two to four place values. For students who are being Primed to learn about adding two-digit numbers, limit variations to the ones and tens place values. Variations can include the hundreds and/or thousands places for students who are being Primed to learn addition with greater numbers.

Select (or create) a variation that focuses on the prior knowledge you are working to shore up for your student group.

Variation 1

Learning Target: Students will solve "combine" addition problems involving regrouping.

Variation Directions:

Begin this variation by organizing students into pairs. Partners will take turns as follows:

1. Partner A builds a number in the top row of the place value chart using place value blocks.

2. Partner B builds a number in the second row of the place value chart using place value blocks.

3. Each partner represents the other partner's number using place value cards.

4. The pair writes an addition expression and the expanded notation of each addend on their adding mat, then they **combine** the two sets of blocks.

5. After combining, partners count the totals for each place value and write the partial sums on their adding mat.

6. Then, together, the partners need to decide if they have the *simplest form* of their number (do they have the fewest possible number of blocks on the place value chart?).

7. Finally, partners will record the sum on their adding mat and compare it to the sum of the partial sums they found before trading.

Figure 5.4.2 Adding It Up Variation 1 example

Partner A builds the number 248 in the top row of the place value chart.

Partner B builds the number 391 in the second row of the place value chart.

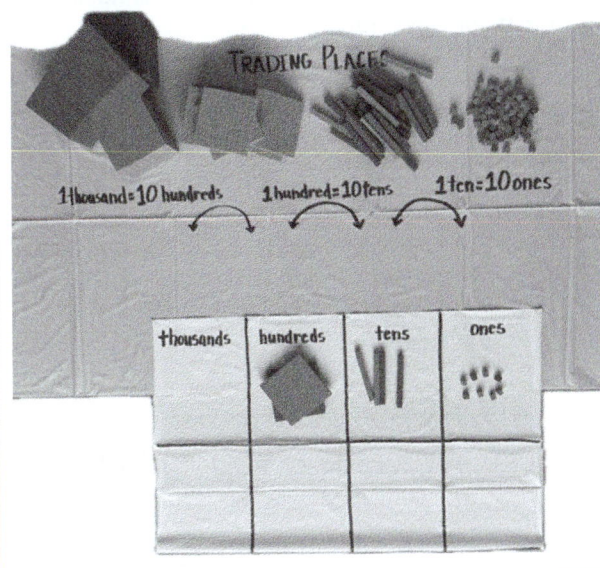
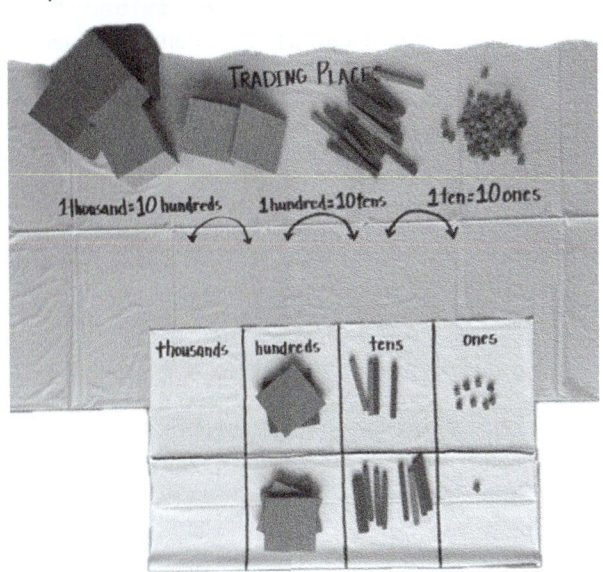

The pair build one another's numbers using place value cards, and they write an addition expression and the expanded form of their addends on their adding mat.

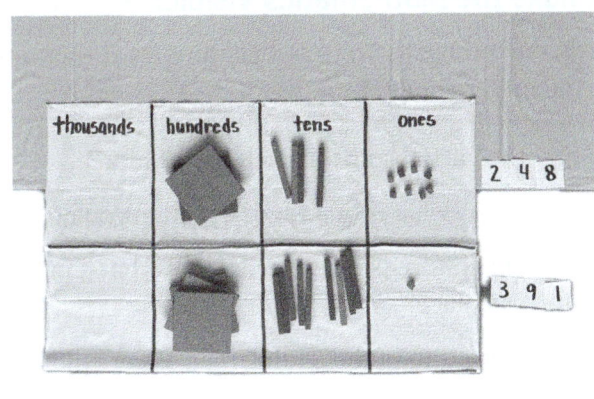
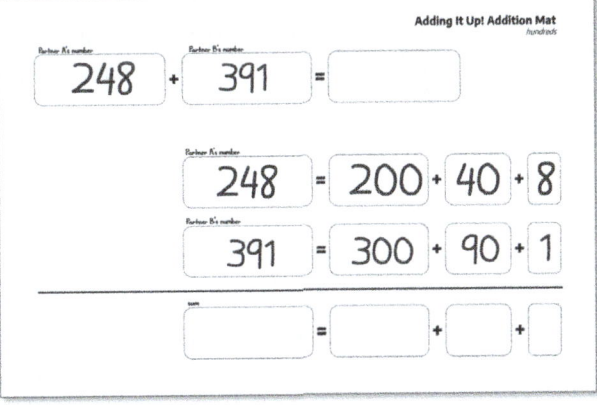

(Continued)

(Continued)

Next, they **combine** their blocks and find the partial sums for each place value.

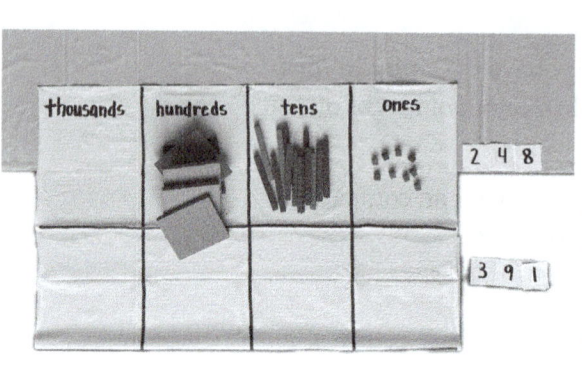
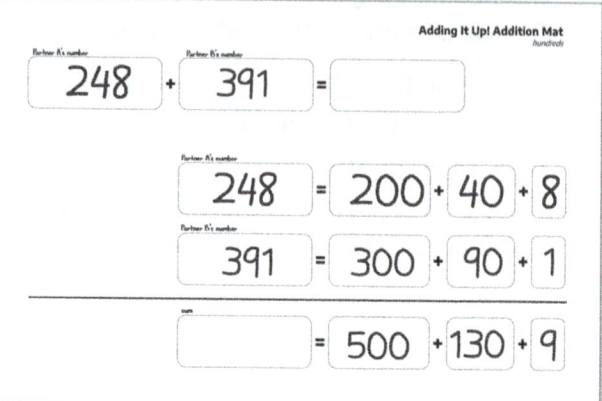

They decide they should group 10 tens and trade them for 1 hundred. They write the sum 639 on their adding mat and discover that it matches the sum of the partial sums from before they traded!

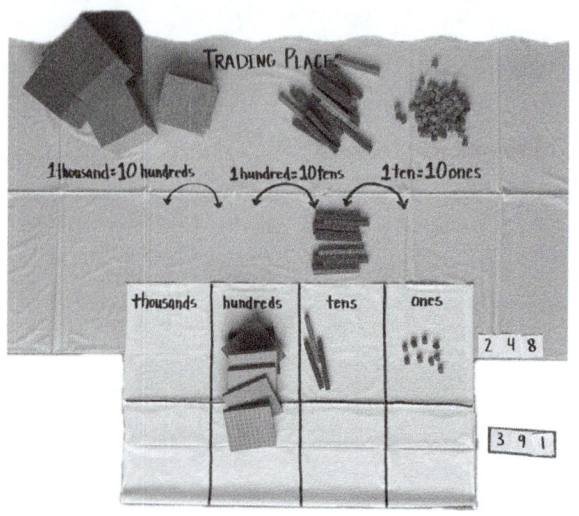
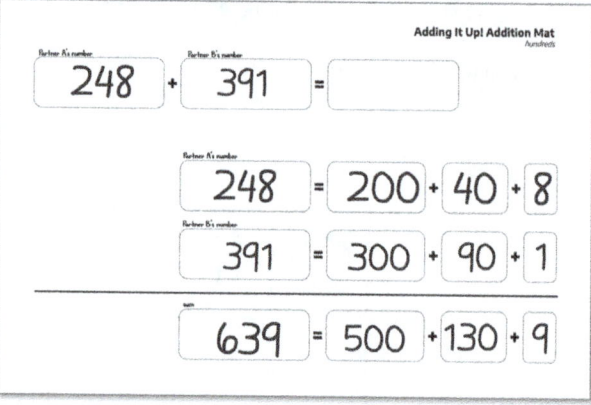

Engage in Math Discourse (Make the Mathematics Visible):

Observation: As students work, encourage them to notice important relationships and push on early and partial understandings.

- How are you counting to find the partial sums in each place value?
- If you add the partial sums together, what is that total? What do you notice?
- When you counted 5 here (in the hundreds place), what does that 5 mean? Is that 5 the same as 5 in the tens place?
- Look at your partial sums at the bottom of the adding mat. Where did that new (hundred) come from?
- Be on the lookout for students who are choosing numbers that do not require regrouping. **Highlight a strength:** Ask what pattern they are noticing that is helping them to be sure they won't have to do any

regrouping. These ideas will be great to lift up for group discussion. Then, challenge the group to show you what they mean by making a problem that breaks their "rule" and requires regrouping.

Interview: After the activity, bring the intervention group together for a discussion. Suggested prompts:

- Did you notice any patterns? Could you predict when you might need to make a trade in a place value? How did you know?

- When you had to make a trade, what happened in your partial sums? How did that change the total?

- Using one or more of the student-generated adding mats, ask students to describe the regrouping that happened. Annotate the regrouping notation using both the expanded and standard form as the students describe each move. For example, you might start the conversation like this . . .

Figure 5.4.3 Sample group conversation

Student(s):	They had too many tens, so they grouped 10 tens to trade for 1 hundred.	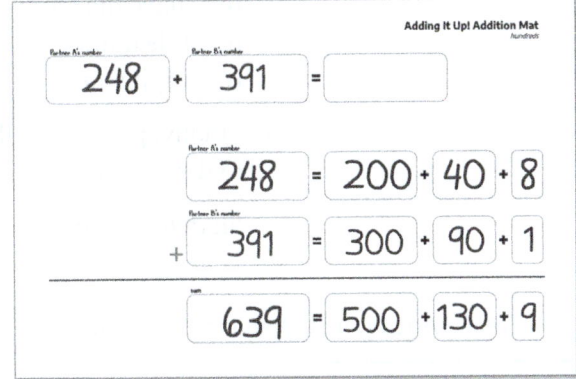
Teacher:	Let's look at the tens on this group's adding mat. Partner A had 4 tens (pointing) and Partner B had 9 tens (pointing). How many tens was that altogether?	
Student(s):	13!	
Teacher:	You said you traded 10 of your tens to make a new hundred. We can write a 1 here above the hundreds place to show that there is one more hundred. Talk with your partners. Where do you see the new one hundred in their partial sums?
Student(s):	(after partner discussion): It's in the 130!
Teacher:	Oh, you said there were 13 tens. How does that compare to 130? That's where the 130 comes from. If this one (point to the standard notation for regrouping) represents the new hundred, where do we see the other 3 tens from the 13 tens in 130?

	Bridging Prompt (Prompt for Classroom Teacher to Use During New Lesson Content):
	• Show the class an addition problem modeled with place value blocks. Ask: "When you are adding two numbers, how do you know you will need to make a new group in a place value?"
	• Provide opportunities for students who participated in the intervention group to share their reasoning.

Variation 2

Learning Target: Students will solve "add to" addition problems involving regrouping.

Variation Directions:

This variation is similar to Variation 1, except instead of **combining** two quantities, students will model **adding to** a quantity. Begin this variation by organizing students into pairs. Partners will take turns as follows:

1. Partner A builds a number on the place value chart using place value blocks. Partner A also represents their number with place value cards.

2. Partner B represents a number using place value cards *only*. This is the amount that will be **added to** Partner A's amount.

3. The pair writes an addition expression and the expanded notation of each addend on their adding mat.

4. Next, the partners **add** Partner B's amount to the mat.

5. After Partner B's amount is **added to** the mat, the partners decide if they have the *simplest form* of their number (do they have the fewest possible number of blocks on the place value chart?). They complete any regrouping needed.

6. Finally, partners will record the sum on their adding mat and compare it to the sum of the partial sums they found.

Switch roles and repeat.

Figure 5.4.4 Adding It Up Variation 2 example

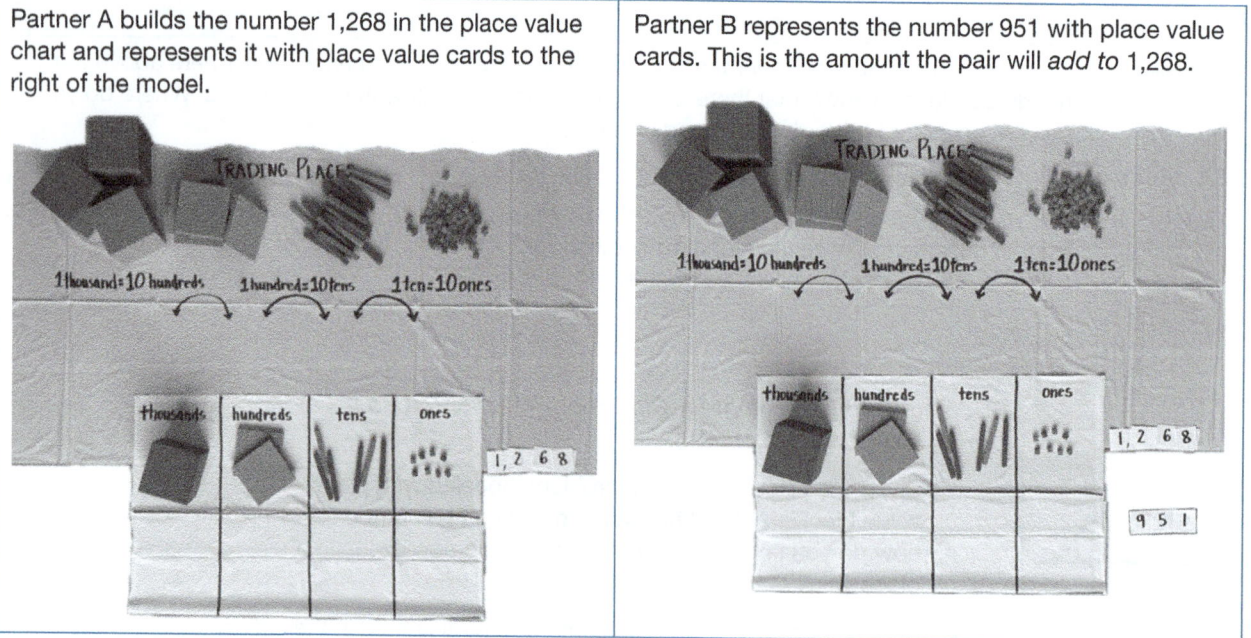

| Partner A builds the number 1,268 in the place value chart and represents it with place value cards to the right of the model. | Partner B represents the number 951 with place value cards. This is the amount the pair will *add to* 1,268. |

The pair writes an addition expression and the expanded form of their addends on their adding mat.

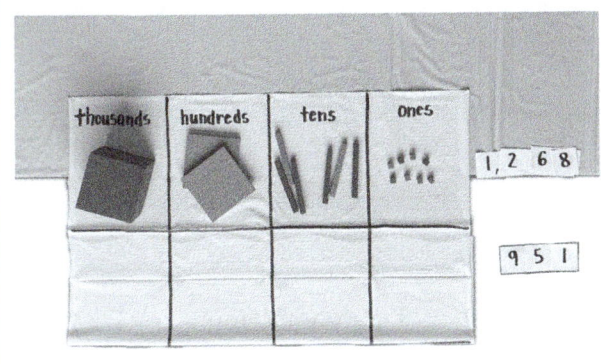 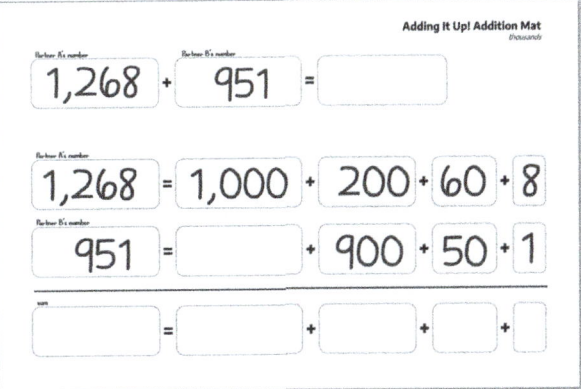

Next, they **add** Partner B's amount (951) **to** Partner A's amount (1,268) and find the partial sums for each place value.

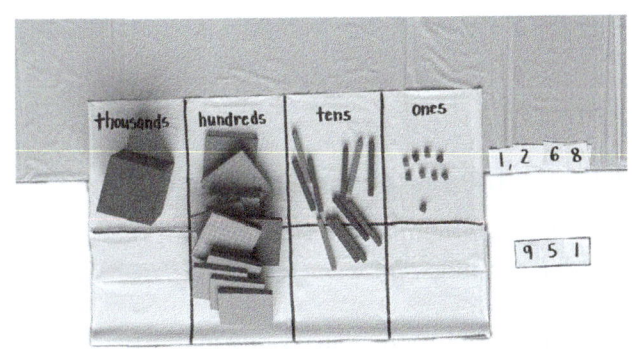 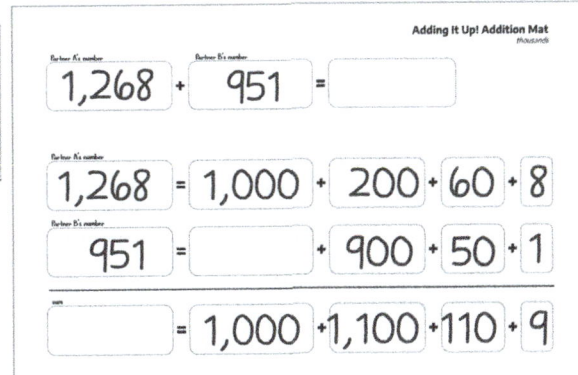

They decide they should group 10 tens and trade them for 1 hundred.

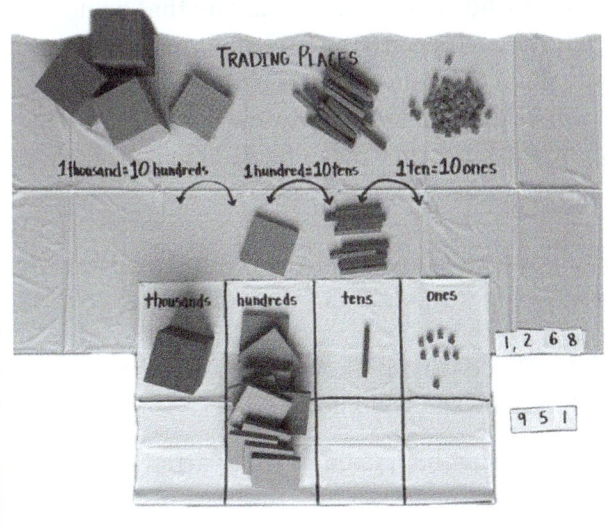

Next, they decide they should group 10 hundreds and trade them for 1 thousand.

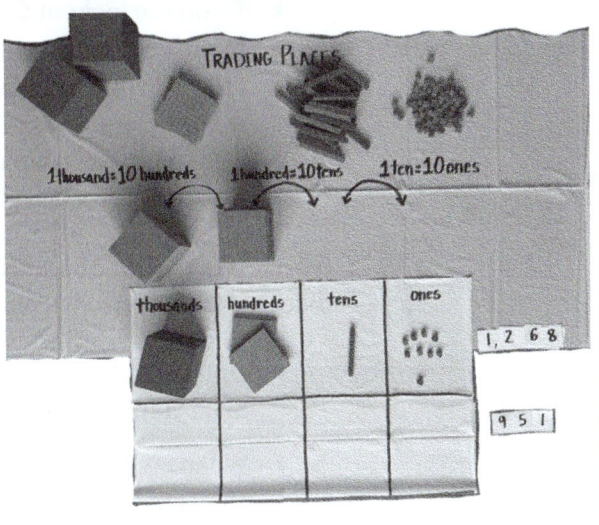

(Continued)

(Continued)

They write the sum 2,219 on their adding mat and discover that it matches the sum of the partial sums from before they traded!

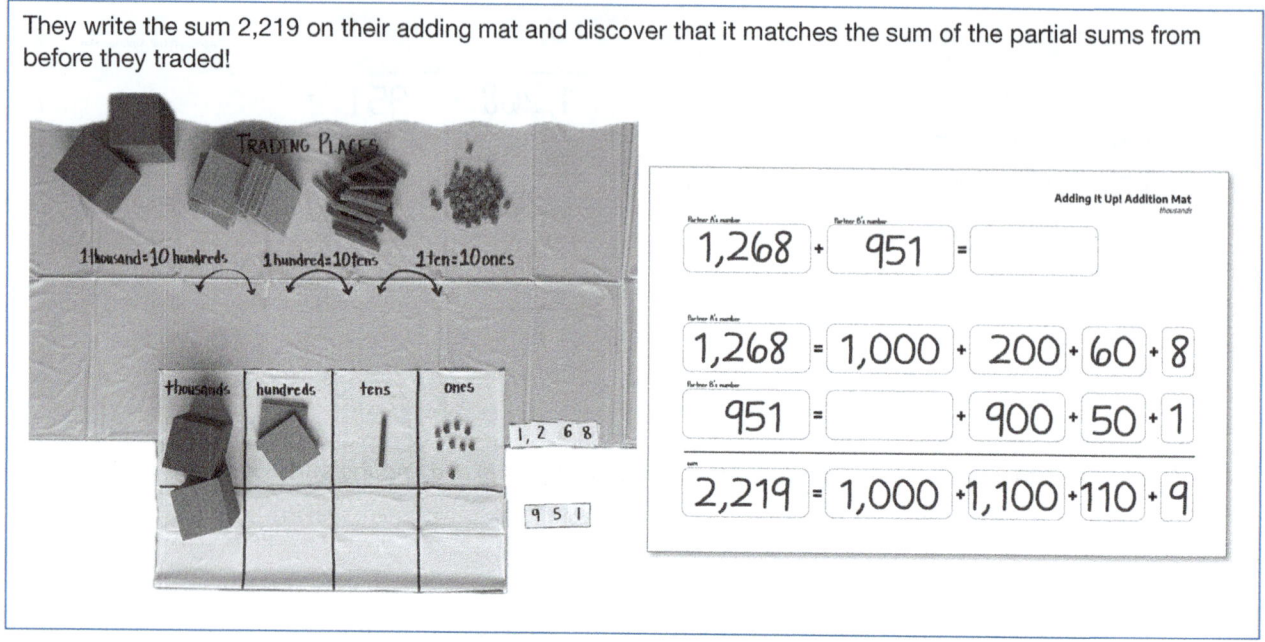

Engage in Math Discourse (Make the Mathematics Visible):

Interview: As students work, encourage them to notice important relationships and push on early and partial understandings.

- When you added Partner B's amount to the place value mat, which place did you add first? Did you add the (hundreds) first? Or the ones? Why?

- How are you deciding when to regroup in a place value? Do you think you could regroup a different way? How?

- Be on the lookout for students who are choosing numbers that do not require regrouping. **Highlight a strength:** Ask what pattern they are noticing that is helping them to be sure they won't have to do any regrouping. These ideas will be great to lift up for group discussion. Then, challenge the group to show you what they mean by making a problem that breaks their "rule" and requires regrouping.

Show Me: After the activity, bring the intervention group together for a discussion. Suggested prompts:

- Did you notice any patterns? Could you predict when you might need to make a trade in a place value? How did you know?

- When you had to make a trade, what happened in your partial sums? How did that change the total?

	**Bridging Prompt (Prompt for Classroom Teacher to Use During New Lesson Content):** • Show the class an addition problem built with place value cards. • Ask students who participated in the intervention group to explain: When you are adding two numbers, how might it help to write each number in expanded notation and then add each place value part?

Task 5

What's the Difference?

General Objective:

Students will develop understanding of regrouping in subtraction problems.

This activity might be used with:

Students who can (prerequisite knowledge):	Students who are Primed to (getting ready to learn):
• describe the relationship between adjacent place values • use models to solve subtractions within 10 • build a number using place value blocks	• subtract with two-, three-, or four-digit numbers • regroup (grouping and ungrouping) in subtraction

Materials:

- Jumbo-sized place value trading chart made on disposable plastic tablecloth or butcher paper (see PRINTABLE directions).
- Place value blocks. If possible, make two colors of place value blocks available for Variation 2.
- PRINTABLE: Place Value Mat
- PRINTABLE: Directions for Making a Jumbo Place Value Chart
- PRINTABLE: Place Value Cards
- PRINTABLE: Take From Mats (suggest laminating so students can use dry-erase markers)
- PRINTABLE: Compare Mat (suggest laminating so students can use dry-erase markers)

 Printables for this task are available for download at **https://companion.corwin.com/courses/ProactiveMathIntervention**.

Recommended Children's Literature:

- *17 Kings and 42 Elephants* – Mahy, 1987
- *Shark Swimathon* – Murphy, 2000
- *Panda Math: Learning About Subtraction from Hua Mei and Mei Sheng* – Nagda, 2009
- *Great Math Tattle Battle* – Bowen, 2006
- *What's the Difference? An Endangered Animal Subtraction Story* – Slade, 2010

Task Overview:

In this activity, students will generate physical models of whole number minuends (for take from/separate problems) and both minuends and subtrahends (for comparison problems) on a jumbo-sized place value chart. They will equate those models to abstract numerical notations, including standard and expanded notation. They will practice taking from and comparing to find a difference and learn how to ungroup when needed to subtract.

Begin each variation with the place value equivalencies left off or covered on the place value trading chart.

Figure 5.5.1 Jumbo Place Value Chart with place value equivalencies covered

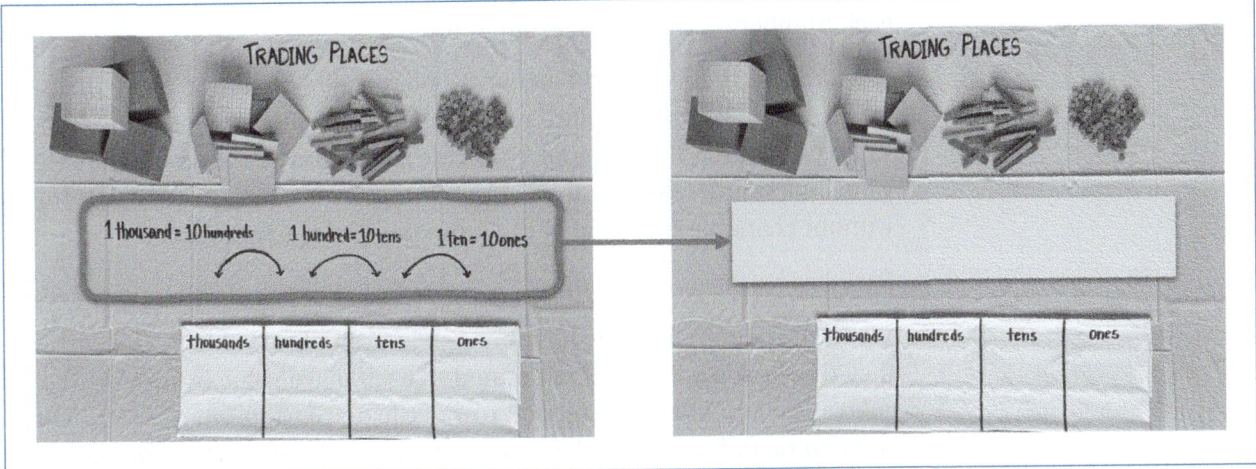

Invite students to explore the base ten manipulatives and ask them to tell you what relationships they notice. As students share (and show) each place value equivalency (e.g., 1 ten = 10 ones), reveal or write the equivalence on the place

value trading chart. Next, show students the place value cards and elicit from them which cards correspond to which place value blocks. Once these correlations have been made, students are ready to begin the variation.

Note: Each variation of this task can be adapted to include anywhere from two to four place values. For students being Primed to learn about subtracting two-digit numbers, limit variations to the ones and tens place values. Variations can include the hundreds and/or thousands places for students who are being Primed to learn subtraction with greater numbers.

Select (or create) a variation that focuses on the prior knowledge you are working to shore up for your student group.

Variation 1

Learning Target: Students will solve "take from" subtraction problems involving regrouping.

Variation Directions:

Begin this variation by organizing students into pairs. Each pair will need a Take From Mat. Partners will take turns as follows:

1. Partner A builds a number on the place value chart using place value blocks.

2. Partner B represents a **lesser** number using place value cards. This is the amount that will be **taken from** Partner A's amount.

3. The pair writes a subtraction expression and the expanded notation of each number on their Take From Mat.

4. Next, the partners expand Partner B's place value cards and place them in the second row of the place value chart under the corresponding place values. Together, the pair look to decide if they have enough of the place value blocks in each place to subtract. If they don't, they need to regroup without changing the value of the minuend (the starting value).

5. Partners record any regrouping they do on their Take From Mat.

6. When the pair is sure they have enough in each place value to subtract, they **take away** Partner B's amount in each place value by moving the subtracted amount to the second row of the place value chart. **Students can choose to begin with the greatest place or the ones place, but direct them to work in order by place value (either least to greatest or greatest to least). They should use Partner B's place value cards to keep track of the parts they have subtracted (see example).*

7. When they have subtracted in all place values, partners record the difference (the amount left in the top row of the place value chart) on their Take From Mat.

Switch roles and repeat.

Figure 5.5.2 What's the Difference? Variation 1 example

Partner A builds the number 518 in the top row of the place value chart.	Partner B represents the number 485 with place value cards next to the second row of the chart.
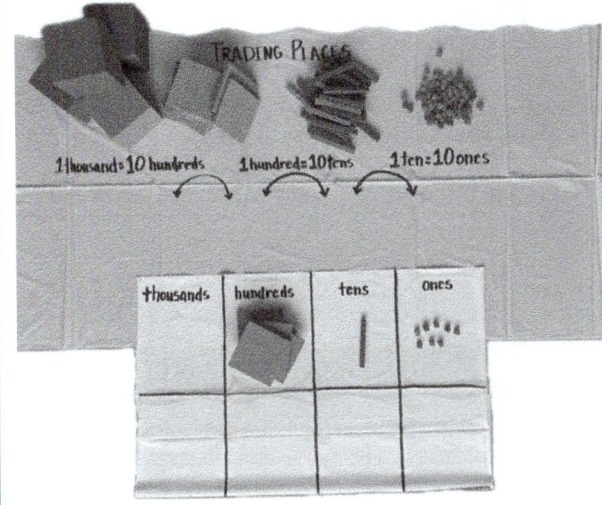	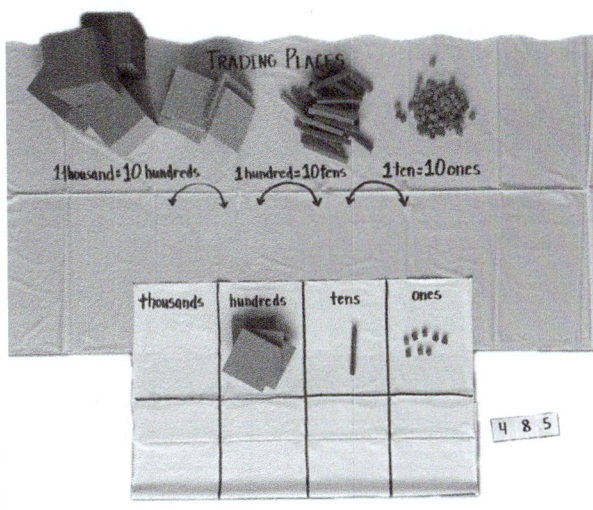

The pair writes a subtraction expression and the expanded form of each number on their Take From Mat.

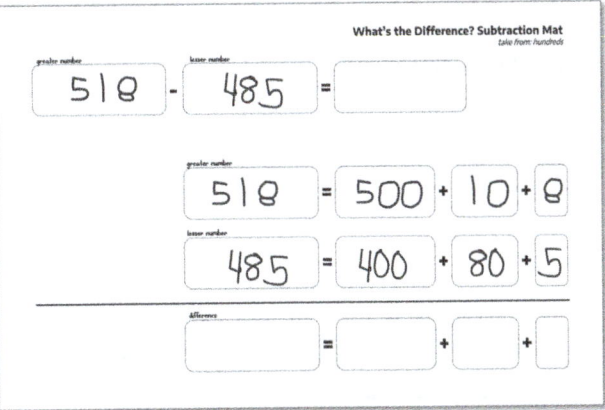

The pair expands Partner B's place value cards and places them in the bottom section of the place value chart under the corresponding place values.

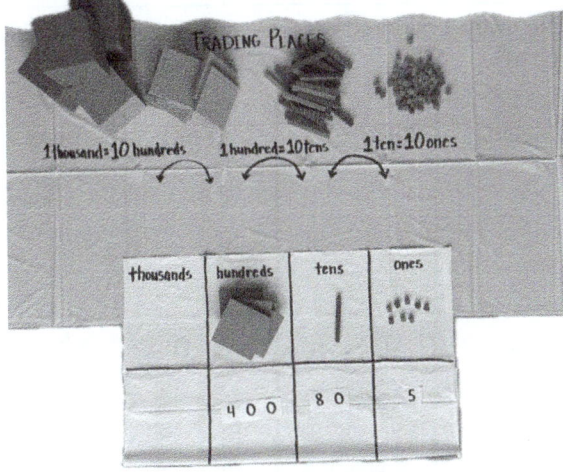

(Continued)

(Continued)

They decide they have enough hundreds and ones to subtract, but there are not enough tens to take away 8 tens. The pair decide to ungroup 1 hundred to get 10 more tens.

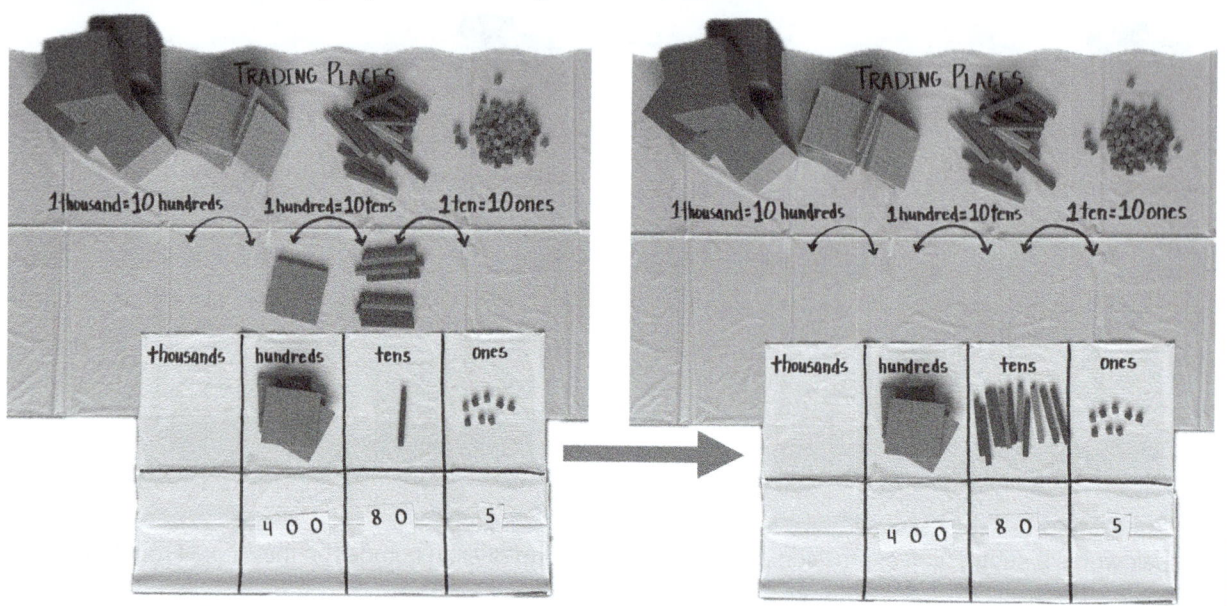

They record this trade on their Take From Mat.

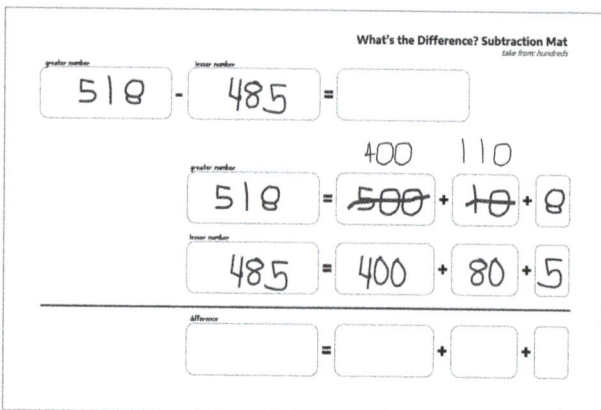

Now they can subtract! They take away 4 hundreds, 8 tens, and 5 ones by moving that amount down to the second row of the place value chart.

They record the final differences for each place value on their Take From Mat (the amount left in the top section).

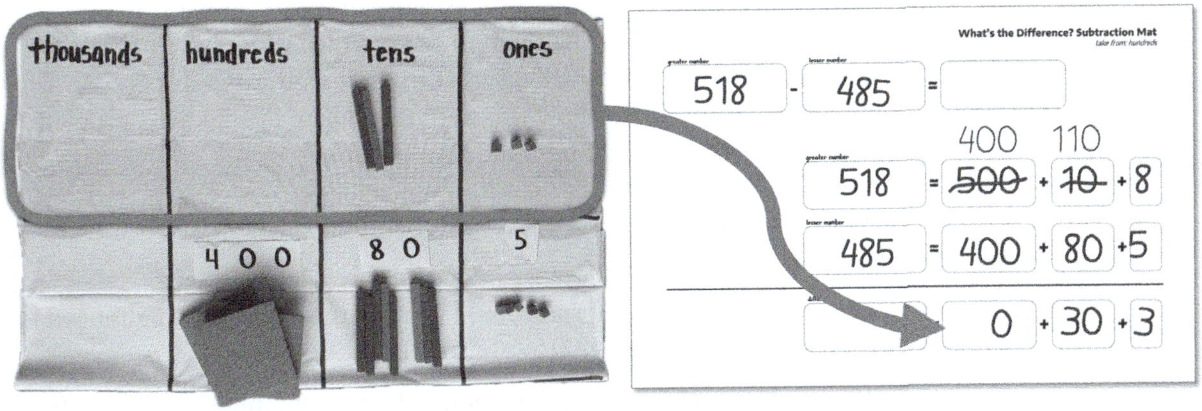

They write the difference 33 on their Take From Mat.

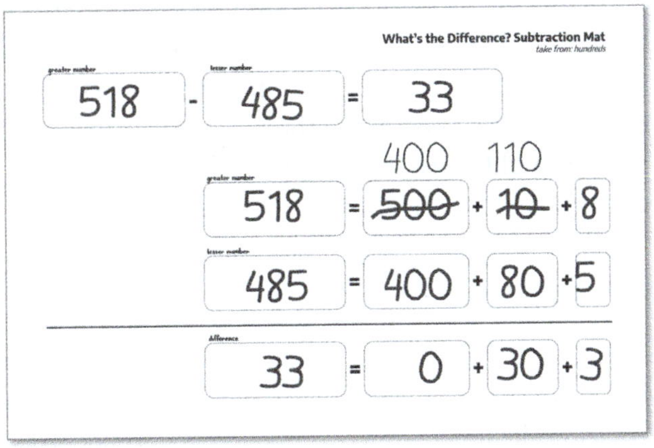

Engage in Math Discourse (Make the Mathematics Visible):

As students work, encourage them to notice important relationships and push on early and partial understandings.

Observation:

- Watch for students who add place value blocks onto their place value mats instead of regrouping. Ask them, "Are we adding? Did you mean to add 10 (tens/ones)?

- How might you get 10 (tens/ones) without **adding** 10 (tens/ones) and changing the total amount? Can you think of another way to represent your original number where you would have more tens than you have now?"

Interview:

- When you started to take Partner B's amount away from the place value mat, which part did you subtract first? Did you subtract the (hundreds) first? or the ones? Why?

Observation:

- Be on the lookout for students who are choosing numbers that do not require regrouping. **Highlight a strength:** Ask what pattern they are noticing that is helping them to be sure they won't have to do any regrouping. These ideas will be great to lift up for group discussion. Then, challenge the group to show you what they mean by making a problem that breaks their "rule" and requires regrouping.

Interview: After the activity, bring the intervention group together for a discussion. Suggested prompts:

- Did you notice any patterns? Could you predict when you might need to regroup in a place value? How did you know?
- How did you get more (tens) without adding something more to your number?
- Using one or more of the student-generated Take From Mats, ask students to describe the regrouping that happened. Annotate the regrouping notation using standard notation as the students describe each move, making connections to their expanded-form annotations. For example, you might start a conversation like this . . .

Figure 5.5.3 Sample group conversation

Student(s):	They didn't have enough tens to take 8 tens away, so they ungrouped 1 hundred to get 10 more tens.
Teacher:	Let's look at this group's regrouping. When these students ungrouped 1 hundred, how many hundreds did that leave them?
Students:	4!
Teacher:	They only had 1 ten before. Now how many did they have?
Students:	11!

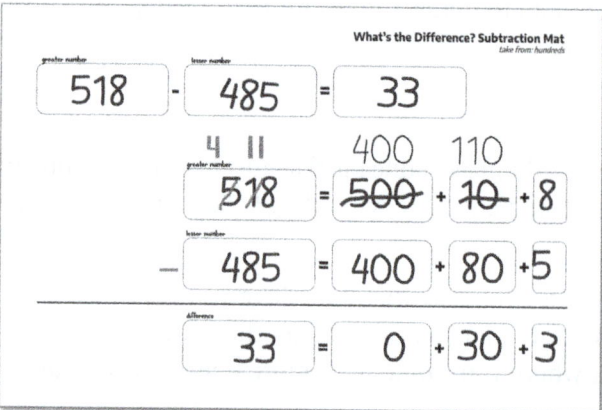

Teacher:	They ungrouped 1 hundred to get 10 more tens. So now they have 4 hundreds. I'll show that here by changing the 5 in the hundreds place to a 4. Where do you see that in their expanded form?
Students:	They changed 500 to 400!
Teacher:	Right, and now they have 11 tens. Let's change that too. I'll change the 1 in the tens place to an 11. Can you see that in their expanded form?

	***Bridging Prompt (Prompt for Classroom Teacher to Use During New Lesson Content):*** • Show the class a minuend built with place value blocks and a subtrahend represented with place value cards. • Ask students who participated in the intervention group to explain: When you are subtracting, how do you know you will need to regroup in a place value?

Variation 2

Learning Target: Students will solve "comparison" subtraction problems involving regrouping.

Variation Directions:

For this variation, unfold your third row or add a third row to the place value chart.

Figure 5.5.4 Place value chart with third row

Begin the variation by organizing students into pairs. If possible, provide each pair with two colors of place value blocks. Each pair will need a Compare Mat. Partners will take turns as follows:

1. Partner A uses one color of place value blocks to build a number in the top row of the place value chart using place value blocks.

2. Partner B uses *the same color* of place value blocks to build a **lesser** number in the bottom row of the place value chart using place value blocks.

3. Each partner represents their number using place value cards.

4. They record the two numbers on the Compare Mat.

5. The pair works together to find the **difference** between the two numbers. To do this, they will use a different color of place value blocks to **add up** from the lesser number to the greater number (e.g., "think addition" to find the difference). The pair will build the **difference** amount in the middle row of the place value chart.

6. As the pair adds up, they will record their moves as lengths on the number line on the Compare Mat.

7. When the adding is complete, students will regroup the **difference** amount as needed so the difference is represented with the fewest place value blocks possible.

8. The partners will then **compare** the numbers. They determine **how much more** Partner A's number is compared to Partner B's number and **how much less** Partner B's number is compared to Partner A's number.

9. Finally, partners will record the difference on their Compare Mat.

Figure 5.5.5 What's the Difference? Variation 2 example

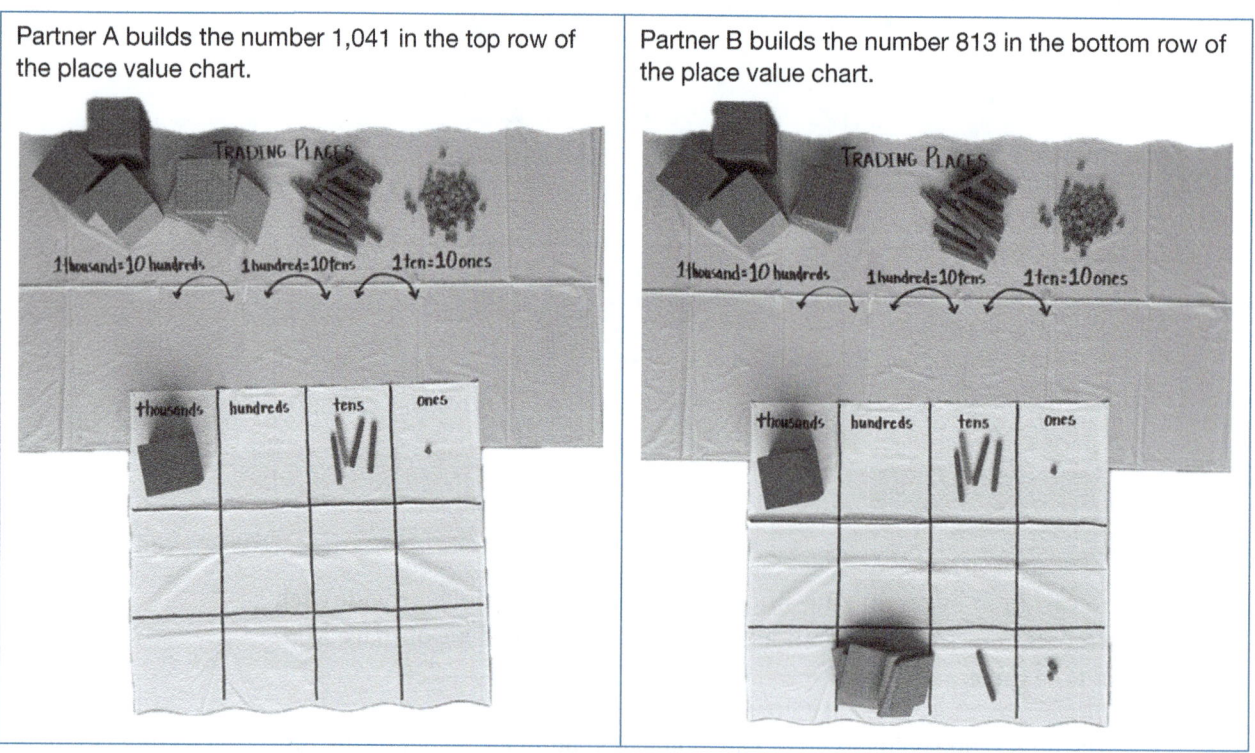

The pair build one another's numbers using place value cards, and they write a subtraction expression and write the numbers in the blanks at each end of the number line on the Compare Mat.

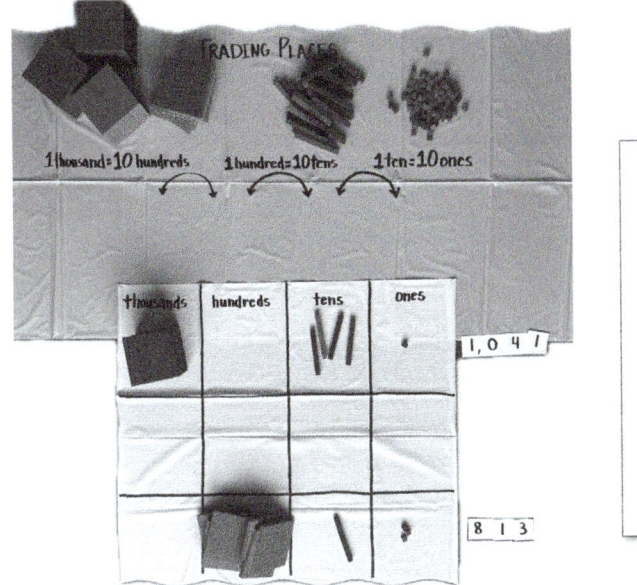

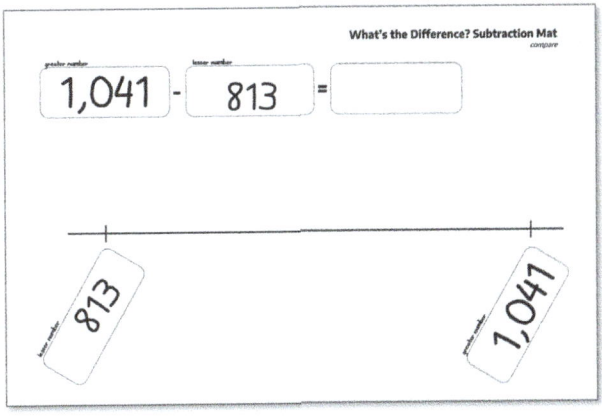

The pair begins adding up from 813 to 1,041 by adding 2 ones to 813. This gets them to 815. They place 2 ones in the middle row of the place value chart and record this on the number line on their Compare Mat.

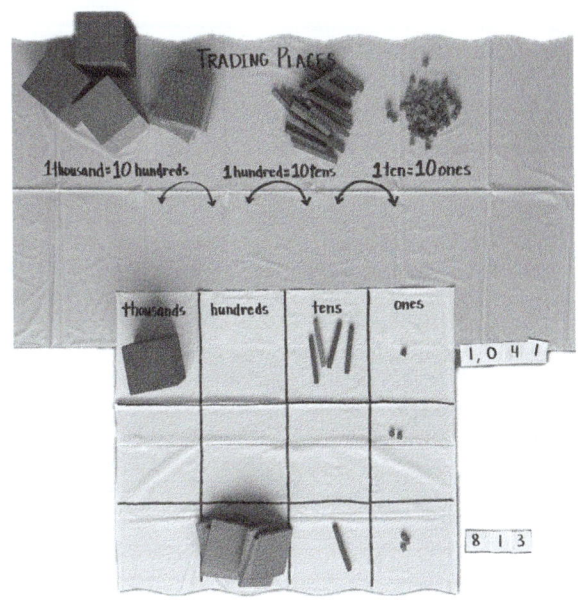

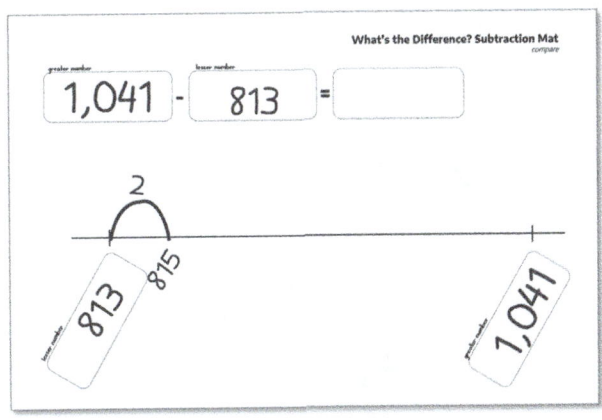

(Continued)

(Continued)

Next, they add 5 ones to 815 to get to 820.

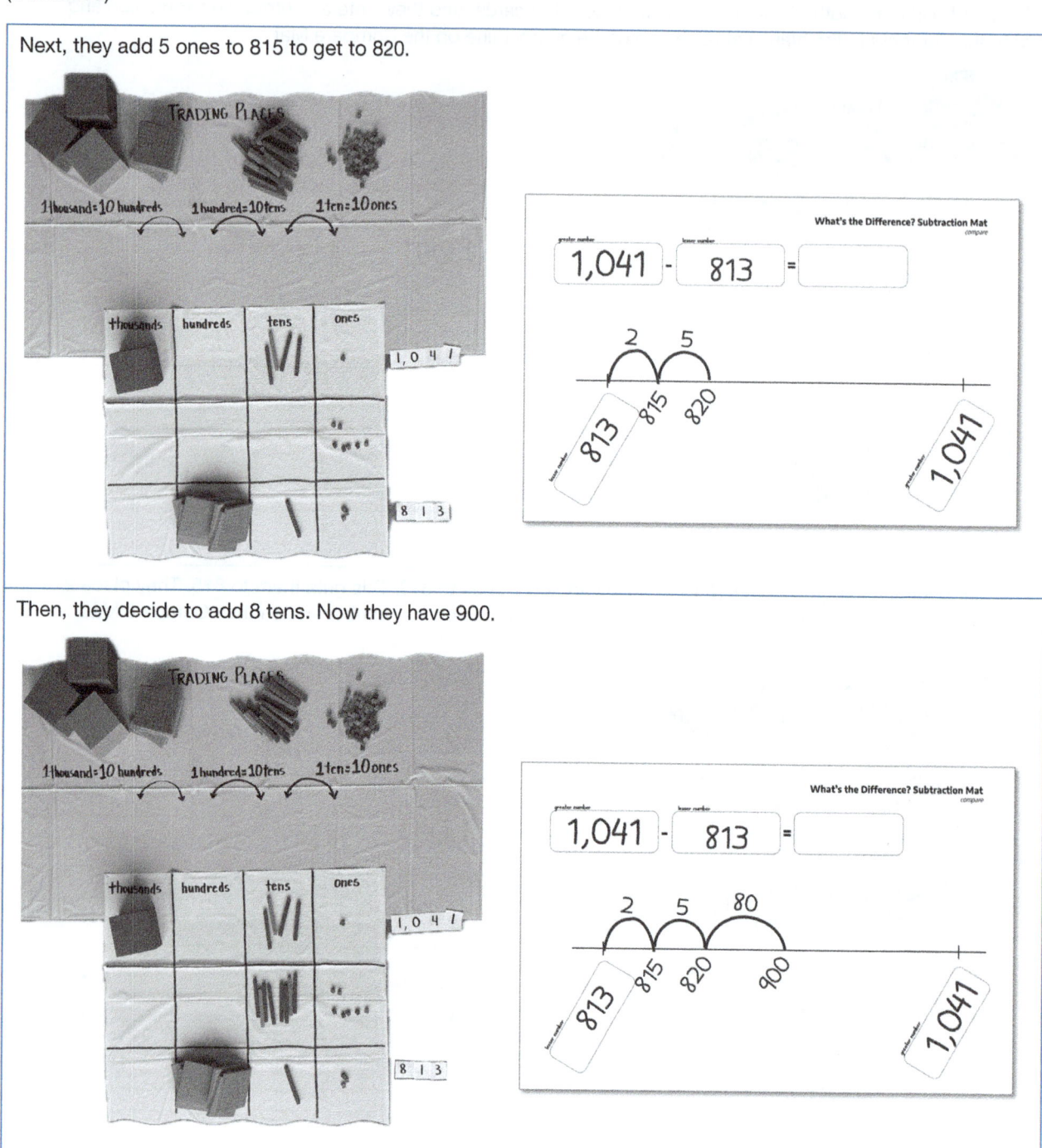

Then, they decide to add 8 tens. Now they have 900.

After some trial and error, the pair decides to add 1 hundred.

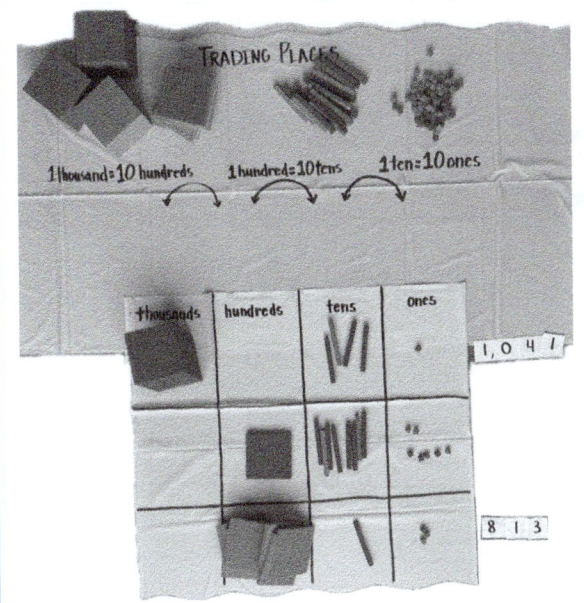
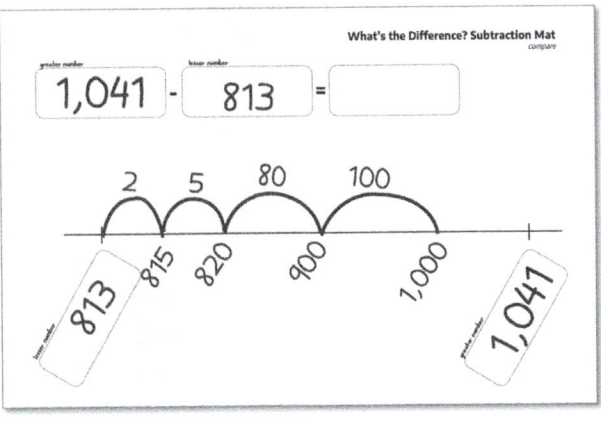

At this point, the pair notices that they can add 41 (4 tens and 1 one) to get to Partner A's number.

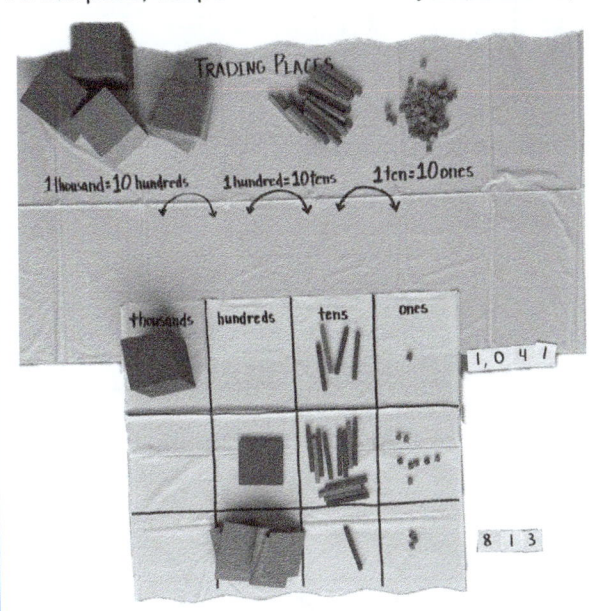
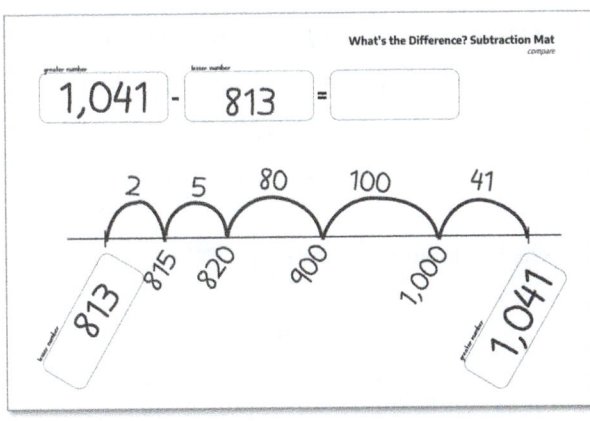

(Continued)

(Continued)

They look at their difference amount and see that they can group 10 tens to make another hundred.

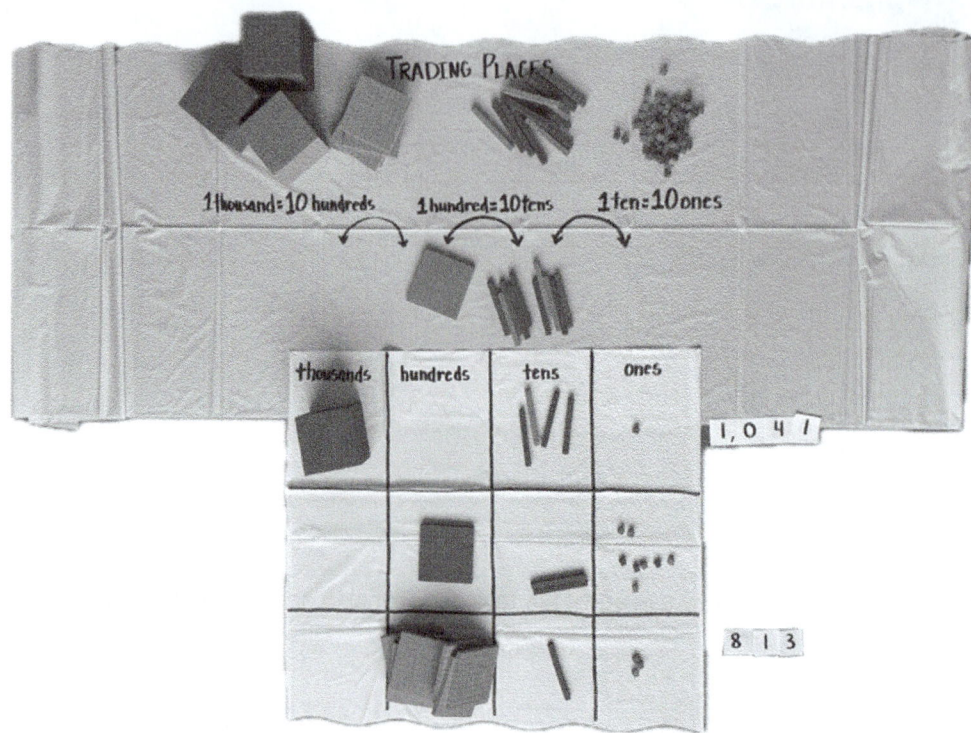

After regrouping, they **compare** their numbers.

- Partner A says, "1,041 is 228 *more than* 813."
- Partner B says, "813 is 228 *less than* 1,041."

They write the difference 228 on their subtracting mat.

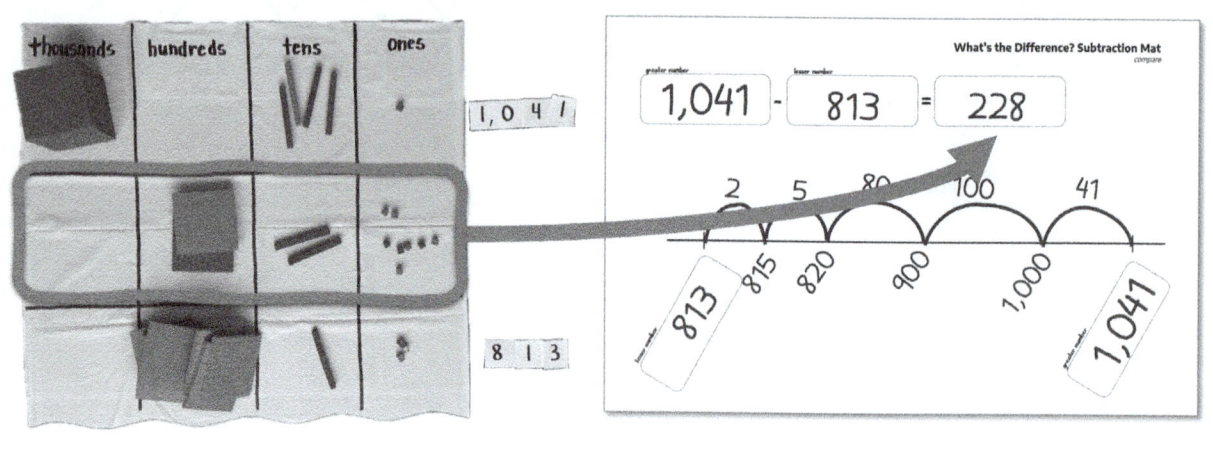

Engage in Math Discourse (Make the Mathematics Visible):

Observation: As students work, encourage them to notice important relationships and push on early and partial understandings.

- Watch for students who add place value blocks onto their place value mats instead of regrouping. Ask them, "Are we adding? Did you mean to add 10 (tens)?
- How might you get 10 (tens) without **adding** 10 (tens) to our number?
- Can you think of another way to represent your original number where you would have some additional tens?"
- Be on the lookout for students who are making a great number of very small lengths on the number line (possibly counting up by ones). **Highlight a strength:** Ask what kinds of numbers they are trying to "get to" on the number line. Celebrate when they talk about wanting to get to useful benchmarks (fives, tens, hundreds, etc.). These ideas will be great to lift up for group discussion. Then, challenge the group to try for a greater benchmark. For example, if students are only counting by ones to the next "five" on the number line, ask them how they might use tens instead or even hundreds.
- When you started to add up from Partner B's amount, how did you decide what to add first? Did you add (hundreds) first? or ones? Why?
- Find the sum of the lengths you made on your number line. What do you notice?

Interview: After the activity, bring the intervention group together for a discussion. Suggested prompts:

- Did you notice any patterns? What kinds of numbers were you trying to reach on the number line? How did those numbers help you? (elicit that benchmark numbers are helpful)
- What did you notice about the sum of all the lengths you made on the number line? How does that compare to the difference?

Bridging Prompt (Prompt for Classroom Teacher to Use During New Lesson Content):
• Show students a subtraction problem represented on the subtraction mat number line.
• Ask students who participated in the intervention group to explain: How could we use "think addition" to find a difference if we don't want to subtract?

TASK 6: Targeted Sum

General Objective:

Students can estimate by rounding and add within 1,000 using models, math sketches, and number strategies based on place value and estimation.

This activity might be used with:

Students who are able to (prerequisite knowledge):	Students who are Primed to (getting ready to learn):
• use estimation and add two-digit numbers using models, math sketches, and number strategies.	• estimate by rounding and use place value structures to add three-digit numbers using models, math sketches, and number strategies. • add and subtract within 1,000 using concrete models or math sketches and strategies based on place value, properties of operations, and/or the relationship between addition and subtraction; relate the strategy to a written method.

Materials:

- Place value blocks
- PRINTABLE: 0–9 Digit Cards

  or

  Index cards with written numerals
- PRINTABLE: Target Sum Challenge Cards
- DIGITAL RESOURCE (number tiles): https://polypad.amplify.com/p#number-tiles

 Printables for this task are available for download at **https://companion.corwin.com/courses/ProactiveMathIntervention**.

Recommended Children's Literature:

- *Betcha!* – Murphy, 1997
- *Counting on Frank* – Clement, 1994

- *Nicky and the Big Bad Wolves* – Gorbachev, 2000
- *Great Estimations* – Goldstone, 2006

Task Overview:

Students will use their reasoning to explain how they arranged the digits to get to a targeted sum as well as explain their addition strategies. In the Targeted Sums activity, students work in pairs to create two three-digit numbers from six drawn 0–9 digit cards, aiming to get a sum close to a target number selected from the target sum challenge cards (in this case 1,000) by using estimation and rounding. They play with another team, sharing their strategies with numbers, math sketches, and place value blocks to explain their thinking on place value and digit placement. As they play, they demonstrate their understanding that in adding or subtracting three-digit numbers, one adds or subtracts hundreds and hundreds, tens and tens, ones and ones; and sometimes, it is necessary to regroup by composing or decomposing ones, tens, or hundreds, etc. After the game, a group discussion helps students reflect on estimation, the impact of digit placement on sums, and visualizing the target number 1,000 using place value blocks. The ability to connect representations is essential.

Select (or create) a variation that focuses on the prior knowledge you are working to shore up for your student group.

Variation 1

Learning Target: Students use estimation by rounding to create two three-digit addends with a sum close to 1,000.

Variation Directions:

For the initial game of Targeted Sums, ask students to pair up and play against a team to make the greatest sum. Each team will take a turn drawing digit cards so that they have six cards in total. Then, both teams will use these same six digit cards to make two three-digit numbers. Students should estimate the sum. They will add these three-digit numbers together to see which team gets the greatest total amount using the same six digits. Then, the other team selects the six digit cards. After several rounds, students can choose another challenge for their learning target sum. See Variation 2.

- Use the six number cards to make two three-digit numbers. First, make an estimate. Then, add to make a sum that is closest to 1,000.
- Pair up with a partner and play against another team.
- To start the game, one team randomly selects six digit cards from the pile.
- Using the six digits, both teams create two three-digit numbers to add to get closest to the target number. Have students show the opposing team the two addends and the sum using numbers, math sketches, and/or place value blocks to explain your addition strategy.

Figure 5.6.1 0–9 digit cards

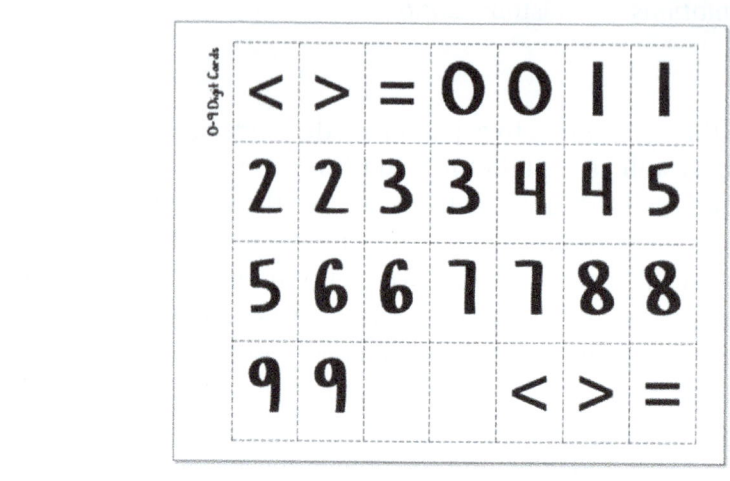

Figure 5.6.2 Targeted Sum Variation 1 example

Kendra and Jeremy pair up to play a game with Emma and Zach. They picked 0, 2, 3, 7, 8, and 9 as the six digit cards. They chose their target sum challenge card and it is "nearest to 1,000." How can they arrange their cards to get closest to 1,000?

Kendra made the numbers 703 and 289 and rounded 703 to 700 and 289 to 300 and estimated that these two addends would get her closest to 1,000. Then she added mentally.

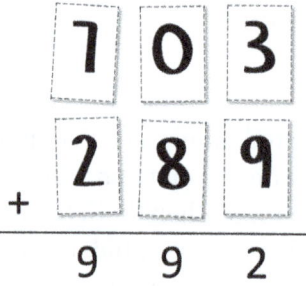

Jeremy made 793 and 280 and estimated by rounding 793 to 800 and 280 to 300 and thought it would be greater than 1,000 but wanted to see how close it was from 1,000. He used partial sums to add and saw that he was over by 73.

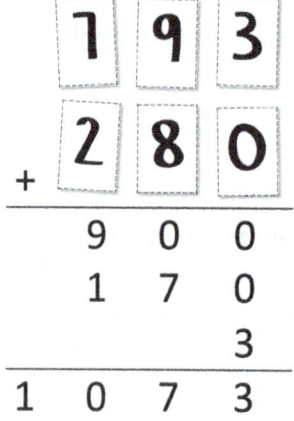

Emma and Zach made the numbers 708 and 293 and rounded, thinking 708 would be close to 700 and 293 close to 300 and thought their number would be closest to 1,000. They represented their numbers with base ten blocks and found a sum of 1,001.

900 + 90 + 11 = 1,001

Engage in Math Discourse (Make the Mathematics Visible):

Observation: As students work, encourage them to notice important relationships and push on early and partial understandings.

- Have students think about place value as they try the Target Challenge Game.
- Using the place value blocks, remind students that the position of the digit determines the place value.

Interview: After the activity, bring the intervention group together for a discussion. Suggested prompts:

- How could you use estimation to get closest to 1,000?
- How is place value important when you add three-digit numbers?
- How does the placement of the digits change the sum?

Show Me:

- How do we show what 1,000 looks like using base ten materials?

Bridging Prompt (Prompt for Classroom Teacher to Use During New Lesson Content):

- Give a problem with two three-digit numbers. Have students create a model for adding the number using the base ten materials. Select students who participated in the intervention group to explain their solution.
- Ask: How is place value important when you add three-digit numbers?

Variation 2

Learning Target: Students will use addition strategies, estimation, and rounding to find sums that meet specific challenges, such as achieving the sum closest to a target number, the greatest sum, or the least sum, and will explain their reasoning to compare results.

Variation Directions:

Students use addition strategies to play variations of the Target Challenge game, which can involve adding two-digit or multidigit numbers with adjusted sums. Different challenges include finding sums nearest to 500, finding the greatest sum, or finding the smallest sum, and students are encouraged to round and estimate before adding. After the activity, teachers should facilitate discussions, use formative assessments, and provide feedback to help students explain their thinking and compare their sums to the target number.

You can play Target Challenge using different challenges such as sum nearest 500, greatest sum, and least (smallest) sum. Encourage students to round before adding to estimate the answer. To find who is the closest, ask students to find how close their sum is to the target number. They can also compare sums of two three-digit addends to see which team is closest.

Figure 5.6.3 Target Sum challenges

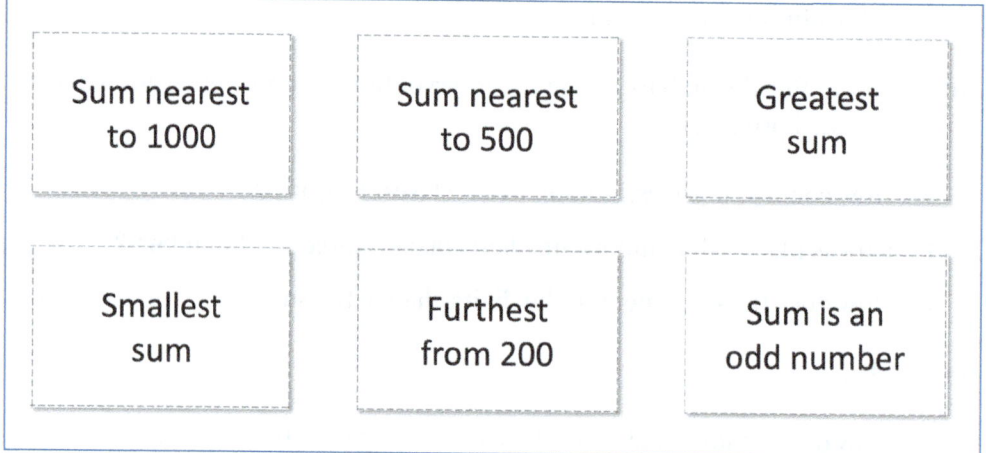

Engage in Math Discourse (Make the Mathematics Visible):

Observation: As students work, encourage them to notice important relationships and push on early and partial understandings.

- Explain when you have to regroup to add.
- How did you compare the sums to see who was closest to the targeted sum?
- Encourage students to fully explain their thinking using models, symbols, and words.

Interview: After the activity, bring the intervention group together for a discussion. Suggested prompts:

> What strategy (or strategies) did you use to determine how you arranged your digits for different versions of the game? Greatest sum? Smallest sum?

	*Bridging Prompt (Prompt for Classroom Teacher to Use During New Lesson Content):* • Play a round of Targeted Sum, explaining the game as a class. Have students who participated in the intervention group explain the game and play a round, explaining their thinking to the class, then have the class play. • Ask: How does the placement of the digit change the sum?

Targeted Differences

General Objective:

Students can estimate and subtract within 1,000 using models, math sketches, and number strategies based on place value and estimation.

This activity might be used with:

Students who are able to (prerequisite knowledge):	Students who are Primed to (getting ready to learn):
• use estimation and subtract two-digit numbers using models, math sketches, and number strategies.	• estimate and use place value structures to subtract three-digit numbers using models, math sketches, and number strategies.

Materials:

- Place value blocks
- PRINTABLE: 0–9 Digit Cards
- PRINTABLE: Target Differences Challenge Cards
- DIGITAL RESOURCE (number tiles): qrs.ly/a5gjcaq

 Printables for this task are available for download at **https://companion.corwin.com/courses/ProactiveMathIntervention**.

Recommended Children's Literature:

- *Shark Swimathon* – Murphy, S., 2000

Task Overview:

Students will use their reasoning strategies to explain how they arranged the digits to get to a targeted difference as well as explain their subtraction strategies. Students will practice their estimation and subtraction skills by using 0–9 digit cards to form two three-digit numbers, aiming for the least possible difference between them. Working in pairs, students will focus on using logical placement of digits and subtraction strategies within 1,000. This activity encourages reasoning about place value, estimation, and comparison to achieve the targeted difference.

Select (or create) a variation that focuses on the prior knowledge you are working to shore up for your student group.

Variation 1

Learning Target: Use the number cards to create two three-digit numbers (a minuend and a subtrahend) using estimation skills to make the least (smallest) difference between the two numbers.

Variation Directions:

1. Pair up with a partner to play with another team.
2. The first team draws six random digit cards from the pile.
3. Using these six digit cards, each team works together to create two three-digit numbers that, when subtracted, produce the smallest possible difference.
4. Show your work by identifying the minuend (the number being subtracted from) and the subtrahend (the number being subtracted) along with the difference, using numbers, math sketches, and/or base ten blocks to explain your subtraction strategy.
5. Then the other team draws cards for the next round.

Figure 5.7.1 Targeted Differences Variation 1 example

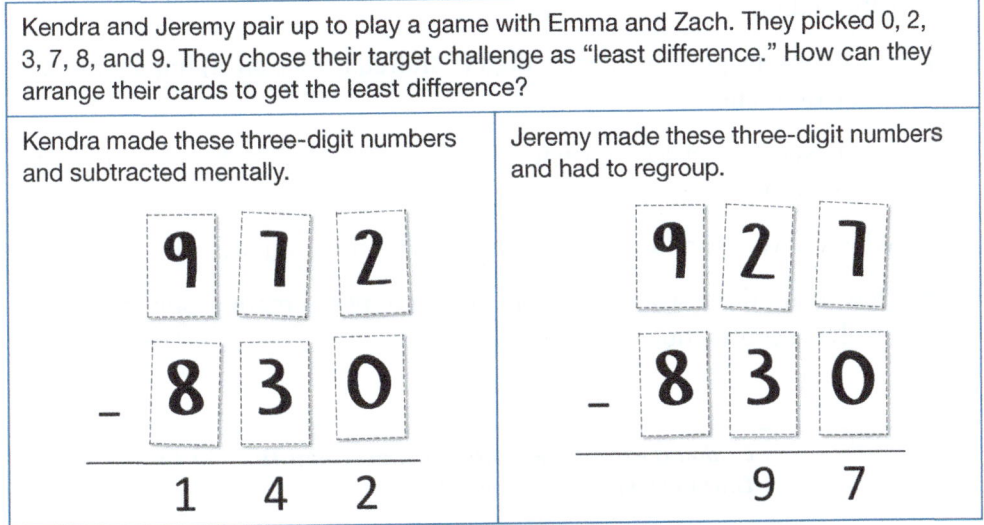

Engage in Math Discourse (Make the Mathematics Visible):

Observation: As students work, encourage them to notice important relationships and push on early and partial understandings.

- What strategies are you using to place the digits in each number?
- How did you decide which numbers should be in the minuend and which should be in the subtrahend?
- Can you estimate the difference before calculating it exactly?

- How does changing one digit affect the difference?
- If you switch the digits in the hundreds or tens place, what happens to the difference?
- Are there other combinations of digits you could try to get a smaller difference?
- How does understanding place value help you arrange the digits?

Interview and **Show Me:** After the activity, bring the intervention group together for a discussion. Suggested prompts:

- How could you use estimation to get the least (smallest) difference?
- How does the placement of the digits change the difference?
- When do you have to regroup when subtracting?
- What was the smallest difference you were able to create?
- What was the most challenging part of choosing the numbers to subtract?
- What strategies worked best to minimize the difference?
- How did using base ten blocks or math sketches help you understand the subtraction?
- How does the position of each digit (hundreds, tens, ones) impact the size of the difference?
- If you were to play again, what strategies might you use to create an even smaller difference?
- What did you learn about subtraction and estimation from this activity?
- How might these strategies apply to other types of math problems involving subtraction or place value?

	**Bridging Prompt (Prompt for Classroom Teacher to Use During New Lesson Content):** • Give the class six digits and ask them to create two different numbers. Then suggest they subtract one from the other. Select students who participated in the intervention group to share their strategies and solutions. • How does the position of each digit (hundreds, tens, ones) impact the size of the difference?

Variation 2

Learning Target: Students will use reasoning, place value knowledge, and subtraction strategies to reach each target. Use the number cards to create two three-digit numbers that achieve the specified target difference.

Variation Directions:

1. Pair up with a partner to work collaboratively.
2. Draw six random digit cards from the pile.
3. Using these six digits, work together to create two three-digit numbers that, when subtracted, achieve the specified target difference.

   Variations of Targets
   - Greatest Difference: Arrange the digits to create the two three-digit numbers with the greatest possible difference when subtracted. Identify the minuend (the number being subtracted from) and the subtrahend (the number being subtracted) and explain your strategy for maximizing the difference.
   - Closest to 200 Difference: Arrange the digits to create two three-digit numbers that, when subtracted, produce a difference as close to 200 as possible. Show your work, and discuss the adjustments made to reach a difference near this target.
   - Even Number Difference: Arrange the digits to create two three-digit numbers that, when subtracted, produce a difference that is an even number. Verify that the difference is even and use numbers, math sketches, or base ten blocks to show your reasoning.
4. Show your work by clearly identifying the minuend and subtrahend along with the difference. Use numbers, math sketches, and/or base ten blocks to explain your subtraction strategy.
5. Reflect on how the target number influenced your choice of digits and discuss any challenges or insights from the activity.

You can play Target Challenge using different challenge cards such as greatest difference, difference closest to 200, and difference is an even number. Encourage students to round before subtracting to estimate. To find who is the closest, ask students to find how close their difference is to the target number. They can also compare the differences to see which team is closest to the target difference.

Figure 5.7.2 Target Differences challenges

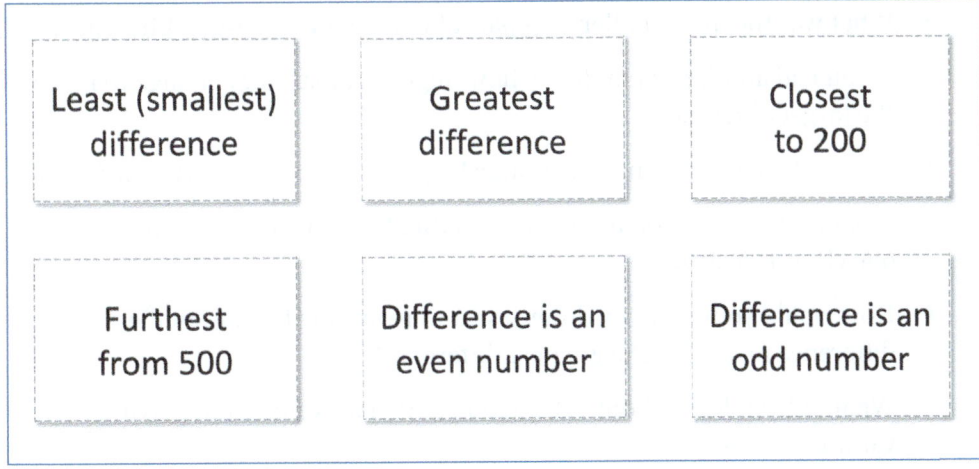

Engage in Math Discourse (Make the Mathematics Visible):

Observation: As students work, encourage them to notice important relationships and push on early and partial understandings.

- Greatest Difference:
  - What placement of digits will give you the greatest difference?
  - How does placing larger digits in the hundreds place affect the difference?
  - Could switching the digits in the tens or ones place make the difference greater?
  - How does your strategy change when you are finding the least difference compared to when you are finding the greatest difference?

- Closest to 200 Difference:
  - How close are you to a difference of 200? What adjustments could you make to get closer to 200?
  - How can estimating help you decide where to place the digits?
  - If your difference is a little over or under 200, what happens when you switch two digits?

- Even Number Difference:
  - How can you ensure your difference is even? What patterns do you notice?
  - What happens to the difference when you change a digit in the ones place? Tens place?
  - Can you predict whether the difference will be even or odd before calculating? How?

Interview and **Show Me:** After the activity, bring the intervention group together for a discussion. Suggested prompts:

- What strategies worked best for achieving your target difference (greatest, closest to 200, or an even number)?
- What was the most challenging part of reaching your target difference?
- If you had another set of digits, how might your strategy change to reach the target difference?
- How did understanding place value help you achieve the target difference?
- What patterns did you notice when trying to create an even difference or one closest to 200?
- How would your strategy differ if you were aiming for the smallest difference instead of the greatest difference?
- What did you learn about subtraction and place value from playing each variation of the game?

Bridging Prompt (Prompt for Classroom Teacher to Use During New Lesson Content):

- Try one of the Target Challenge Activities with the whole group. Bring the students who participated in the intervention group up to show a sample problem (can be one they used during intervention time). Then, ask those students, "What strategies worked best for achieving your target difference (greatest, closest to 200, or even)?" Then give the rest of the class a chance.

TASK 8

Valuable Digits

General Objective:

Students will be able to confidently read, write, and compare multidigit numbers.

This activity might be used with:

Students who are able to (prerequisite knowledge):	Students who are Primed to (getting ready to learn):
• identify the place and value of each digit in a three-digit whole number.	Whole Number Place Value • read, write, and identify the place and value of each digit in a multidigit whole number. • compare and use >, =, and < symbols to record the results of comparisons. Decimal place value • compare and order three decimals to the thousandth place from least to greatest or greatest to least.

Materials:

- Base ten manipulatives
- PRINTABLE: Valuable Digits Game Boards
- PRINTABLE: 0–9 Digit Cards

   or

   Deck of cards using digit cards (no picture cards)
- DIGITAL RESOURCE (number tiles): qrs.ly/a5gjcaq

 Printables for this task are available for download at **https://companion.corwin.com/courses/ProactiveMathIntervention**.

Recommended Children's Literature:

- *How Much Is a Million? (¿Cuánto Es Un Millón?)* – Schwartz, D., 2001
- *House With 100 Stories* – Iwai, 2023

- *Thousand Star Hotel* – The Okee Dokee Brothers, 2017
- *Piece = Part = Portion: Fraction = Decimal = Percent* – Gifford & Thaler, 2008

Task Overview:

By having concrete and/or virtual manipulatives, students can see the structure of the base ten system, where a digit moving to the left one place value represents 10 times more and moving to the right one place value (yes, even to decimals) represents one-tenth as much. This game is called Valuable Digits because as students play this game and place digits on their place value game board and compare numbers, we want them to say the value of the digit (being determined by its place).

Note that this game can be used for a wide range of learners working with two-digit to multidigit whole numbers and decimals.

The goal of the game is to create the largest possible number from the digits that each student picks from a stack of digit cards. This is a game that students can play many times in each of its many variations.

Select (or create) a variation that focuses on the prior knowledge you are working to shore up for your student group.

Variation 1

Learning Target: Students will build and compare multidigit whole numbers and decimals by creating numbers on a game board, reading them aloud, writing them using expanded form, and explaining how place value impacts the size of each number in the context of our base ten system.

Variation Directions:

Organize students into pairs or teams of two. Each group needs one set of digit cards per player/team and an extra set of digit cards for the "draw pile."

Players take turns drawing a digit card and placing it on their game board to build a multidigit number without showing their opponent. Before each round begins, the pair will decide on a goal: (1) build the greatest number, (2) build the least number, or (3) build the number nearest in value to "a number of your choice" (selected by the players). Once all digits are placed, players reveal their numbers, read them aloud, and use comparison symbols (> , = , <) to determine who has the greater or lesser number or closest, depending on the goal for that round. Players need to build their numbers with place value blocks (if playing with Whole Number Game Board A), write them in expanded form, and read the number aloud. They should also be able to explain how place value changes the value of a digit with each position.

Rules of the game:

1. Students choose a goal for the round:
    a. build the greatest number
    b. build the least number

c. build the number nearest in value to "a number of your choice" (for this goal, the players need to agree on a target number before play begins)

2. Student A and Student B place their game board hidden from view.

3. Students A and B take turns picking the digit cards or cards from a deck (remove all picture cards or jokers) where ace = 1 and ten = 0.

4. As each digit is picked, both Student A and B place the digit on their game board, hiding where they have placed it from their opponent.

5. When the game board is filled with digits, students reveal their game board and compare who had the greater number (or whatever the goal). The first person to reach the goal must read it and use the >, =, and < symbols to make a winning statement as they compare with the other player.

6. They play another round, and this time, they can choose a different goal.

Figure 5.8.1 Valuable Digits Variation 1 example (playing "the greatest number" with Whole Number Game board A)

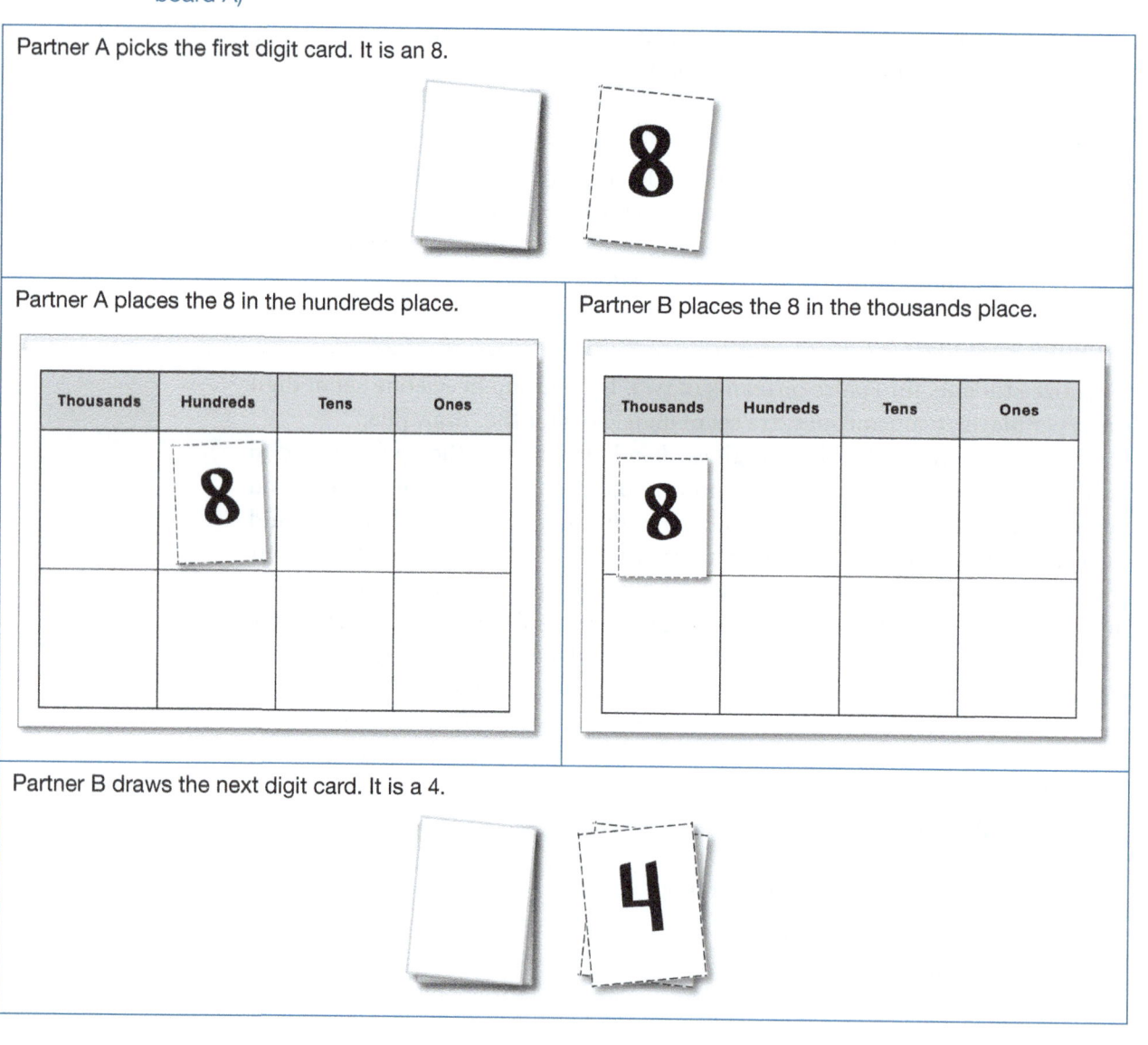

Partner A places the 4 in the tens place.

Partner B places the 4 in the ones place.

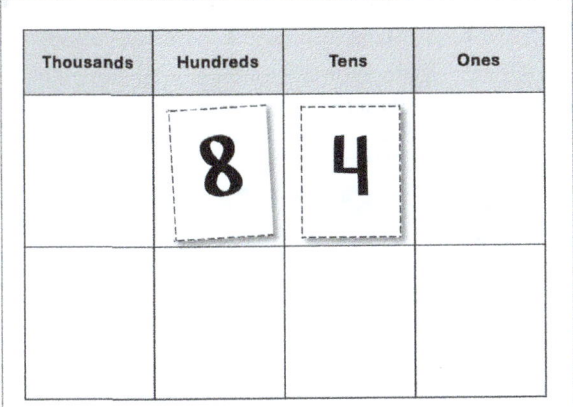

Partner A draws the next digit card. It is another 8.

Partner A places this 8 in the thousands place.

Partner B places this 8 in the hundreds place.

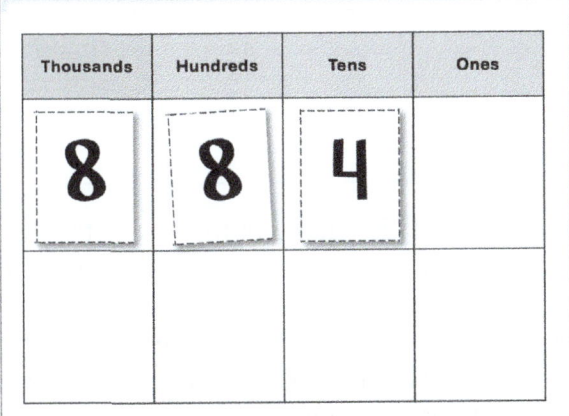

Partner B draws the last digit card. It is a 2.

(Continued)

(Continued)

Partner A places the 2 in the ones place.	Partner B places the 2 in the tens place.
Thousands / Hundreds / Tens / Ones: 8 8 4 2	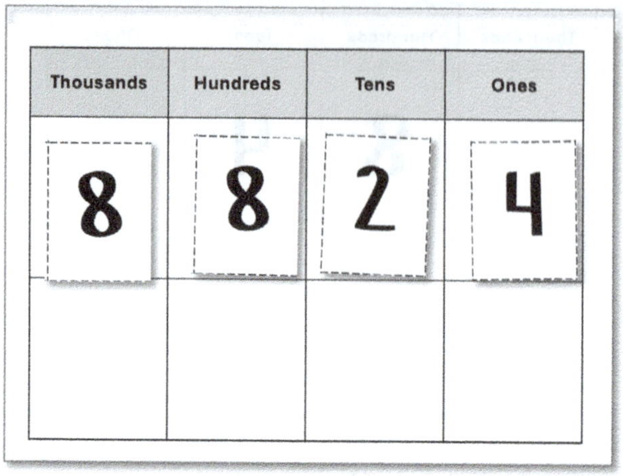 Thousands / Hundreds / Tens / Ones: 8 8 2 4

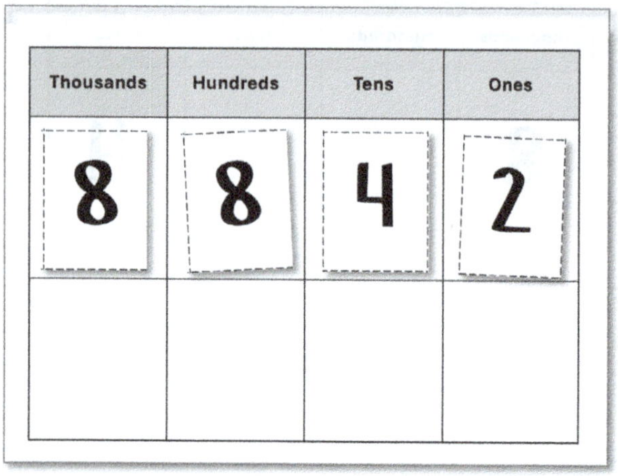

The partners reveal their boards and compare their numbers.

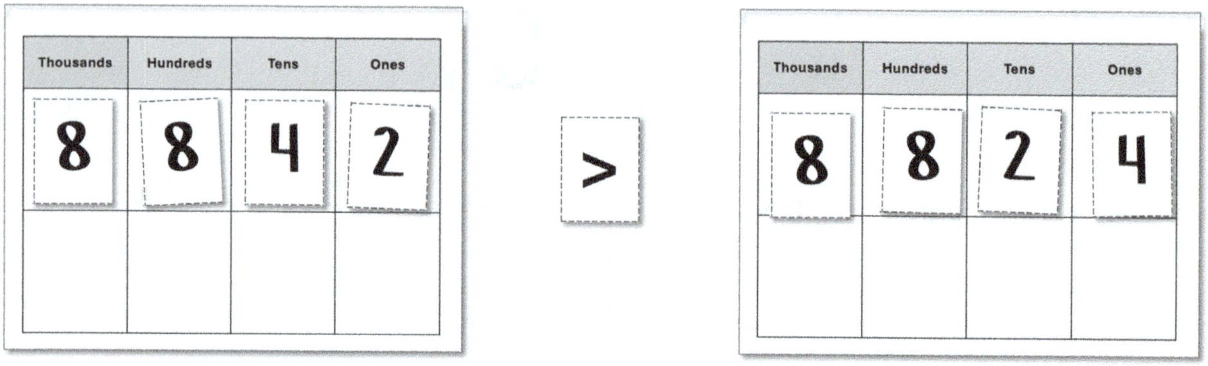

8 8 4 2 > 8 8 2 4

Partner A has won, so partner A says, "8,842 is greater than 8,824."

Other Adaptations—One way to adapt this game is to start with Whole Number Game Board A. Work up confidence reading three-digit numbers then to four digits and so on. An adaptation that supports the CSA progression is to have students build the number using place value blocks. After building it, they then read it with the written form matching the number. This would also be a great opportunity to use the expanded form stating the equivalence:

Figure 5.8.2 Equivalent representations of 2,346

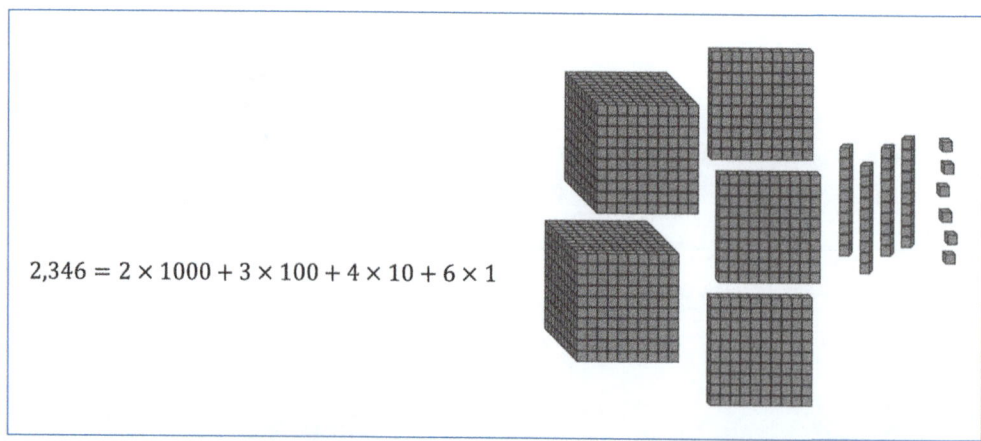

$2{,}346 = 2 \times 1000 + 3 \times 100 + 4 \times 10 + 6 \times 1$

Use the whole numbers to a thousand or a million place value game, or decimal place value game to relate the importance of the structure of our base ten system. This game can be played to preview grade-level standards of reading, writing, and comparing whole numbers and decimals.

Figure 5.8.3 Valuable Digits Whole Number Game Board B

Millions	Hundred Thousands	Ten Thousands	Thousands	Hundreds	Tens	Ones

Engage in Math Discourse (Make the Mathematics Visible):

Observation: As students are playing the game, they need to think about the structure of the base ten system and the importance of "the place of a digit tells us its value." This phrase can be written down and repeated verbally throughout the game. As students work, encourage them to notice important relationships and push on early and partial understandings.

- Observe how students position digits like 0, 1, and 2 and digits 8 and 9. Ask them how the goal they selected impacts their placement decisions.

- Remind students that we don't know what card might come next, so there is an element of chance. There are also times when one can think about patterns to be strategic. For example, if they know that two 9s are in the pile, the chance of another 9 coming up is impossible if we have already used those two cards.

Interview and **Show Me:** After the activity, bring the intervention group together for a discussion. Suggested prompts:

- How did the placement of the digits change the value of the number? What was a strategy that you used in the game to win?

- What's a digit you would want if you were trying to create the greatest number? Where would you put it? What digit would you want if you were creating the least number?

- How does playing different versions of the game such as building the greatest number and building the least number change your strategy in placing your digits?

- How does the concept of base ten help you understand the value of each digit in a number?
- Why is it important to understand that each place value represents a power of ten?

Bridging Prompt (Prompt for Classroom Teacher to Use During New Lesson Content):

- Play a quick round of the game with several students who participated in the intervention group in front of the whole class where it is teacher versus students. As the numbers are called, ask the students to share their thinking. Ask, "Where might you position digits like 0, 1, and 2 and digits 8 and 9 when playing the version of the game of building to the greatest number versus building to the least number?"

Variation 2

Learning Target: Students will be able to read and write decimals using base ten numerals, number names, and expanded form for whole numbers less than 1 million. Compare two decimals based on the value of the digits in each place, using >, =, and < symbols to record the results of comparisons.

Variation Directions:

This variation follows the same game rules as Variation 1, except students will use the Decimal game board. You can add an element of excitement with building one's own number by picking their own digits for the game board and giving students one opportunity to discard one card.

Figure 5.8.4 Valuable Digits Decimal Game Board

Ones	Tenths	Hundredths	Thousandths

- Similar to the previous example, a great visual for decimals that use the CSA progression is to say to the students, "Let's assume this big place value cube that we used for a thousand is now one."

Figure 5.8.5 Using place value blocks to represent decimals

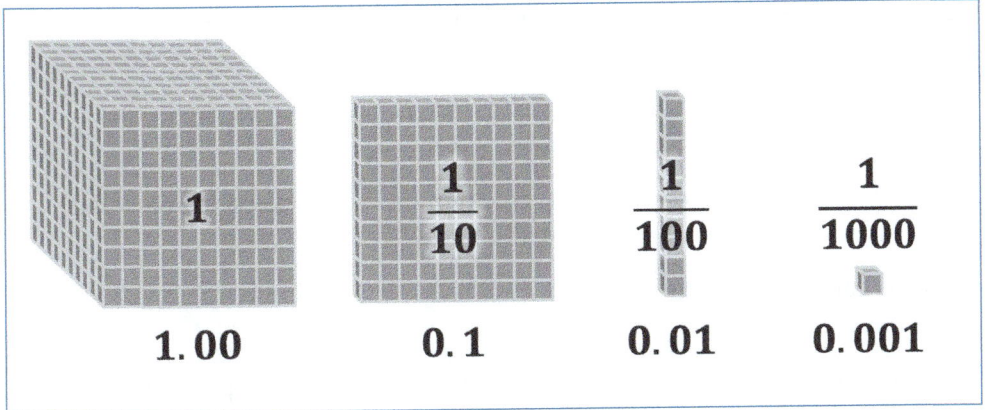

- In this intervention, we want them to generalize that the value of the digit is dependent on where it is positioned on the place value chart. We also want them to recognize that moving to the left in each place value represents a value 10 times greater, and moving to the right in each place value (yes, to decimals) represents the number's value as one-tenth of the previous place value.

Engage in Math Discourse (Make the Mathematics Visible):

Observation: As students work, encourage them to notice important relationships and push on early and partial understandings.

- Between your number and your partner's, which one is greater? Can you explain why?

- How do the symbols >, =, or < help you show which number is greater or smaller in the equation or inequality you wrote?

- How does the position of a digit affect its value when comparing two numbers?

- How does playing with decimals change your strategy for where you will place a number? Or is it the same?

- When we add a decimal point to your number, how does the place value of each digit change?

- If you add a digit in the tenths place, how does that compare to adding that same digit in the hundredths place?

- What do you notice when you move the decimal point one place to the right in a number? What about when you move it one place to the left in a number?

 **Interview** and **Show Me:** After the activity, bring the intervention group together for a discussion. Suggested prompts:

- What strategy did you use to compare numbers? How did the position of each digit influence your decisions?
- How does the concept of base ten help you understand the value of each digit in a number?
- Why is it important to understand that each place value represents a power of ten?
- In what ways are comparing whole numbers and comparing decimals similar? In what ways are they different?
- What did you learn about how decimal points affect the value of a number?
- Can you give an example of a number in which a digit's position changes its value dramatically?
- Where might you position digits like 0, 1, and 2 and digits 8 and 9 when playing the version of the game of building to the greatest number versus building to the least number?

> **Bridging Prompt (Prompt for Classroom Teacher to Use During New Lesson Content):**
>
> - Play a quick round of the decimal version of the game with several of the students who participated in the intervention in front of the whole class where it is teacher versus students. As the numbers are called, ask the students to share their thinking. Ask, "Where might you position digits like 0, 1, and 2 and digits 8 and 9 when playing the version of the game of building to the greatest number versus building to the least number?"

Picture This

TASK 9

General Objective:

Students will learn how to draw mathematically useful representations (math sketches) to interpret story problems.

This activity might be used with:

Students who can (prerequisite knowledge):	Students who are Primed to (getting ready to learn):
• read and write math expressions and equations • draw pictures	• draw math representations (math sketches) • match math representations to story contexts • interpret math story contexts using multiple representations

Materials:

- Collections of stories and representations from your curriculum materials
- Index cards
- PRINTABLE: Math Story Cards (Additive Situations A: no question)
- PRINTABLE: Math Story Cards (Additive Situations B: with question)
- PRINTABLE: Math Story Cards (Multiplicative Situations A: no question)
- PRINTABLE: Math Story Cards (Multiplicative Situations B: with question)
- PRINTABLE: Math Representations Game Cards

 Printables for this task are available for download at **https://companion.corwin.com/courses/ProactiveMathIntervention**.

Recommended Children's Literature:

- *The Girl Who Made a Million Mistakes* – Li, 2023 (growth mindset and building confidence)
- *How to Solve a Problem: The Rise (and Falls) of a Rock-Climbing Champion* – Shiraishi, 2020 (keep trying new strategies—learn from mistakes)

Task Overview:

In this task, students will explore math story contexts with a focus exclusively on representations. Students will not be asked to solve the math stories but will engage in building concrete models, drawing semiconcrete representations as math sketches, generating abstract expressions and equations, and matching those representations to story contexts.

Select (or create) a variation that focuses on the prior knowledge you are working to shore up for your student group.

Variation 1

Learning Target: Students will determine characteristics of representations that are mathematically useful for problem-solving.

Variation Directions:

Begin this variation with students together in a whole group. Show and read a brief math story (without a question) to the group. For example: "Miss Andrews has 12 markers. Mrs. Scorrano has 8 markers."

Show the group two images (one image should be a "math sketch" that highlights the mathematical relationship described (Image B), and the other image should be related to the story, but not a math sketch (Image A). See examples in Figure 5.9.1.

Figure 5.9.1 Two example images—one related to the story (A) and one a mathematical sketch of the story (B)

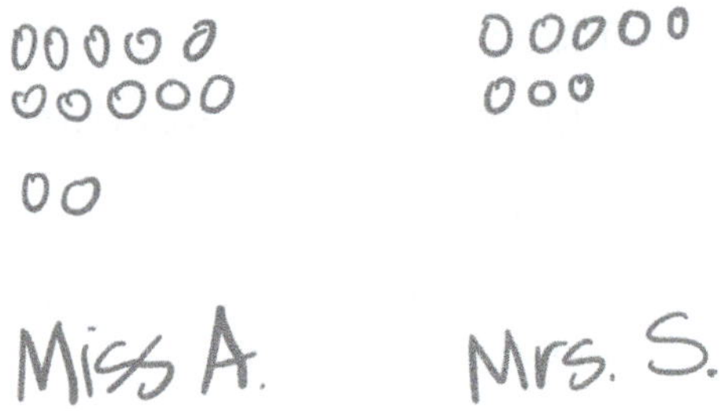

Ask the group, "What do you think these two images might have to do with this math story?" Elicit student ideas about both images. Students should notice that both images represent something from the story. Encourage them to describe what the images tell you about the story—what story details can they see in each image? For example, with the images shown here, students might notice:

CHAPTER 5. Intervention Tasks

Figure 5.9.2 Student ideas about images

Image A	Image B
• it's a marker, and the story was about markers • there's only one, but the story was about more than that • this marker is blue, but the story doesn't tell what color markers either person has	• the circles represent markers • the circles are organized in groups of 5, that makes them easy to count • Miss A has 12 circles because the story says she has 12 markers • Miss A's circles are organized into a group of 10 and then 2 more—that equals 12 • Mrs. S has 8 circles because the story says she has 8 markers • Mrs. S's circles are organized into a group of 5 and 3 more—that equals 8 • Miss A has more markers than Mrs. S • Mrs. S has fewer markers than Miss A

After discussing the images with students, tell them, "One of these drawings is a **math sketch** (Image B), and one of these drawings is the kind of drawing we might do in art class as an illustration (Image A). Talk to your partner. Which drawing do you think is the **math sketch**? Which drawing is the kind of drawing you might do in art class? Which kind of drawing can help you solve a story problem?"

Discuss this together with the group and create a chart together about characteristics of a **math sketch**. Use the following questions to elicit key characteristics for the chart:

Figure 5.9.3 Whole-group chart prompts

Suggested prompt:	Related idea to elicit:
Does an art class drawing or a math sketch give us more information about the math story?	Math sketches show what happened in the story.
Would an art class drawing or a math sketch take more time to make?	Math sketches are *efficient*.
Did the art class drawing or the math sketch show us who had more markers in this story?	Math sketches show *relationships* from the story.
Could we use the art class drawing or the math sketch to find out how many markers there were in all?	Math sketches can help us answer questions about the story.

Next, organize students into pairs and provide each pair a math story problem. (Use one or more sets of math story cards included here, or source math stories from your own curriculum.) Note that it will be helpful to omit the question from each story at first and ask them to make a math sketch to represent their story. Bring the whole group together to discuss students' representations. Add samples of stories and math sketches to the whole-group chart.

Figure 5.9.4   Sample whole-group chart

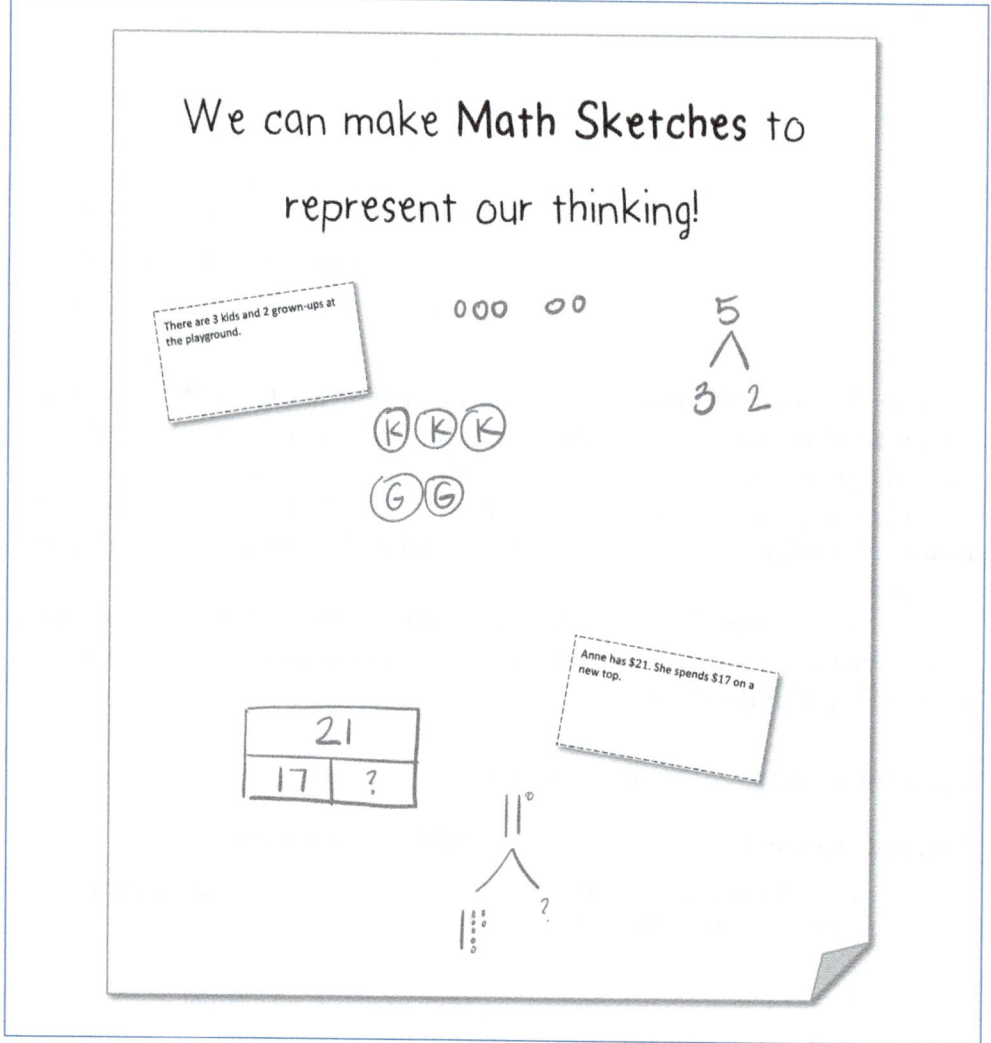

Engage in Math Discourse (Make the Mathematics Visible):

Observation and **Show Me**: As students work, encourage them to notice important relationships and push on early and partial understandings.

- What math representations do you know that might help you represent your math story? (elicit models used in your curriculum—e.g., ten frames, place value materials, area models, models of equal groups, bar diagrams, number lines, etc.)

- How does your math sketch show ___? (pick a detail from the story)
- (indicate a part of the math sketch) What part of the story does this represent? How might you label your sketch so someone else will know that?
- Why is it important to focus on the math relationships rather than just drawing a picture of something from the story problem?

Interview and **Show Me**: After the activity, bring the whole group together for a discussion. Suggested prompts:

- Let's look at some of the different ways we know to represent math stories! What representations do you see?
- Show the intervention group a specific math sketch and read the related story. Ask, "Turn and tell your partner how you see ___ (indicate a detail from the story) in the math sketch."
- What are some questions we could ask about this math story that this sketch could help us answer?

Bridging Prompt (Prompt for Classroom Teacher to Use During New Lesson Content):

- *Show two images, one a math sketch and one an art drawing (can be Figure 5.9.1). Then ask, "How are math sketches different from the drawings we do in art class?"*
- *Consider bringing some of the math sketches and related stories from this activity with the intervention group to the grade-level class to use as examples to create a similar "Characteristics of Helpful Math Sketches" wall chart together with the grade-level class.*

Variation 2

Learning Target: Students will match math stories to corresponding math sketches.

Variation Directions:

Begin this variation with students together in a whole group. Show the group a story card and a math sketch card from the collection of Math Representations Game Cards. Have students partner talk to decide if the math sketch matches the story or not. Elicit student ideas. Encourage students to describe specific aspects of the math sketch and how they do (or do not) match specific details from the math story. Repeat with another set of story and math sketch cards.

When students have a clear understanding of how to identify a "matched set" of cards, organize the group into pairs for the task. Give each pair a set of math representations game cards. The printable game cards provide 12 sample stories with common quantities but varying situations. Representations provided in the printable cards show one possible way to represent each of the 12 stories with a math sketch.

For initial rounds of the game, teachers may wish to have partner pairs play with smaller decks of game cards (perhaps only three stories with three corresponding math sketches). When students are comfortable with the game, they can add to their deck.

Students should lay out their cards face down in an array. Partners will take turns as follows:

1. Partner A flips over one story card and one math sketch card.

2. Partner A decides if the math sketch *matches* the story or *does not match* the story and explains why.

    a. If the math sketch *matches* the story, Partner A gets to keep both cards.

    b. If the math sketch *does not match* the story, Partner A turns both cards back over.

    Either way, partner A's turn ends.

3. Partner B flips over one story card and one math sketch card.

4. Partner B decides if the math sketch *matches* the story or *does not match* the story and explains why.

    a. If the math sketch *matches* the story, Partner B gets to keep both cards.

    b. If the math sketch *does not match* the story, Partner B turns both cards back over.

    Players continue alternating turns until all stories have been matched and removed from play.

Note that while students will have the goal of collecting the most matches (winning the game!), which will keep them engaged, the (teacher) goal for this task is the discussion students will have around *both matches and mismatches*. As partner pairs play, be sure to listen in on their conversations to hear how they are analyzing stories and math sketches.

Teachers may elect to use these preprinted cards initially, but then, have students create their own matching games by adding student-generated representations and/or story contexts from your own curriculum materials. For an added challenge, include multiple representations for each story and challenge students to search for multiple matches.

Figure 5.9.5 Picture This Game example

Players arrange the cards in an array. *Note that story cards have a thought bubble on the back and math sketch cards have a picture frame on the back. Students will turn over one of each for each turn in the matching game.*	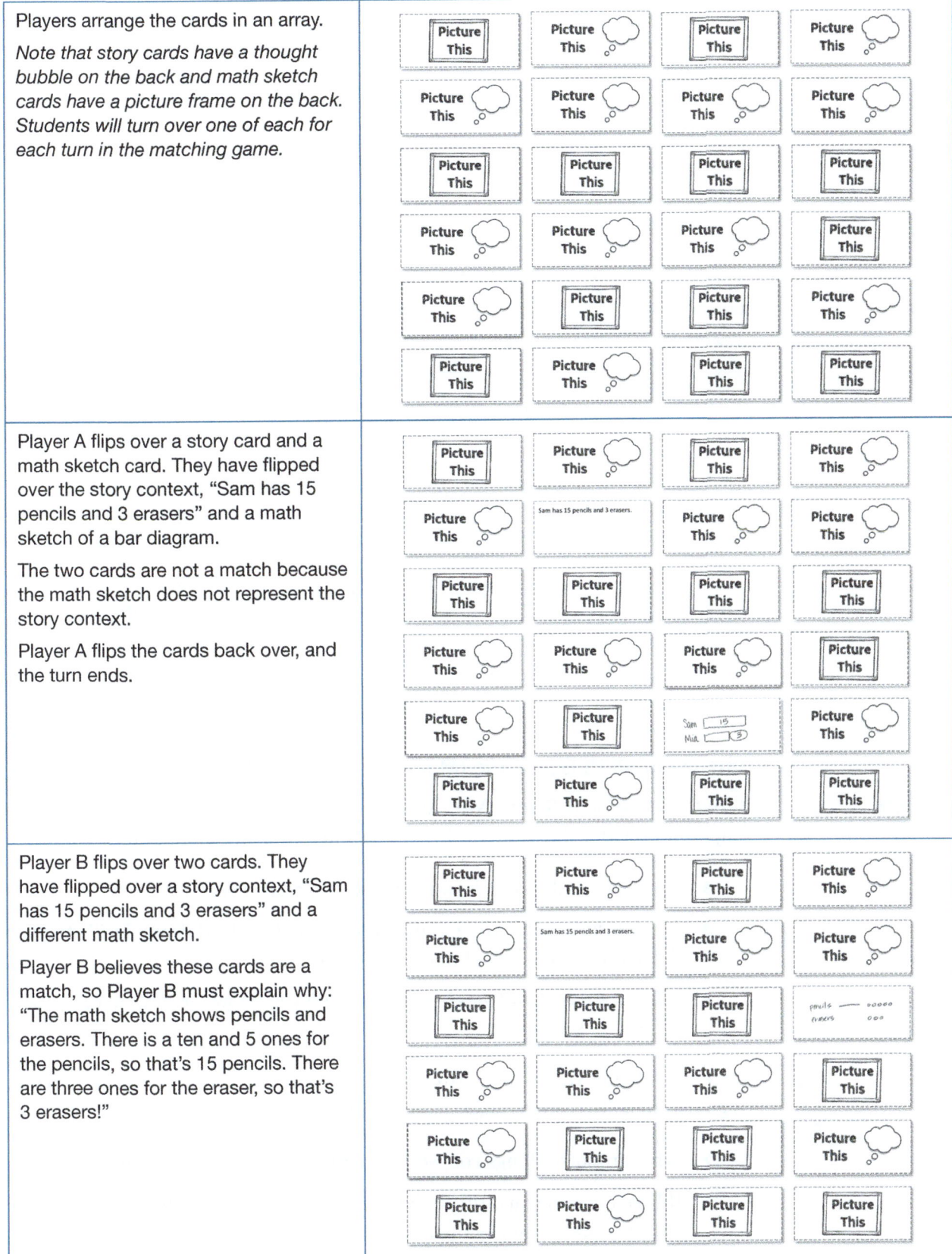
Player A flips over a story card and a math sketch card. They have flipped over the story context, "Sam has 15 pencils and 3 erasers" and a math sketch of a bar diagram. The two cards are not a match because the math sketch does not represent the story context. Player A flips the cards back over, and the turn ends.	
Player B flips over two cards. They have flipped over a story context, "Sam has 15 pencils and 3 erasers" and a different math sketch. Player B believes these cards are a match, so Player B must explain why: "The math sketch shows pencils and erasers. There is a ten and 5 ones for the pencils, so that's 15 pencils. There are three ones for the eraser, so that's 3 erasers!"	

(Continued)

(Continued)

Player B claims the match and gets to keep those cards in their match pile. 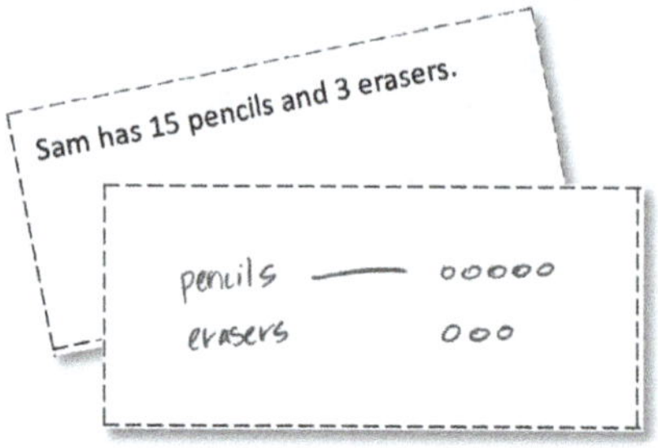 *Sam has 15 pencils and 3 erasers.* *pencils ——— ooooo* *erasers ooo*	
Player A takes another turn, and play continues . . .	(grid of "Picture This" cards with one card showing "Sam has 15 pencils. He has 3 times as many pencils as erasers." and another showing "Sam 15 / Mia 3")

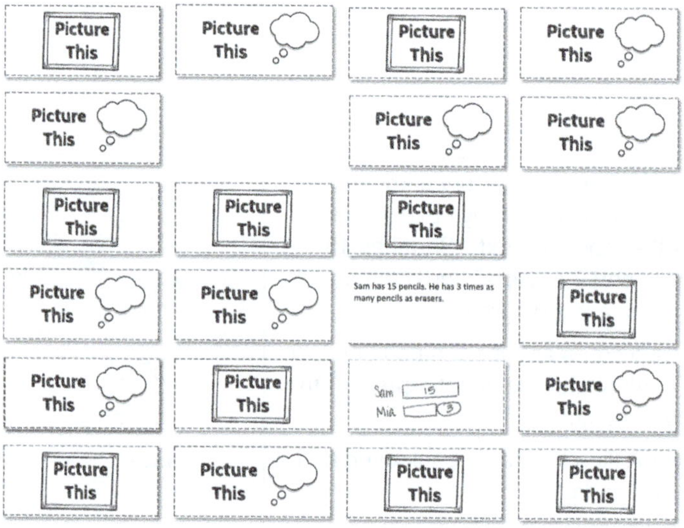

Engage in Math Discourse (Make the Mathematics Visible):

Observation and **Show Me**: As students work, encourage them to notice important relationships and push on early and partial understandings.

- How are you deciding if two cards are a matched set?
- *When a student decides a pair is not a match:* Why did you decide that this math sketch did not match the story?
- *When a student matches a pair:* How did you know for sure that this was a matched set?

Interview and **Show Me**: After the activity, bring the intervention group together for a discussion. Suggested prompts:

- What are some things you were noticing as you played the game?
- If I see the same quantities (the same numbers) in a math sketch and a story, does that make them a matched set?
- Show two cards and ask students if they are a matched set. Ask how they know.

Bridging Prompt (Prompt for Classroom Teacher to Use During New Lesson Content):

- Show a quick version of the game with four cards face down (different from the game cards that will be used). Have two of the students who participated in the intervention come flip two cards and talk about why the two cards match or why there is not a match to the rest of the class.

- Ask students to talk about what parts of the math sketches helped them connect the representations to the stories.

TASK 10: Balancing Act

General Objective:

Students will learn the meaning of the equal sign (=) and recognize the use of a variety of equation formats as options for recording mathematical ideas.

This activity might be used with:

Students who are able to (prerequisite knowledge):	Students who are Primed to (getting ready to learn):
• construct models for computational situations • understand the relational concepts of greater than, less than, and equal to with concrete materials	• accurately interpret the meaning of the equal sign and develop symbol sense • show relational understanding of equivalence of fractions • show procedural fluency with all four operations using whole numbers, fractions, and decimals • use related fact strategies to derive unknown quantities • use reasoning to find answers

Materials:

- One-inch cubes or unit cubes
- Collections of Pattern Blocks – yellow hexagons, red trapezoids, blue rhombuses, and green triangles
- PRINTABLE: Pattern Block Cutouts
- PRINTABLE: Balancing Act Game Board
- PRINTABLE: Balancing Act Game Cards and Relationship Cards
- PRINTABLE: Balancing Expressions
- PRINTABLE: Balancing Shapes
- PRINTABLE: Is It True?

 Printables for this task are available for download at **https://companion.corwin.com/courses/ProactiveMathIntervention**.

Recommended Children's Literature:

- *Equal Shmequal* – Kroll, 2005
- *How Many Elephants in a Blue Whale?* – Weeks, 2010 (Fun comparisons of equality such as the weight of 25 elephants equals one blue whale. Great for creating problems.)

Task Overview:

There is no more important symbol to communicate algebraic thinking in the elementary grades than the equal sign. Knowledge of the meaning of this symbol has been a first-grade standard in almost all U.S. states since 2010, yet there is still evidence that it is poorly understood (Blanton et al., 2018). Avoiding this common hiccup in students is important as early as the intervention can take place. As with all the tasks in this book, these variations can be repeated and can also be used over multiple days.

Select (or create) a variation that focuses on the prior knowledge you are working to shore up for your student group.

Variation 1

Learning Target: Students will move from a focus on seeing the equal sign as a signal to find answers to understanding that the equal sign (and signs for inequalities) is about a focus on relationships between quantities.

Variation Directions:

To begin this variation, students must learn the meaning of the equal sign in conjunction with other relational symbols such as greater than and less than rather than connecting it to operational symbols such as the addition and subtraction signs. From the beginning, do not use the language "makes" or "is" when reading an equation with an equal sign—just say "equals" or "is equal to." We are also starting with concrete materials prior to using math sketches and more abstract problems. Organize students into pairs. Each team needs a Game Board Balance, Balancing Act Game Cards, Pattern Blocks, and one set of Relationship Cards.

Figure 5.10.1 Relationship cards

Start by sharing with the group a situation with 1-inch cubes or unit cubes in which an image of the Balancing Act Game board is projected from a document camera. Set up an inequality in which there are three cubes on one side and one cube on the other (see Figure 5.10.2).

Figure 5.10.2   Cubes on Balancing Act Game board

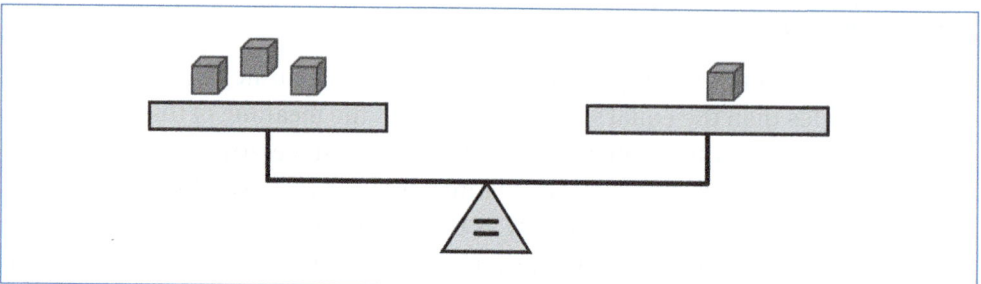

Ask students, does this image show two quantities that will balance? How do you know? What does it show? Who can come up and fix the amount of cubes on the balance so that it is balanced correctly?

Teams of two each get one Balancing Act Game Board. Students will use pattern blocks to model the problems written on the Balancing Act Game Cards. Start by showing students a set of pattern blocks (either concrete manipulatives or see included printable option).

Figure 5.10.3   Pattern Block pieces

Ask students, "Are these pieces the same as the cubes?" [No, they are different]. "Are the pattern blocks all the same size or are there different sizes?" Ask students what pattern block pieces balance or are the same as (equal to) the red trapezoid piece if you match them. Have students model and discuss possible equivalencies on the projected Balancing Act Game board [a trapezoid is the same as another red piece; two blues and a green, etc.].

Figure 5.10.4 Two of the possible matches for the red trapezoid pattern block

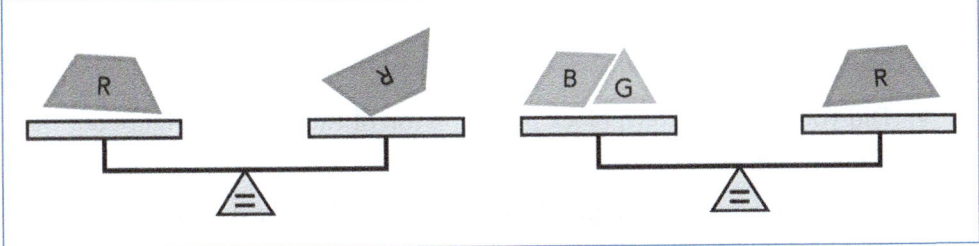

Then, students take turns as follows:

1. Partner A draws a game card. They use the pattern block pieces indicated by the cards and the first letter of the color (Y = yellow hexagon; R = red trapezoid; B = blue rhombus; and G = green triangle) to represent the two parts of the equation. They set the pieces on the balance. Then, they compare the pieces and select the appropriate relationship card – either Greater Than, Less Than, or Equal to. Partners A and B talk to see if their choice is correct. Then they need to say the entire inequality or equation aloud. Such as "1 red trapezoid is greater than 2 green triangles." Each player gets 1 point for each correct answer. Expect the common hiccup that students will be challenged by the similarities of the greater than and less than symbols. As a strategy, ask them to circle the lesser quantity then put in the relational notation. A teacher we collaborate with developed this approach and reports that it works!!!

2. Next, Partner B draws another game card and carries out the same process.

3. Play continues until both players have tried all of the game cards or the instructional time period is 5 minutes from ending.

4. Ask students for an example of each of the three relationships (greater than, less than, or equal to) to be explained to the group.

Engage in Math Discourse (Make the Mathematics Visible):

Observation: As students work, encourage them to notice important relationships and push on early and partial understandings.

- Look for the pattern block materials to be overlaid upon each other to make actual comparisons. Avoid conversations about alligators eating the larger amount or other "tricks" to remember the greater than and less than symbols. Just keep repeating the name of the symbol and focusing on the use of the inequalities to reinforce the understanding of the two symbols.

- Watch for students who may not initially recognize that we are comparing the first amount on the game card to the second. Have students read out the equation or inequality, for example, "1 red trapezoid is equal to 3 green triangles." Ask them to explain their thinking.

Interview and **Show Me:** After the activity, bring the intervention group together for a discussion. Suggested prompts:

- Can you show me what strategies you used to compare the two amounts?
- Did you set the pattern block pieces on top of each other to figure out the relationship each time? How did you see it?
- Choose one or two examples from student work and ask, "Is there a way we could think about the pattern blocks as numbers?"

> **Bridging Prompt (Prompt for Classroom Teacher to Use During New Lesson Content):**
>
> - Show students one of the game card questions and, using a projected Balancing Act Game Board, ask a team of students who participated in the intervention group to come up and explain how they know which relationship symbol to use.
>
> **Figure 5.10.5** Balancing Act Game Board with shapes and symbols
>
> [Balance diagram: 3 R (trapezoids) on left; 1 Y (hexagon) + 1 G (triangle) on right; fulcrum labeled =]
>
> - Ask, "How do the red trapezoid pieces compare to the combination of the yellow hexagon and green triangle pieces? Are they greater than, less than, or equal to? How did you figure that out?"
> - Bring up two students who participated in the intervention group to explain a strategy for how you can accurately decide between the greater than and less than signs.

Variation 2

Learning Target: Students will use illustrations of shapes to continue to explore the relationship of equality.

Variation Directions:

In this variation, students will use semiconcrete models of equal quantities on either side of the Balance to figure out the relationships between the values of the different shapes. You may want to provide the Balancing Act Game Board in sheet protectors so they can use dry-erase markers to mark what they notice.

To begin, go back to the pattern blocks and, on a document camera, show 1 blue rhombus on the left and 2 green triangles on the right sides of the Balancing Act Game Board. Explain that the sides are balanced. Ask, "What does that mean?" What is the comparison? [One blue rhombus is equal to two green triangles.] That can be represented by 1B = 2G.

Figure 5.10.6   Balancing Act Game Board using pattern blocks

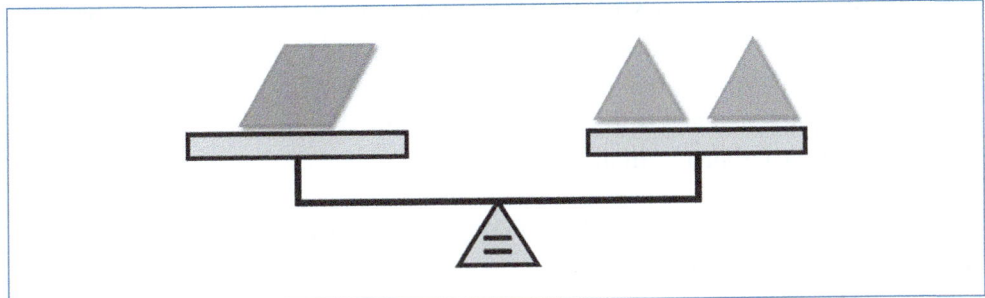

Then, shift to this more abstract variation with an image of different shapes (not pattern blocks). Show the students this image of a Balancing Act Game Board on a projection device:

Figure 5.10.7   Image of balanced shapes

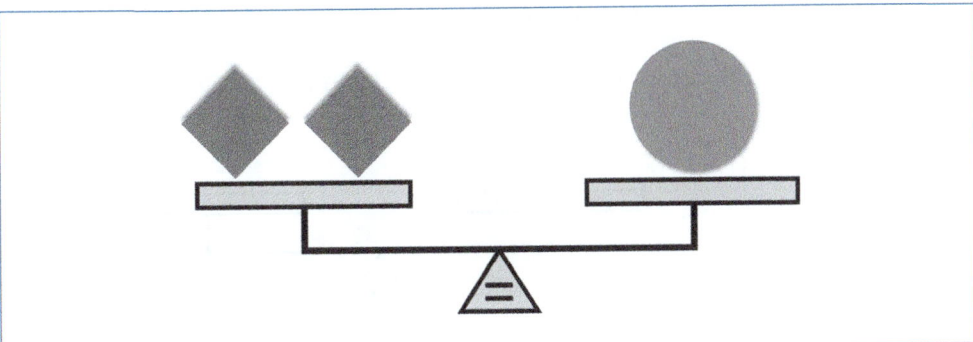

Ask again what it means when the sides are balanced. What does it mean about the weights of the shapes on the two sides of the balance? [The weight of the 2 squares is equal to the weight of the circle.] Discuss what we know about the relationship between the weight of the circle and the weight of the two squares. [The squares weigh half of the weight of the circle, or 2 squares = 1 circle.] Then, share the Balancing Shapes activity sheet for students to work on in pairs. If you need, let them try one more and discuss their thinking. Then, let them move on to the rest of the Activity Sheet.

Engage in Math Discourse (Make the Mathematics Visible):

As students work, encourage them to notice important relationships and push on early and partial understandings.

Observation: Observe whether students are associating values with the shapes, such as saying if the squares each are 1, then the circle is 2.

Show Me:

- Ask students, "If you add another circle on the right side, can you show me what needs to happen to the left side to stay balanced?"

- If you have a balance with the shapes on the scale [show a new problem or select one of the problems on the sheet] and add the same shape (weight) to both sides of the balance, what will happen? What if you take off the same shape on each side—what will happen? Will the amounts on each side of the balance remain equal?

Interview: After the activity, bring the intervention group together for a discussion. Suggested prompts:

- What strategies did you use to figure out how the different shapes compare in value?

	Bridging Prompt (Prompt for Classroom Teacher to Use During New Lesson Content):
	• Show students a set of shapes representing various weights as used in the intervention on the balance, such as this problem:
	Figure 5.10.8 Balancing Act Game Board with shapes
	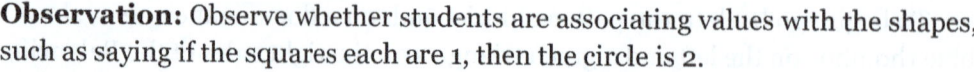
	• Ask: "If the balance is balanced with these shapes, what can you tell about the relationship of the weight of the arrows to the weight of the hearts? [e.g., 1 arrow = 2 hearts, so 2 arrows = 4 hearts; the weight of 1 arrow is > the weight of 1 heart]. How did you figure it out?"

Variation 3

Learning Target: Students will make connections between the concrete, semi-concrete, and symbolic versions of thinking about the meaning of the equal sign.

Variation Directions:

To begin this variation, organize students into pairs. Each team will need a copy of the Balancing Expressions Activity Sheet. Ask them what they notice about the balance. [That the balance has an equal sign on it in this position.] Students will collaborate to fill in the boxes representing unknown values. Let them use counters or base ten materials to model the problems if needed.

As an extension, try the Is it True? Activity Sheet to see if the students can begin to use relational thinking to solve problems.

Engage in Math Discourse (Make the Mathematics Visible):

Observation and **Show Me**: As students work, encourage them to notice important relationships and push on early and partial understandings.

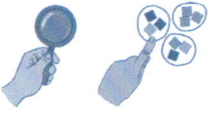

- Listen for language from students that suggests they are beginning to notice that the amount on one side should be the same as the quantity on the other side if they are equal.

- Challenge students to show you how they would now figure out $8 + 4 = ? + 5$. Let them have materials available to model the two sides of the balance.

After the activity, bring the intervention group together for a discussion. Suggested prompts:

- What strategies did you use to find the unknown amount(s)?

- What do you notice about the relationship between one side of the balance and the other?

> **Bridging Prompt (Prompt for Classroom Teacher to Use During New Lesson Content):**
>
> - Show students one of the balances from the Activity Sheet or create a new problem. For example:
>
> Figure 5.10.9 Balancing Act Game Board with expressions
>
> [Balance diagram: left side shows "12 + 2", right side shows "☐ + 5", with an equal sign as the fulcrum]
>
> - Allow students to use materials if needed. Ask, "What is the first step you take to solve this?" [Possible answer: Start with the side of the balance I can calculate and work toward figuring out the unknown side to make both sides equal.]

Task 11: Escape Room

General Objective:
Students will practice problem-solving using word problems.

This activity might be used with:

Students who are able to (prerequisite knowledge):	Students who are Primed to (getting ready to learn):
• understand the meaning of the operations of addition, subtraction, multiplication, and division (can also be adapted for fraction problems) • carry out operations with multidigit numbers	• recognize the importance of understanding the context in problem-solving • focus on patterns in the meaning of the operations • emphasize persevering to be able to complete work within a given period of time

Materials:

- Tape measures (optional)
- Bathroom scale (optional)
- PRINTABLE: Student Task Sets—Animals vs. Humans; Insect Data
- PRINTABLE: Escape Room Teacher Recording Sheet and Codes
- PRINTABLE: Translation Task Template

 Printables for this task are available for download at **https://companion.corwin.com/courses/ProactiveMathIntervention**.

Recommended Children's Literature:

(Some are books of data you can use to create problems for an escape room, and others are about people who solve codes, puzzles, and ciphers.)

- *If You Hopped Like a Frog* – D. Schwartz, 1999
- *Animals by the Numbers* – Jenkins, 2016
- *Disasters by the Numbers* – Jenkins, 2024
- *Insects by the Numbers* – Jenkins, 2020

CHAPTER 5. Intervention Tasks   167

- *Actual Size* – Jenkins, 2009
- *Can You Crack the Code: A Fascinating History of Ciphers and Cryptography* – E. Schwartz, 2009
- *Molly and the Mathematical Mysteries* – Cheng, 2021

Task Overview:

This is an adaptation of a lesson by Dawn Pilotti of Nashville, Tennessee.

This activity focuses on the importance of working on problems that involve students in multiple mathematical ideas and computational skills. We share here ideas with whole-number operations, but the activity of an Escape Room can be used with topics such as fractions, decimals, measurement, and so on. The activity can take place potentially using one intervention period to solve the problems and figure out the code. This task also builds the needed mathematical practices and processes of understanding the problem and persevering. Time frames for the escape are guided by student needs.

Learning Target: The students will use their knowledge of addition, subtraction, multiplication, and division to think about problem-solving of one- and two-step problems. They will also identify a pattern to do a self-check of exiting the "escape room."

Variation Directions

This variation can be delivered over several days during any upcoming topics related to computational situations. We share problem sets for two different "escape rooms," but any set of five problems from the curriculum materials used at your school can be adapted for this task.

In your intervention group, organize students into teams of two or three. Then, select one of the included task sets, or use your own school's curriculum for a set of five word problems. The teacher will need to read the problem (possibly multiple times) to remove any barriers with reading and language. Let's look at one of the options for the escape room—Animals Versus Humans. If students use their own measurements, make sure they round to the nearest unit (you will also need a bathroom scale for weight measures). You may need to have students discuss the main points of careful measurement such as no gaps or overlaps in lengths. The Task set can be delivered as a page or cut into cards that can be placed in different parts of the space like learning centers so teams can work on the problems. Stagger their starting points so they all have time to work without overlapping conversations. If you use stations, you may even want two stations with copies of the same problem to spread people around.

As students solve a problem, they bring their work (not just an answer) to the teacher to explain their thinking. If they are correct, use the Escape Room Teacher Recording Sheet and Codes (need to fill in correct solutions, card numbers, and codes page) and give them the code number that goes with the activity card they completed. Students write the code number next to the problem on their team task set sheet. When all code numbers are collected, they then try to find a pattern. Once identified, they add two more numbers in the sequence to ESCAPE! There can even be a box to open where students can get a sticker for an escape room code-breaker collection.

Figure 5.11.1 Sample of teacher key to assess student work and give code number for Animals Versus Humans

Activity Card Number	Solution	Code Number
1	116 inches	6
2	3.650 pounds	2
3	8 inches	4
4	1,160 inches	10
5	4,930 inches	8

When students have found all of the code numbers, they should rearrange them to make a pattern (2, 4, 6, 8, and 10). Once they discover a pattern and use the pattern to identify the next two numbers (12 and 14), they escape!

Engage in Math Discourse (Make the Mathematics Visible):

Observation: As students work, encourage them to notice important relationships and push on early and partial understandings.

- Listen for language from students that suggests they are working together to understand the problem situation. They might say something like, "I know we are looking at a comparison where one part is so many times larger than another. It seems that is multiplication."

- Watch for students who may try to look for key words or just take the numbers and "do something" with them. They need to be able to describe their reasoning for selecting a particular operation based on the meaning of the operation—such as combining or joining, separating, comparing or taking away amounts, equal groups, multiplication comparisons of "how many times more," and fair sharing or repeated subtraction of equal-groups situations.

- Also, look for ways the students are using the code numbers to organize them into patterns. Ask them about strategies—does it help to put the numbers in some numerical order [let them tell you that they should try from greatest to least or least to greatest as a starting point]?

Interview and **Show Me:** After the activity, bring the intervention group together for a discussion. Suggested prompts:

- Which problem was the most challenging? Can your group come up and show me the way you made sense of the problem?

- What makes these problems interesting?

- What is another number pattern we could use for another escape room code?

Bridging Prompt (Prompt for Classroom Teacher to Use During New Lesson Content):

- Ask two groups of students to come up to the front of the class and give each team the same problem that they already solved from the intervention Escape Room Session (the whole class should solve as well). When the students who participated in the intervention group get an answer, they need to go to the teacher and give an explanation of their answer, modeling that for the whole class. Ask the class as a whole, "What are the features of a good explanation?" and jot down those ideas for students to refer to. Then use the Escape Room format to solve five problems in the grade-level curriculum.

Task 12: Things That Come in Groups

General Objective:
Represent a story involving equal groups with math sketches.

This activity might be used with:

Students who are able to (prerequisite knowledge):	Students who are Primed to (getting ready to learn):
• identify things that come in groups • work with equal groups of objects to represent repeated addition	• describe how a math sketch represents equal groups • represent stories involving equal groups with math sketches • use equal groups of objects to represent multiplication situations

Materials:

- Counting cubes or counters (at least 20 per student)
- Chart paper; label each with 1s, 2s, 3s, 4s, 5s
- Chart paper
- One number cube per group
- Marker, crayon, or pencil per group
- PRINTABLE: Group Images
- PRINTABLE: Representing Equal Groups Recording Sheet (1 per pair of students)
- PRINTABLE: Boxes and Candy Recording Sheet (1 per group)
- PRINTABLE: Translation Task Template

 Printables for this task are available for download at **https://companion.corwin.com/courses/ ProactiveMathIntervention**.

Recommended Children's Literature:

- *Each Orange Had Eight Slices* – Giganti, 1999
- *Amanda Bean's Amazing Dream* – Neuschwander, 1998

- *What Comes in 2s, 3s, & 4s?* – Aker, 1993
- *How Many Feet in the Bed?* – Hamm, 1994

Task Overview:

In this task, students build on their work with equal groups of objects from previous grades. In this task, students focus on the meaning and representation of multiplication as equal groups and the ways in which multiplication situations can be represented.

Select (or create) a variation that focuses on the prior knowledge you are working to shore up for your student group.

Variation 1

Learning Target: Students will identify, represent, and describe stories involving equal groups.

Variation Directions:

Begin with a whole-group discussion. Show the winter mice image and say, "Find all the groups of two on the winter mice." (Arms, legs, eyes, hands, ears, buttons or pockets on the coat.) List the ideas on a whole-group chart labeled 2s.

Figure 5.12.1 Winter mice

Source: istock.com/Reginast777

Next, show the hiking path image and ask students to tell you what they notice about the picture (items with groups of three including fish, rowboats, bridges, windows on the house, branches on the tree). List their ideas on the whole-group chart under 3s. They can also add to the 2s page (cable cars, direction arrows, kayaks), 4s page (mushrooms, berry bushes), and 5s page (snow-capped mountains, waves in the pond).

Figure 5.12.2 Hiking path

Source: istock.com/Lexi Claus

Next, show the Food Store image (see Figure 5.12.3) or another image of your choice and ask students to tell you what they notice about groups of 2s, 3s, 4s, and 5s in the picture. Have them list their ideas on the equal-group charts.

Figure 5.12.3 Food store shelves

Source: istock.com/Zentangle

Organize students in groups of three and distribute one piece of paper to each group. Pose the question to students, "What are some things that come in groups of five?" (Possible responses: five fingers on one hand, five toes on one foot.) Provide students with at least 30 seconds of individual "think" time and then provide students with 1 minute to draw a math sketch of their idea. Then discuss as a group and record student responses on chart paper.

Select one example from the whole-group charts (mouse has two buttons). Pose the question to students: "Sariah mouse has three winter coats. Each coat has two buttons. How would you represent this story?" Provide students with at least 30 seconds of individual think time. Then, provide paper, markers, or crayons and ask them to create their representation. Discuss with the students that the drawings should be math sketches that represent mathematical ideas, not illustrations. Provide students with 2 to 3 minutes to work. Share and record student responses, focusing on how the representations connect to the problem.

Write 3 coats × 2 buttons = _____ buttons and ask students how many total buttons.

Next, divide the students into partners to write other math stories about something from the 2s list and 3s list. Distribute a large piece of paper with 2 Story on one-half and 3 Story on the other half (see Figure 5.12.4).

Figure 5.12.4 Story chart for groups of 2s and 3s

2 Story	3 Story

Say, "Now, you and your partner are going to create representations of other equal-groups stories." Allow time for partners to represent each story and then share with the whole group.

Engage in Math Discourse (Make the Mathematics Visible):

Interview: As students work, encourage them to notice important relationships and push on early and partial understandings.

- What are the equal groups in your story? How many objects are in each group?
- How does your math sketch show the number of groups and the objects in each group?
- What equation can you write to represent your story? How does it match your math sketch?
- How did you represent the (four chairs, three tables)? How did you represent the (three petals on each flower, five chairs at each table, 10 pieces of candy in each box)?

Interview: After the activity, bring the intervention group together for a discussion. Suggested prompts:

- How did you decide what to include in your 2 story and 3 story?
- How were the stories and representations from different groups similar or different?
- How do your story and representation show equal groups? What does it tell us about multiplication?

Bridging Prompt (Prompt for Classroom Teacher to Use During New Lesson Content):

- Share an equal-groups image for one of the stories written by the students who participated in the intervention group.
- Ask the class:
  - What equation could we write to represent this story?
  - How does the equation help us understand the total number of objects?
  - How would you use a math sketch to show this story?

Variation 2

Learning Target: Students will identify, represent, and describe stories involving equal groups.

Variation Directions:

Gather the students in a whole group and introduce the book *Each Orange Had 8 Slices* by Paul Giganti Jr. Begin by reading the title page aloud: "Each orange had eight slices, and each slice had two seeds. How many seeds are there in all?" Ask students:

- What is the problem asking us to figure out?
- How could we represent this problem?

Discuss how the book represents mathematical stories and connect the examples to multiplication and equal groups. Elicit story details from the students.

- Each orange had eight slices.
- Each slice had two seeds.

Have a student come up to model the situation, possibly creating a math sketch of one orange with eight slices, labeling the seeds in each slice. Next, record on the chart:

8 slices × 2 seeds = _____ seeds.

Divide the students into small groups and assign each group a scenario from the book (e.g., groups of chairs, rows of cars, pots of blue flowers). Provide students with paper, crayons, and markers to represent their assigned problem using math sketches. Encourage students to include labels and equations in their math sketches to connect their representation to multiplication. Consider using the Translation Task Template to organize their work.

Show Me: Bring the intervention group back together and have students share and discuss their math sketches and equations.

Engage in Math Discourse (Make the Mathematics Visible):

Observation: As students work, encourage them to notice important relationships and push on early and partial understandings.

- How does your math sketch represent the multiplication problem?
- What strategies are you using to solve your problem?
- Explain your equation and how it matches your math sketches.
- How do the representations help us understand multiplication as equal groups?

Interview: After the activity, bring the intervention group together for a discussion. Suggested prompts:

- What did you notice about the problems in the book? How were they similar?
- How does using math sketches help us understand multiplication problems?
- What other stories can you think of that involve equal groups?

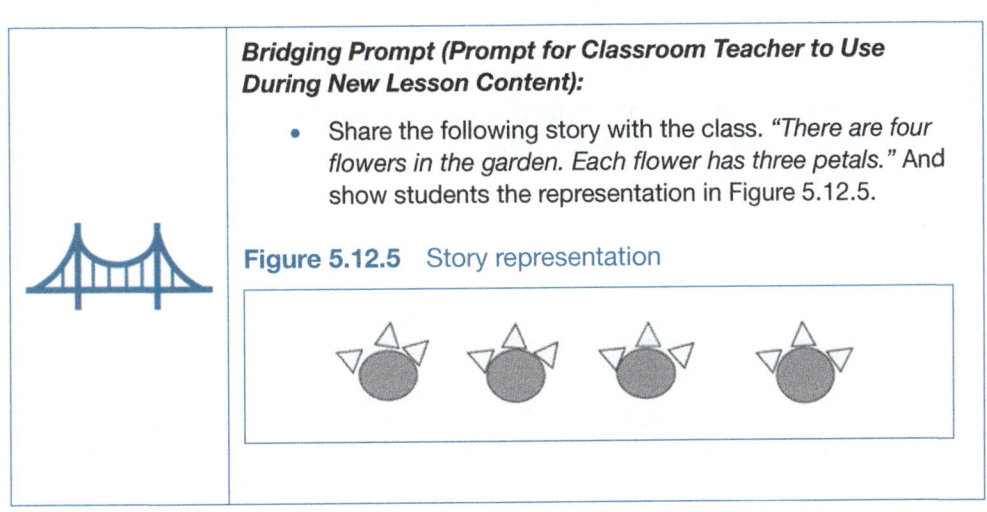

Bridging Prompt (Prompt for Classroom Teacher to Use During New Lesson Content):

- Share the following story with the class. *"There are four flowers in the garden. Each flower has three petals."* And show students the representation in Figure 5.12.5.

Figure 5.12.5 Story representation

(Continued)

(Continued)

	• Show the class this image and say, "Charles made this math sketch to match the representation and story. How does the math sketch represent the number of flowers and the number of petals on each flower?" Engage the students who participated in the intervention group in answering this question. A sample student response might be, "The math sketch shows four circles, one for each flower, and the three triangles on each circle represent the three petals on each flower."

Variation 3

Learning Target: Students will use equal groups of objects to represent multiplication situations.

Variation Directions:

Gather the students together for a whole-group discussion. Begin by saying, "We are going to play the Candy Factory Game! Candy is packed in boxes for the candy factory, and they need your help. Today, you and your partner will practice making representations of boxes of candy and math sketches to show what you know about equal groups. Let's walk through how this activity works together." Use a document camera or chart paper to demonstrate the directions:

Figure 5.12.6 Candy Factory Game directions

1. Roll the number cube to determine how many boxes to draw. Write the number in the correct space on the activity sheet and draw the boxes.

   First roll is a 3. Record the 3 on the recording sheet and draw three boxes:

   ☐ ☐ ☐

2. Roll the number cube again to determine how many pieces of candy to place in each box. Record the number of candies. Draw that number of candy pieces inside each box.

   Second roll is a 2:

   [three boxes each containing 2 dots]

3. Write an equation to represent the total number of candies (e.g., 3 × 2 = 6).

4. Divide the students into partners and distribute the Candy Box Recording Sheet and remind them to work together to take turns rolling the number cube. They should draw boxes and candies to create a matching math sketch and equation.

After the game, bring the students back together for a whole-group discussion.

Engage in Math Discourse (Make the Mathematics Visible):

Observation: As students work, encourage them to notice important relationships and push on early and partial understandings.

- How did you decide how many boxes to draw?
- What does your math sketch show about the total number of candies?
- How is the number of candies in each box related to your equation?

Interview: After the activity, bring the intervention group together for a discussion. Suggested prompts:

- What strategies did you use to represent your candy boxes and math sketches?
- How did your math sketches help you visualize the total number of candies?
- How is your equation connected to your math sketch?
- How does the math sketch you created represent the number of groups and the number of objects in each group?
- How do the representation and math sketch show equal groups?

Bridging Prompt (Prompt for Classroom Teacher to Use During New Lesson Content):
• Imagine the snack shop is preparing a special order of candy boxes for a school event. If each candy box must hold the same number of candies, what are some different ways you could pack 24 candies into boxes? How many boxes would you need? How many candies would go into each box? Use cubes, math sketches, or numbers to show your thinking. Ask students who participated in the intervention group to describe one way of packing the candies.

Variation 4

Learning Target: Students will represent multiplication situations using equal groups and create math sketches to model their thinking.

Variation Directions:

Begin with a whole-group game. Say, "Today, we're going to use our own bodies to create equal groups and learn more about multiplication." Have all students stand in an open space where they can move freely. Play music and encourage students to move, dance, or walk around the room. Stop the music suddenly and call out a number (e.g., "Form groups of 3!"). Students must quickly organize themselves into groups with the specified number of people. Once the groups are formed, ask:

- How many groups did we make? How many people are in each group?
- What is the total number of people in all the groups?

Write the matching multiplication equation on the board (e.g., two groups of three students, 2 × 3 = 6). Repeat the process varying the group sizes (e.g., groups of two, four, or five, depending on the size of your intervention group). Each time, record the equation, emphasizing the group size and number of groups formed.

Divide the students into pairs and provide each pair of students with the Representing Equal Groups Recording Sheet. Read the first story together: "There are four flowers in the garden. Each flower has three petals. Instruct students to draw a math sketch to represent the situation (e.g., flowers and their petals) and write a multiplication equation to match their drawing (e.g., 4 × 3 = 12). Have the students repeat this process for the other stories on the recording sheet. Bring students back together for a group discussion, asking pairs to present their work for one story, explaining their math sketch and equation.

Engage in Math Discourse (Make the Mathematics Visible):

Observation: As students work, encourage them to notice important relationships and push on early and partial understandings.

- What do the groups in your story represent?
- How many objects are in each group? How did you show this?
- What multiplication equation matches your math sketch?

Interview: After the activity, bring the intervention group together for a discussion. Suggested prompts:

- What similarities and differences did you notice in the representations?
- How does the math sketch help us understand the total number of objects?

Bridging Prompt (Prompt for Classroom Teacher to Use During New Lesson Content):

- Show the class one of the math sketches that the intervention group created to represent an equal-groups story. Ask the students to explain the story they wanted to tell.
- Ask the class to think about something they see every day in equal groups—chairs, books, trees, or something else.

- Organize students in pairs and have them brainstorm their own equal-groups story (e.g., "There are five lunch tables. Each table has four chairs.").

- Provide the Translation Task Template for students to:
  - Write their story.
  - Draw a representation/math sketch.
  - Write a multiplication equation.

- Ask a few pairs to present their original stories, representations/math sketch and equations.

- Encourage classmates to discuss what they notice and suggest alternative ways to represent the story.

- Look for opportunities to lift up the reasoning of students who participated in the intervention group.

Task 13

Garden Spaces

General Objective:

Students will develop understanding of arrays and partitioning rectangles by rows and columns to find the total.

This activity might be used with:

Students who are able to (prerequisite knowledge):	Students who are Primed to (getting ready to learn):
• skip count • engage in repeated addition • have a collection of known multiplication facts	• connect skip counting to factors and multiples • engage in early multiplicative thinking • know that area is a measure of covering space

Materials:

- Dice
- Inch tiles or centimeter cubes for concrete representations
- PRINTABLE: 1-centimeter grid paper (printable or blank paper will work) to draw rectangles with rows and columns
- PRINTABLE: Garden Area Game Board

 Printables for this task are available for download at **https://companion.corwin.com/courses/ProactiveMathIntervention**.

Recommended Children's Literature:

- *In Our Garden* – Miller & Crowton, 2019
- *City Green* – DiSalvo, 1999
- *Bigger, Bigger, Best* – Murphy, 2006

Task Overview:

The objective of the Grow Your Garden game is for students to engage in hands-on activities that reinforce multiplication, arrays, and area calculation. By rolling dice and creating arrays to represent dimensions of garden plots, students will apply their understanding of multiplication, use visual models to represent areas, and compare the areas of their gardens.

Select (or create) a variation that focuses on the prior knowledge you are working to shore up for your student group.

Variation 1

Learning Target: Students will solve array problems by drawing rows and columns and finding the total square units.

Variation Directions:

Before starting, define rows and columns for arrays.

Figure 5.13.1 Rows and columns

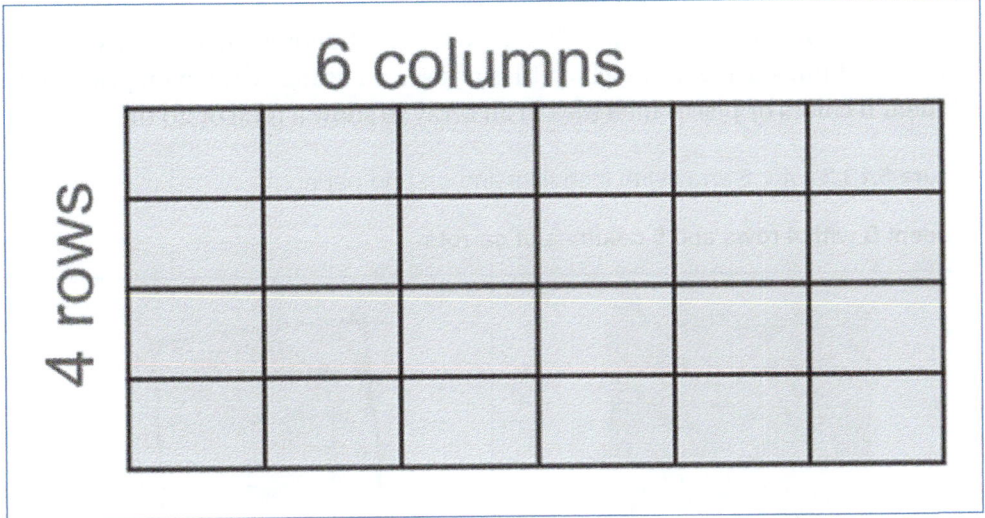

Start by setting the context of growing a garden. Either telling a story about your own backyard or using a piece of children's literature, give students a situation in which multiple vegetables are going to be planted.

Objective: Students will grow vegetables by rolling dice and creating garden spaces using inch tiles or crayons with grid paper. The goal is to practice multiplication and area calculation.

Each student receives a piece of 1-inch grid paper and a set of colored inch tiles or crayons.

Let students decide which vegetable they would like to grow:

Student A: Cucumbers (green inch tiles or crayon)

Student B: Carrots (orange inch tiles or crayon)

How to Play:

Roll the dice:

Each student rolls two dice. The first roll will determine the number of *rows* and the second will determine the number of *columns* in their garden space.

Student A: Rolls a 4 and another 4. Then, the student builds an array with four rows and four columns using green tiles or uses a green crayon with grid paper. Student A colors or places inch tiles in an array to show a total of 16 tiles.

Figure 5.13.2 4 × 4 array with inch tiles and on grid paper

Student A with 4 rows and 4 columns of cucumbers

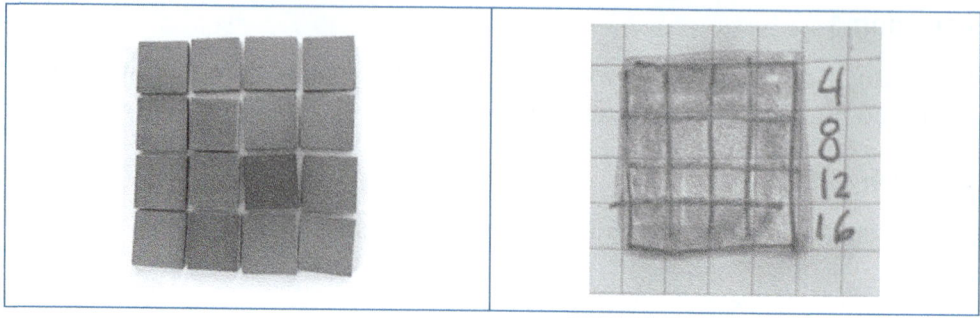

Student B: Rolls a 4 and a 5. Then, the student builds an array with four rows and five columns using orange inch tiles or uses an orange crayon on grid paper. Student B colors or places inch tiles in an array to show a total of 20 tiles.

Figure 5.13.3 4 × 5 array with inch tiles and on grid paper

Student B with 4 rows and 5 columns of carrots

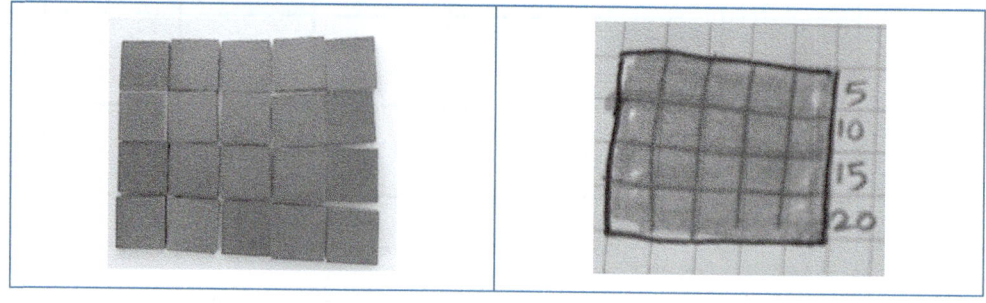

Record the area:

Each student records the total number of square units (inch tiles) on their graph paper.

Student A writes "16 square units" for cucumbers.

Student B writes "20 square units" for carrots.

Repeat: Students can continue to roll the dice and build new garden plots, recording the area each time.

The game can be played for a set amount of time or until a certain number of garden plots are created. The student with the most total square units of garden space at the end of the game wins. This game helps students practice multiplication, understand the concept of area, and have fun modeling with CSA.

Engage in Math Discourse (Make the Mathematics Visible):

Observation: As students work, encourage them to notice important relationships and push on early and partial understandings.

- How is skip counting faster than counting each square?

- How do the rows and columns help you chunk numbers when counting?

- How does drawing the arrays on the grid paper help you see the total square units?

- Check that students are following the convention that the first factor represents the number of rows and the second factor represents number in each row (e.g., 2 × 4 represents two rows of four). *Note that this is a convention in the United States. Other countries have different conventions for how to interpret multiplication equations. Therefore, some students may have experiences in which an opposite convention was used (for example, they may have interpreted 2 × 4 as two columns of four). As needed, remind students of the agreed-upon convention. Sharing a common convention allows students to engage in effective discussion and build meaning around this concept.*

Interview and **Show Me:** After the activity, bring the intervention group together for a discussion. Suggested prompts:

- How did you decide where to place each digit you rolled in your array? What was your strategy for building your garden?

- How do the rows and columns help you calculate the area of your garden plot?

- Can you explain how you found the total number of square units in your garden?

- How is multiplication connected to the number of rows and columns in your garden?

- Did you use skip counting to help you count the inch tiles in your garden? If so, how did that make counting faster or easier?

- How can you use skip counting in the future to make multiplication or area problems easier to solve?

Bridging Prompt (Prompt for Classroom Teacher to Use During New Lesson Content):

- Show these images on a projection to the class. Invite two of the students who participated in the intervention group to come up and respond to this question: "What is similar and different in the ways Jeremy and Zach counted the total square unit of carrots in their garden?"

(Continued)

(Bridging Prompt Continued)

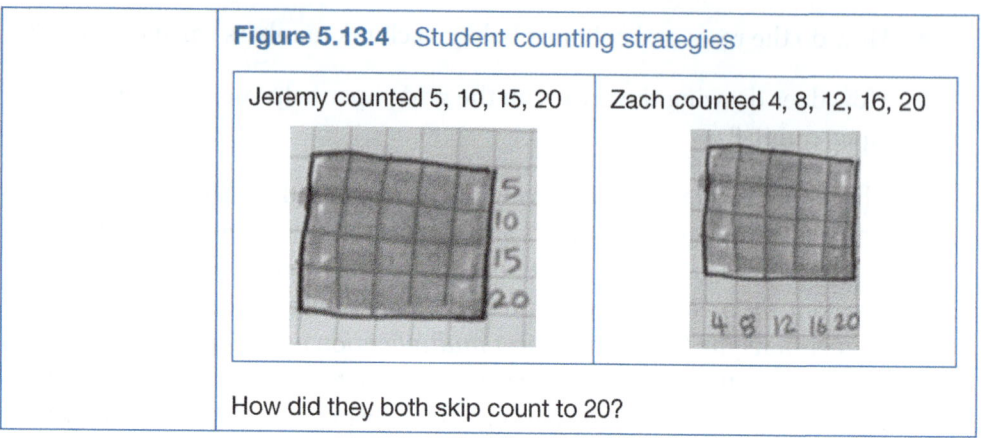

How did they both skip count to 20?

Variation 2

Learning Target: Students will solve array problems by drawing rows and columns (arrays) and finding the total square units.

Variation Directions:

The goal is for each student to fill their side of the neighborhood garden with rolled dice arrays. Organize students into teams with two players on each team. Each team will need a sheet of 1-cm grid paper (portrait layout), two six-sided dice (or other numbered dice of your choice), and colored pencils or crayons (one color per player).

How to Play:

1. Take the sheet of grid paper and divide it in half by drawing a horizontal line across the middle of the page. You can also use the Garden Area Game Board. Each player will have one half of the grid for their garden.

2. Players sit across from each other, each with their own garden. The first player rolls the dice to determine the size of the array they will try to fit into their side of the garden.

3. Based on the numbers rolled, draw an array corresponding to the numbers. So rolling a 3 and a 5 will create an array that is three rows of five.

    Once the roll and shape are decided, color in the squares on your side of the garden (the half you are working on). The array must fit entirely within your garden, and it cannot overlap the boundary line into your neighbor's garden.

4. The second player takes their turn, rolling the dice and drawing their shape onto their half of the garden.

5. Continue taking turns, filling your garden with arrays according to your rolls. Also write in the equation. The goal is to fit the rectangular arrays

on your side of the grid without going over the line in the middle. When the time allotted for this activity is almost over, the game should stop, and students figure out their total area of their garden.

6. The goal is for the player to successfully design their garden without crossing over to the other side (see Figure 5.13.5). If they can't place a roll such as 6 × 6 for Player B on the bottom half of this figure, they would lose their turn.

Figure 5.13.5 Students' Garden Area Game Board (Player A's numbers on top are upside-down because students sit across from each other.)

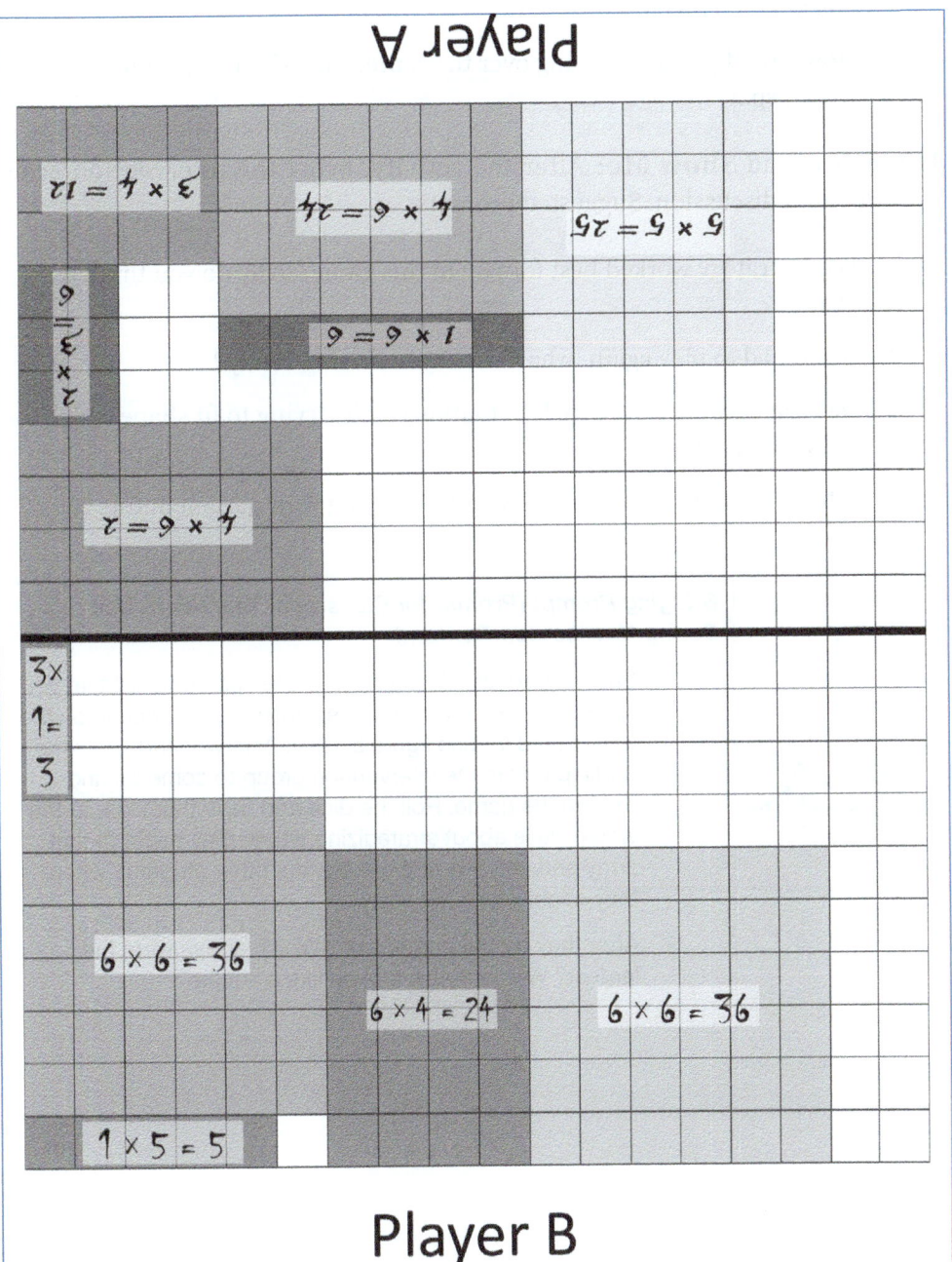

Engage in Math Discourse (Make the Mathematics Visible):

Observation: As students work, encourage them to notice important relationships and push on early and partial understandings.

- How did you decide where to place your array after your roll?
- Did you plan ahead for where to place your next arrays?
- What's your strategy when you have limited space left?
- How are you managing the space in your garden? Are you trying to fill it up without gaps or leave space for bigger shapes?
- When you get a larger array, how do you make sure it fits in without blocking too much space?
- How would you avoid going over the center line when you get to a tight spot?

Interview and **Show Me:** After the activity, bring the intervention group together for a discussion. Suggested prompts:

- What strategy worked best for you in this game? Why do you think it worked?
- If you had to play again, what would you do differently?
- Did you encounter any tricky situations while trying to fit shapes in? How did you solve them?
- What's the most difficult part about designing your garden in this game?

Bridging Prompt (Prompt for Classroom Teacher to Use During New Lesson Content):

- Show a partially filled garden design game board that you saved from the intervention session as a projection on the whiteboard (or use figure 5.13.4). Ask two students who participated in the intervention group to come up and explain the game. Roll the dice and have them talk to the whole class about strategizing where they might fit that array and why. Have them ask the other students where they might place the array.
- After they try the game, ask, "Which arrays were the largest? Which arrays looked like a square?"

Build It Bingo

TASK 14

General Objective:

Students will generate representations to interpret missing-factor equations and use those representations to determine the missing value.

This activity might be used with:

Students who can (prerequisite knowledge):	Students who are Primed to (getting ready to learn):
• understand the meaning of the equal sign • multiply • make and draw an array • recognize the commutative property of multiplication	• interpret multiplication and division equations with missing factors • interpret multiplication and division equations with variables • apply the commutative property to multiply • use the strategy "think multiplication" to divide • use known facts to find related unknown facts

Materials:

- Connecting cubes
- Strips of grid paper (precut)
- Marker boards and dry-erase markers
- Counters to use as bingo markers
- PRINTABLE: 1-centimeter or 1-inch grid paper
- PRINTABLE: Bingo Cards
- PRINTABLE: Equation Cards (multiplication)
- PRINTABLE: Equation Cards (multiplication with variables)
- PRINTABLE: Equation Cards (division)
- PRINTABLE: Equation Cards (division with variables)

 Printables for this task are available for download at **https://companion.corwin.com/courses/ProactiveMathIntervention**.

Recommended Children's Literature:

- *Safari Park* – Murphy, 2001 (solving for unknowns)
- *Ready, Set, Hop!* – Murphy, 1996 (division)

Task Overview:

In this activity, students will be presented with a missing-factor equation. Using stacked connecting cubes or strips of grid paper, students will construct a representation to match the equation and identify the missing factor. The different variations of this task provide students with varied levels of abstraction in their representations.

Select (or create) a variation that focuses on the prior knowledge you are working to shore up for your student group.

Variation 1

Learning Target: Students will use stacked cubes to generate concrete arrays to represent equations with missing factors.

Variation Directions:

To prepare for this variation, create collections of cube stacks of equal lengths (e.g., 10 to 12 cube stacks that are all eight cubes long; 10 to 12 cube stacks that are all five cubes long, etc.). Each pair will need a set of equation cards, and each student will need a bingo card. Use the multiplication Build It Bingo card set at first, and then add in the division bingo card set.

To begin this variation, organize students into partner pairs. Partners take turns. Each turn consists of:

1. drawing a card
2. reading the equation aloud
3. building a model to represent the equation
4. identifying the missing factor and marking that factor (if available) on the bingo card.

If the player draws a "FREE SPACE" card, the player can mark any space on the bingo card and end their turn without building an equation.

The first player to get a "bingo" (five counters in one row, five counters in one column, or five counters on a diagonal) wins. If you need to speed the game up, have both players mark their bingo cards with each player's turn.

CHAPTER 5. Intervention Tasks 189

Figure 5.14.1 Build It Bingo Variation 1 example

Player A	Table Space	Player B
Build It BINGO Card A 3 7 4 9 5 7 5 6 4 8 6 1 8 3 1 2 1 8 9 4 5 9 6 2 7	Build It BINGO (card deck)	Build It BINGO Card B 5 3 8 1 2 8 7 5 2 2 9 1 6 6 5 3 9 4 7 8 6 4 3 9 1
Player A draws a card and reads the equation aloud. "Two times some number equals ten."	Build It BINGO $2 \times \square = 10$	Player B listens to provide feedback on how the equation was read. "I agree!"
Player A builds a model to represent the equation. "I think two groups of 5 will equal 10."	(two rows of 5 boxes each)	Player B observes to provide feedback. "Yeah, you can skip count! 5, 10!"
Because the missing factor is 5, Player A marks any open 5 spot on the bingo card. *If there are no open 5s, Player A would not be able to place a counter on this turn.* Build It BINGO Card A 3 7 4 9 5 7 ● 6 4 8 6 1 8 3 1 2 1 8 9 4 5 9 6 2 7	Build It BINGO $2 \times \square = 10$	Now, Player B's turn will begin by drawing a new card. Build It BINGO Card B 5 3 8 1 2 8 7 5 2 2 9 1 6 6 5 3 9 4 7 8 6 4 3 9 1

Engage in Math Discourse (Make the Mathematics Visible):

Observation: As students work, encourage them to notice important relationships and push on early and partial understandings.

- Watch for students who might still be counting by ones. Encourage these students to think about how they might use skip counting or multiplication to find totals of equal amounts more efficiently. Point to one group/row at a time as students count.

- Watch to see if students are leveraging the commutative property when it might be useful (for example, if modeling the equation $8 \times \square = 24$, does a student who doesn't know their 8s facts spend time guessing and checking to find 8 groups of some amount, or do they recognize that they could also skip count by 8s to model the related fact $\square \times 8 = 24$? As you observe, make note of these examples to highlight during the group discussion.

- Ask students:
  - How are you deciding what length of cube stack to use for your array?
  - What part of this equation represents the number of cubes in each stack?
  - How did you know when to stop adding stacks to your model?
  - What part of this equation represents the number of stacks you used?
  - What does this value (indicate a given part of the equation) represent in your model?

Interview: What strategies are you using to figure out the missing factors? Are you using the same strategy every time? Do the numbers you are given for the equation influence what strategy you use? Can you give an example?

Interview and **Show Me:** After the activity, bring the intervention group together for a discussion. Suggested prompts:

- Did you discover any patterns as you were working to find missing factors?

- Did it make a difference which factor was missing? How did that change your strategy?

> ***Bridging Prompt (Prompt for Classroom Teacher to Use During New Lesson Content):***
>
> - Show a multiplication card with a model and ask two of the students who participated in the intervention to explain it to the group. Then, show the full class a Build It Bingo multiplication card and challenge them to construct a cube stack model to represent the equation. Share models created by students who participated in the intervention.

Variation 2

Learning Target: Students will use strips of grid paper to mimic the stacks they used and generate semiconcrete arrays to represent equations with missing factors.

Variation Directions:

To prepare for this variation, cut strips of grid paper in equal lengths (e.g., 10 to 12 strips that are all 1×8 square units; 10 to 12 strips that are all 1×5 square units, etc.). Each pair will need a set of equation cards, and each student will need a bingo card. Use the multiplication and division Build It Bingo card sets at first. Or, if it fits your instructional goals, use the bingo cards with variables or unknowns.

Before using the bingo cards with variables, conduct a quick mini-lesson to introduce them:

Figure 5.14.2 Mini-lesson

Write an equation with a box for the unknown on the board.	$5 \times \square = 30$
Ask students to explain what the box (□) means. Some students will say it means "6." Listen for students to say that it is the missing number or "where you put the answer" or something along those lines. Tell students, "Yes, this box is what teachers sometimes write when we want students to figure out what number belongs in this spot. It's like a **mystery number**." It is used to mark the place for an unknown value." Write a question mark on the box in the equation.	$5 \times \boxed{?} = 30$
Tell students, "When mathematicians are writing an equation with a mystery number, they have a different way to show it. Look at this equation. What part do you think represents the mystery number in this equation?" When students suggest that the "n" represents the mystery number, introduce the vocabulary word "variable."	$5 \times n = 30$
Tell students, "Mathematicians use letters to represent unknown quantities (like our mystery numbers) in expressions and equations. We can use any letter." Show another example.	$5 \times y = 30$
Discuss with students that the multiplication sign they've been using can be confusing when we start using variables because it can be hard to tell if we mean to write a multiplication sign or a variable "x," so mathematicians often use a • to represent multiplication when they write expressions or equations with variables. Notice that all letter variables are written in italics.	~~$5 \times x = 30$~~ $5 \cdot x = 30$

This variation follows the same structure as Variation 1, only students construct their models with grid paper strips instead of concrete cube stacks. Partners take turns. Each turn consists of:

1. drawing a card
2. reading the equation aloud
3. building a model to represent the equation
4. identifying the missing factor and marking that factor (if available) on the bingo card

If the player draws a "FREE SPACE" card, the player can mark any space on the bingo card and end their turn without building an equation.

The first player to get a "bingo" (five counters in one row, five counters in one column, or five counters on a diagonal) wins. To speed the game up, have both players mark their bingo cards with each player's turn.

Figure 5.14.3 Build It Bingo Variation 2 example (mid-game)

Player A	Table Space	Player B
Build It BINGO — Card B 5 3 8 1 2 8 7 5 2 2 9 1 6 6 ● 3 9 4 7 8 6 4 3 9 1	Build It BINGO $6 \div \square = 2$	Build It BINGO — Card A 3 7 4 9 5 7 5 6 ● 8 6 1 8 3 1 2 ● 8 9 4 5 9 6 2 7
Player A draws a card and reads the equation aloud. "Six divided by some number equals 2."		Player B listens to provide feedback on how the equation was read. "I agree!"
Player A builds a model to represent the equation. "I'm going to make groups of two until I use all six cubes."		Player B observes to provide feedback. "I see three groups of two! What do you put in the blank?"

Player A	Table Space	Player B
"It took three groups, so 3 goes in the blank."		
Because the missing divisor was 3, Player A marks any open 3 spot on the bingo card.	Build It BINGO $6 \div \square = 2$	Now, Player B's turn will begin by drawing a new card.

Engage in Math Discourse (Make the Mathematics Visible):

Observation: As students work, encourage them to notice important relationships and push on early and partial understandings.

- Watch for students who might still be counting by ones. Encourage these students to think about how they might use skip counting or multiplication to find factors/divisors more efficiently.

- Watch to see if students are leveraging the inverse relationship between multiplication and division to find missing divisors (for example, if modeling the equation $21 \div \square = 7$, does a student who has fragility with division concepts spend time "guessing and checking" to find a way to make 21 with random groups, or do they recognize that they could use the related multiplication equation $7 \times \square = 21$ and skip count by 7s until they reach 21? Make note of these examples to highlight during the group discussion.

Interview: Ask students:

- How are you deciding what length of cube stack to use?
- What part of this equation represents the number of cubes in each stack?
- How did you know when to stop adding stacks to your model?
- What part of this equation represents the number of stacks (or groups) you used?
- What does this value (indicate a given part of the equation) represent in your model?
- What strategies are you using to figure out the missing divisors? Are you using the same strategy every time?

Show Me: Do the numbers you are given for the equation influence what strategy you use? Can you give an example?

Interview: After the activity, bring the intervention group together for a discussion. Suggested prompts:

- Did you discover any patterns as you were working to find missing factors?
- Did it make a difference which factor was missing? How did that change your strategy?

	**Bridging Prompt (Prompt for Classroom Teacher to Use During New Lesson Content):** • Show students a Build It Bingo division card and challenge them to construct a grid paper model (to support their representation of stacks) to match the equation. Look for opportunities to have students who participated in the intervention share their thinking.

Variation 3

Learning Target: Students will sketch semiconcrete arrays including "quick arrays" to efficiently represent equations with missing factors.

Variation Directions:

Begin this variation with a brief mini-lesson on sketching quick arrays:

Figure 5.14.4 Mini-lesson

Show students a cube stack or strip of grid paper that is seven units long and ask them to make a "quick math sketch" of the stack/strip. They can also use grid paper if they have challenges with fine motor skills.	(colored strip of 7 units)
Encourage students to share and discuss their sketches. Notice that students (likely) marked out each individual unit in their images and that they may not be proportional.	(strip of 7 empty units)
Show students the following sketch. Tell them another student made this sketch to represent the cube stack. Ask students to talk about how this sketch could represent the stack/strip too. Ask how this quick array is similar to or different from the original model and their original sketches. Ask them why there is a 7 written on this sketch of an array. Elicit from students that this quick array represents one "group of 7."	(7)
Next, show students this image and ask them to work in pairs to determine what total value is represented. Ask them to explain how they know.	(7) (7) (7)

To prepare for the variation, provide each student with a marker board and dry-erase marker. Each pair will need a set of equation cards, and each student will need a bingo card. Use the multiplication and division Build It Bingo card sets at first. Or, if it fits your instructional goals, use the bingo cards with variables.

Follow the same structure as Variation 1, only students will sketch models on marker boards using quick arrays instead of constructing them with cube stacks or grid paper.

Engage in Math Discourse (Make the Mathematics Visible):

Observation: As students work, encourage them to notice important relationships and push on early and partial understandings.

- Watch for students who are struggling with the quick arrays. Provide access to grid paper strips and cube stacks for students to use as a scaffold to support sense-making.

- Watch to see if students are leveraging the related facts to find missing factors/divisors without needing to skip count every group (for example, if modeling the equation $\square \times 6 = 42$, does a student leverage a known fact like 5×6 to know that there are more than five groups of six in 42? Do they start with $5 \times 6 = 30$ and count on two more groups of six to find that $7 \times 6 = 42$? Make note of these examples to highlight during the group discussion.

Interview: Ask students:

- How are you deciding how to label your quick array?

- What part of this equation represents the number of groups in your quick array?

- What part of this equation represents the amount in each group in your quick array?

- How did you know when to stop adding groups to your model?

- What does this value (indicate a given part of the equation) represent in your model?

- What strategies are you using to figure out the missing factors/divisors? Are you using the same strategy every time?

Show Me: Do the numbers you are given for the equation influence what strategy you use? Can you give an example?

Interview: After the activity, bring the intervention group together for a discussion. Suggested prompts:

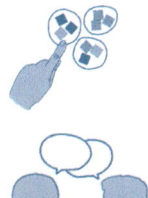

- Did you discover any patterns as you were working to find missing factors/divisors/quotients?

- Did it make a difference which factor was missing? How did that change your strategy?

	**Bridging Prompt (Prompt for Classroom Teacher to Use During New Lesson Content):** • Show students a Build It Bingo multiplication or division card and challenge them to draw a quick array model to represent the equation. Listen and look for students who participated in the intervention to share ideas. Lift up their thinking to the class.

Rearrange It

TASK 15

General Objective:
Students will model and apply properties of multiplication.

This activity might be used with:

Students who can (prerequisite knowledge):	Students who are Primed to (getting ready to learn):
• make equal groups • skip count using a model	• learn about the Commutative Property of Multiplication • learn about the Associative Property of Multiplication

Materials:
- Marker boards
- Counters
- Inch tiles
- Markers or colored pencils
- Small paper plates
- Large construction paper or pieces of butcher paper (or flat trays if available)
- PRINTABLE: 1-centimeter or 1-inch grid paper
- PRINTABLE: Place Value Cards (only the "ones")
- PRINTABLE: Array Cards

Array Cards can also be made from scrap pieces of cardstock or old file folders (see figure 5.15.1).

Figure 5.15.1 Array cards

- PRINTABLE: Associative Expression Mat

 Printables for this task are available for download at **https://companion.corwin.com/courses/ProactiveMathIntervention**.

Recommended Children's Literature:

- *Bats on Parade* – Appelt, 1999 (arrays)
- *Arctic Fives Arrive* – Pinczes, 1998 (counting by fives)
- *Spunky Monkeys on Parade* – Murphy, 1999 (skip counting in groups of 2s, 3s, and 4s)
- *Each Orange Had 8 Slices* – Giganti, 1999
- *Grandpa's Quilt* – Franco, 1999

Task Overview:

In this activity, students will use manipulatives to model multiplications of two or three single-digit numbers. They will use rearrangement to understand and represent the commutative property and the associative property of multiplication.

Select (or create) a variation that focuses on the prior knowledge you are working to shore up for your student group.

Variation 1

Learning Target: Students will represent the commutative property of multiplication using an equal-groups model.

Variation Directions:

Begin this variation by organizing students into pairs. Each group will need a collection of small paper plates, counters, and a set of place value "ones" cards (each set should have at least two of each "ones" digit).

Each partner will draw one digit card. Their job is to create an equal-groups model using the two cards as factors. Each partner will use the digit card *they* drew as their first factor *(number of groups)*. They will use the card *their partner* drew as their second factor *(number of items in each group)*. After their models are built, students should each write multiplication equations to represent their models (equations should be written in the form: *number of groups × number of items in each group*), compare, and discuss observations.

After discussion, students should shuffle cards again and repeat the activity.

Figure 5.15.2 Rearrange It! Variation 1 example

Partner A draws a 4. 	Partner B draws a 2. 
Partner A lays out four plates to represent four equal groups and puts two counters on each plate to show two in each group. 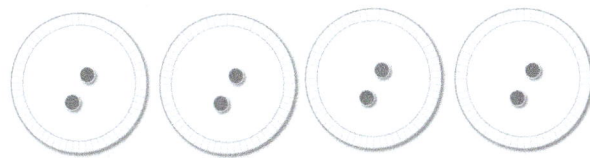	Partner B lays out two plates to represent two equal groups and puts four counters on each plate to show four in each group. 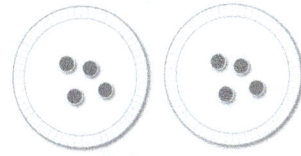
Partner A writes an equation to represent the equal groups: $4 \times 2 = 8$	Partner B writes an equation to represent the equal groups: $2 \times 4 = 8$

The students compare their models and equations.
- Partner A: I have more plates than you do!
- Partner B: Yeah, but my plates have more counters on them.
- Partner B: We have the same amount of counters—we both have 8!
- Partner A: I have 4 times 2 and you have 2 times 4.

Engage in Math Discourse (Make the Mathematics Visible):

Observation: As students work, encourage them to notice important relationships and push on early and partial understandings.

- What did you notice about your models?
- Where is the "4" in Partner A's model? What about Partner B's model? Where is the "2" in each model?
- What did you notice about the equations you wrote?
- Is there a way you could change Partner A's model to make it match Partner B's model?

Interview: After the activity, bring the students together for a discussion. Suggested prompts:

- Show the group some of the equations students wrote and ask them to describe the models that match. For example . . .
  - How is the model for 4×2 different from the model for 2×4?

Show Me: If $4 \times 2 = 8$ and $2 \times 4 = 8$, could we say that $4 \times 2 = 2 \times 4$? How could we show that with the plate models?

- Build 4×2. How could we change this model to show 2×4?

	**Bridging Prompt (Prompt for Classroom Teacher to Use During New Lesson Content):** • Show an image of plates with counters and ask students to write a corresponding expression. Ask how the image would change if the factors switched places (were commuted). Call on the students who participated in the intervention group to share their thinking.

Variation 2

Learning Target: Students will represent the commutative property of multiplication using an array model.

Variation Directions:

Begin this variation by organizing students into pairs. Each group will need two sets of Rearrange It Array Cards or a collection of inch tiles and a set of place value "ones" cards (each set should have at least two of each "ones" digit).

Each partner will draw one digit card. Their job is to create an array model using the two cards as factors. Each partner will use the digit card *they* drew as their first factor *(number of rows)*. They will use the card *their partner* drew as their second factor *(number of unit squares in each row, or number of columns)*. After their models are built, students should write multiplication equations to represent their models (equations should be written in the form: *number of rows × numbers of columns*), compare, and discuss observations.

Figure 5.15.3 Rearrange It! Variation 2 example (with array cards)

Partner A draws a 4.	Partner B draws an 8.
Partner A chooses an array card with four rows and then lays an array card with eight columns on top of it. The open spaces that show up create an array that represents four rows of eight.	Partner B chooses an array card with 8 rows and then lays an array card with four columns on top of it. The open spaces that show up create an array that represents eight rows of four.

CHAPTER 5. Intervention Tasks

Partner A writes an equation to represent the array: $4 \times 8 = 32$	Partner B writes an equation to represent the array: $8 \times 4 = 32$
The students compare their models and equations. • Partner A: I could turn my cards and make them match yours! • Partner B: Me too! • Partner B: I wonder if it would be the same if I put the 8 card on top instead of the 4 card. • Partner A: Let's try it. Look—it stayed the same!	

Figure 5.15.4 Rearrange It! Variation 2 example (with inch tiles)

Partner A draws a 4.	Partner B draws an 8.
Partner A arranges inch tiles into four rows of eight.	Partner B arranges inch tiles into eight rows of four.
Partner A writes an equation to represent the array: $4 \times 8 = 32$	Partner B writes an equation to represent the array: $8 \times 4 = 32$
The students compare their models and equations. • Partner A: I think we have the same number of squares! • Partner B: Me too! • Partner B: Your rectangle looks like mine, it's just sideways. • Partner A: Yeah, it's like they are turned sideways.	

Engage in Math Discourse (Make the Mathematics Visible):

Observation: As students work, encourage them to notice important relationships and push on early and partial understandings.

- What did you notice about your arrays?
- Where is the "4" in Partner A's array? What about the "4" in Partner B's array? Where is the "8" in each array?
- What did you notice about the equations you wrote?
- Is there a way you could change Partner A's array to make it match Partner B's array?

Interview: After the activity, bring the students together for a discussion. Suggested prompts:

- Show the group some of the equations that students wrote and ask them to describe the models that match. For example . . .
  - How is the array for 4 × 8 different from the array for 8 × 4?

Show Me: If 4 × 8 = 32 and 8 × 4 = 32, could we say that 4 × 8 = 8 × 4? How could we show that with the array cards/inch tiles?

- Build 4 × 8. How could we change this array to show 8 × 4?

> **Bridging Prompt (Prompt for Classroom Teacher to Use During New Lesson Content):**
>
> - Use the Array Cards to represent several multiplications. Ask students who participated in the intervention group to tell what expression is being represented. Rotate the cards 90° and ask how the expression will change. Will the total amount change?

Variation 3

Learning Target: Students will represent the commutative property of multiplication using an area model.

Variation Directions:

Begin this variation by organizing students into pairs. Each group will need grid paper, markers or colored pencils, and a set of place value "ones" cards (each set should have at least two of each "ones" digit).

Each partner will draw one digit card. Their job is to sketch a rectangle area model on grid paper using the two cards as factors. Each partner will use the digit card *they* drew as their first factor *(length of the rectangle/how many rows)*. They will use the card *their partner* drew as their second factor *(width of the rectangle/how many unit squares in a row)*. After their models are drawn, students should write multiplication equations to represent their models (equations should be written in the form: *number of rows* × *number of columns*), compare, and discuss observations.

Figure 5.15.5 Rearrange It! Variation 3 example

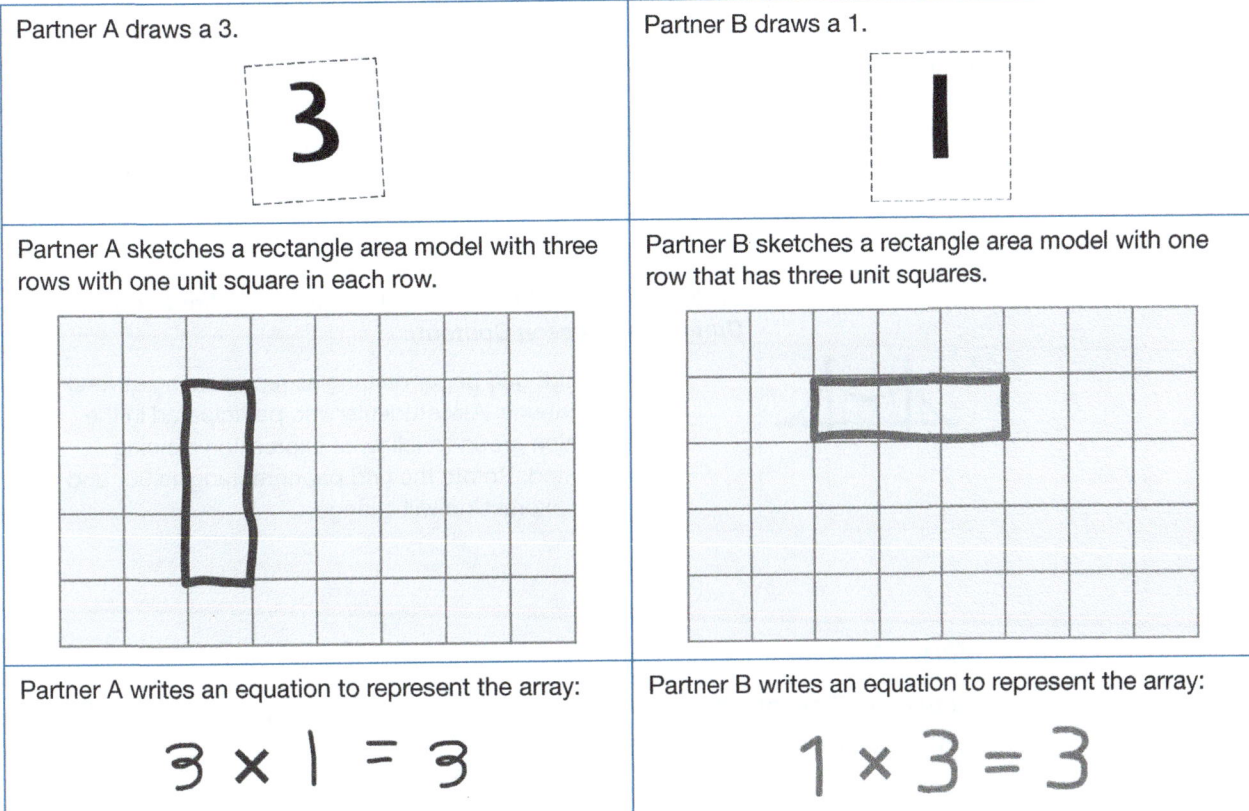

The students compare their models and equations.

- Partner A: We both have three squares.
- Partner B: Our rectangles are the same.
- Partner B: I think that will always happen.
- Partner A: Let's try another card.

Engage in Math Discourse (Make the Mathematics Visible):

Observation: As students work, encourage them to notice important relationships and push on early and partial understandings.

- What did you notice about your area models?
- Where do you see the "3" in Partner A's area model? What about Partner B's area model? Where do you see the "1" in each area model?
- What did you notice about the equations you wrote?
- Is there a way you could change Partner A's area model to make it match Partner B's area model?

Interview: After the activity, bring the students together for a discussion. Suggested prompts:

- Show the group some of the equations that were written and ask them to describe the models that match. For example . . .
  - How is the area model for 3×1 different from the area model for 1×3?

Show Me: If $3 \times 1 = 3$ and $1 \times 3 = 3$, could we say that $3 \times 1 = 1 \times 3$? How could we show that with an area model?

- Build 3×1. How could we change this area model to show 1×3?

	Bridging Prompt (Prompt for Classroom Teacher to Use During New Lesson Content):
	• Show large grid paper rectangles representing several multiplications. Ask students who participated in the intervention group to tell what expression is being represented. Rotate the grid paper rectangles 90° and ask how the expression will change.

Variation 4

Learning Target: Students will represent the associative property of multiplication using an equal-groups model.

Variation Directions:

Begin this variation by organizing students into groups of four (partnered pairs). Each group will need:

- an Associative Expression Mat (cut apart: Pair A gets top half; Pair B gets bottom half)
- a collection of small paper plates
- counters
- trays/construction paper mats
- two sets of place value "ones" cards (one set per pair; each set should have at least two of each "ones" digit). *It may be helpful to initially limit the cards to 1 through 5 in order to manage the amount of space students will need for their models.*

The group will work in partner pairs. Pair A will draw three digit cards and place them on their half of the Associative Expression Mat. Pair B will match the cards and the order of the cards on their half of the Associative Expression Mat. Each pair will build equal-groups models to represent their expression.

- The top half of the Associative Expression Mat groups the first two factors: $(a \times b) \times c$.
- The bottom half of the Associative Expression Mat groups the second two factors: $a \times (b \times c)$.

To build their models, students will *first* create an equal-groups model representing the expression *inside the parentheses* using plates and counters. *Next*, they will iterate that model the number of times indicated by the third factor. Each iteration will be shown on a separate tray/construction paper mat. After models are built, pairs should find each product, compare models and equations, and discuss observations.

Figure 5.15.6 Rearrange It! Variation 4 example

Pair A draws the cards 3, 4, and 2, and places them on their part of the Associative Expression Mat.	Pair B finds the cards 3, 4, and 2, and places them in the same order on their part of the Associative Expression Mat.

Pair A makes a model showing three groups of four (three plates with four counters each) iterated two times.	Pair B makes a model showing three groups (three trays) with four groups of two on each tray.
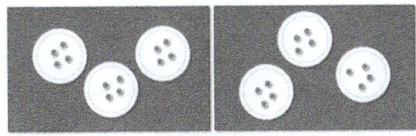	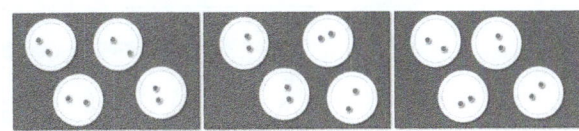

The students compare their models and equations.

- Student 1: You have more trays and more plates than we do.
- Student 3: Yeah, but there are more counters on your plates.
- Student 2: But I think you have more counters than we do.
- Student 3: I don't think so.
- Student 4: Let's count!
- Student 1: We have 24. How many do you have?
- Student 4: We have 24 too!

Engage in Math Discourse (Make the Mathematics Visible):

Observation: As students work, encourage them to notice important relationships and push on early and partial understandings.

- What did you notice about your models?
- Where is the "3" in Pair A's model? What about the "3" in Pair B's model? Where is the "4" in each model? Where is the "2"?

- What is the total amount of counters for each of your models? What equation could you write to describe your model?
- What is an expression you might write to describe just this one tray? *Have students label each tray with an expression.*

Figure 5.15.7 Trays labeled with partial products showing $(3 \times 4) \times 2 = (3 \times 4) + (3 \times 4)$

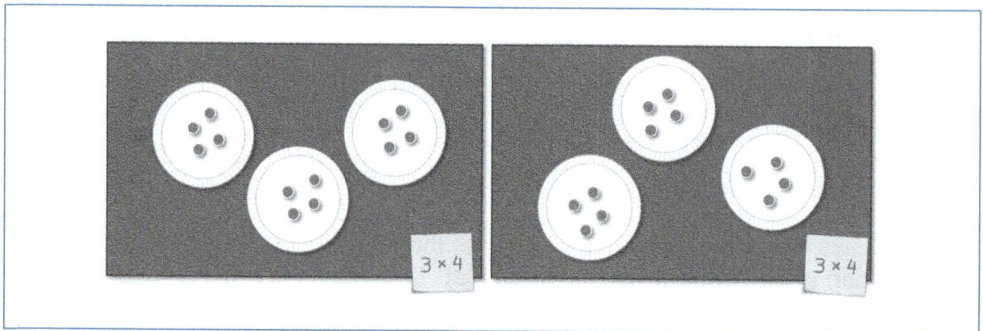

- What is the total number of counters on each tray? *Have students label each tray with the partial product.*

Figure 5.15.8 Trays labeled with partial products showing $3 \times (4 \times 2) = 8 + 8 + 8$

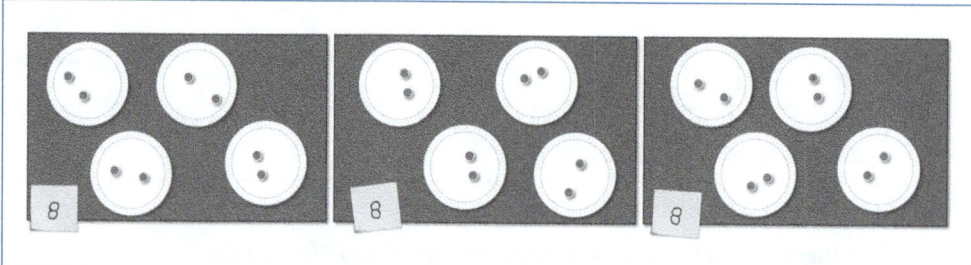

- What other equations might you write to describe the total amount?
- How might you change Pair B's model to make it match Pair A's model without changing the total number of counters?

Interview and **Show Me:** After the activity, bring the students together for a discussion. Suggested prompts:

- Share with the group some of the equations that were written and ask students to describe the models that match. For example . . .
    - How is the model for $(3 \times 4) \times 2$ different from the area model for $3 \times (4 \times 2)$?
- If $(3 \times 4) \times 2 = 24$ and $3 \times (4 \times 2) = 24$, can we say that $(3 \times 4) \times 2 = 3 \times (4 \times 2)$? How could we show that with a model?
- Show a model that students have labeled with expressions to represent each tray.

Figure 5.15.9 Trays labeled with partial products showing $(3 \times 4) \times 2 = (3 \times 4) + (3 \times 4)$

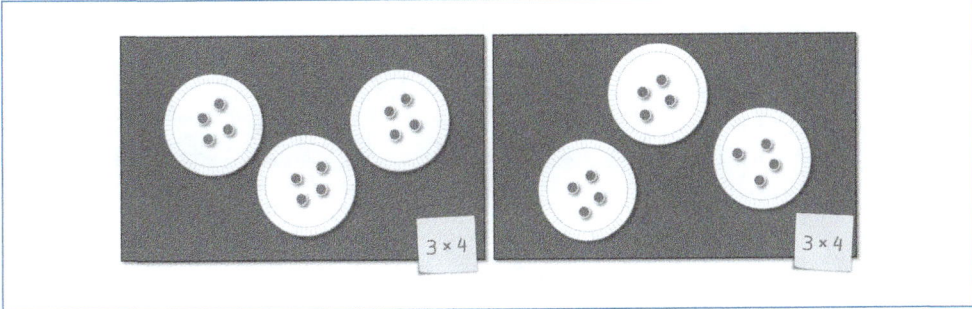

- This model was built to represent the expression $(3 \times 4) \times 2$. This group noticed that each tray represents 3×4. How many groups of 3×4 are there? How does that match the expression $(3 \times 4) \times 2$?

▶ Show a model that students have labeled with partial products to represent each tray.

Figure 5.15.10 Trays labeled with partial products showing $3 \times (4 \times 2) = 8 + 8 + 8$

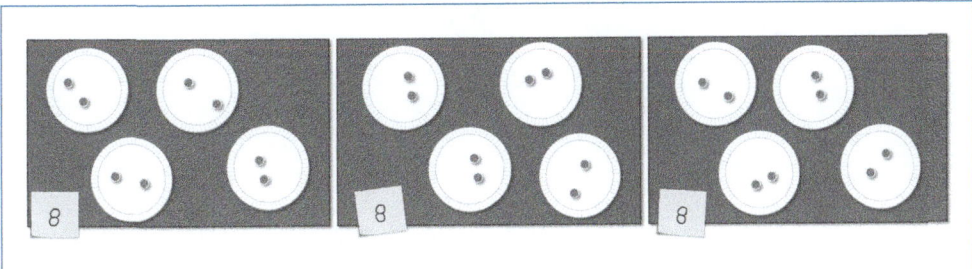

- This model was built to represent the expression $3 \times (4 \times 2)$. This group noticed that each tray represents 8. How many groups of 8 are there? How does that match the expression $3 \times (4 \times 2)$?

Bridging Prompt (Prompt for Classroom Teacher to Use During New Lesson Content):

- Show a picture of trays, plates, and counters. Ask students to write an expression to describe the total number of counters. Accept and discuss expressions that show multiplication of three factors, repeated addition of expressions, and multiplication of partial products. Call on the students who participated in the intervention group to share their thinking. For example:

(Continued)

(Continued)

	Figure 5.15.11 Trays representing 3 × 4 × 2
	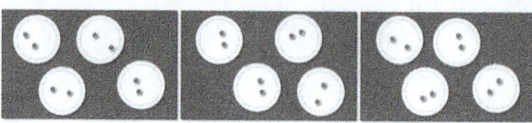
	- 3 × (4 × 2) - (4 × 2) × 3 - (4 × 2) + (4 × 2) + (4 × 2) - 8 × 3 - 3 × 8 - 8 + 8 + 8 - Etc.

CHAPTER 5. Intervention Tasks   209

Multiplication Chains

TASK 16

General Objective:

Students will recognize and leverage related facts to build fluency with whole-number multiplication and division facts.

This activity might be used with:

Students who can (prerequisite knowledge):	Students who are Primed to (getting ready to learn):
• accurately recall *some* basic multiplication facts • represent basic multiplication problems using arrays, equal groups, or area models • state the definition of a factor and a product • understand the similarities and differences between addition and multiplication	• use known facts to derive unknown facts • understand and apply the distributive property as a strategy to multiply

Materials:

- Counters or square-inch tiles
- Scissors
- PRINTABLE: 1-centimeter or 1-inch grid paper
- PRINTABLE: Multiplication Chain Number Cards
- PRINTABLE: Multiplication Chain Expression Frame
- PRINTABLE: Multiplication Chains Activity Page

 Printables for this task are available for download at **https://companion.corwin.com/courses/ ProactiveMathIntervention**.

Recommended Children's Literature:

- *3 × 4* – Brunelli, 2018

Task Overview:

In this activity, students will generate chains of related facts, creating models to show how each fact in the chain can be used to derive the next.

Select (or create) a variation that focuses on the prior knowledge you are working to shore up for your student group.

Variation 1

Learning Target: Students will create chains of related multiplications by adding to or subtracting from equal-groups models.

Variation Directions:

To begin this variation, organize students into pairs. Each pair will need a full set of multiplication chain number cards, a collection of counters, a multiplication chain sentence frame page, a multiplication chain activity page, and table or floor space to create their model.

Student pairs will generate chains of related facts as follows:

1. Partner A draws two multiplication chain number cards and places them in the expression frame.

2. Partner A builds an equal-groups model to represent the fact.

3. Partner A explains how their model represents the multiplication fact.

4. Partner B draws a new multiplication chain number card and places it over one of the two factors in the expression frame.

5. Partner B changes the model to make it match the new fact.

6. Partner B explains how the new fact is related to the original fact and records the addition/subtraction equation to get the new fact.

7. Next, Partner A takes a turn by drawing a new multiplication chain number card to change the expression (and model) again.

Continue taking turns as time permits.

Figure 5.16.1 Multiplication Chains Variation 1 example

Partner A draws 3 and 8 and places the number cards in the expression frame to make the expression 3 × 8.	Multiplication Chain EXPRESSION FRAME: 3 × 8

Partner A builds an equal-groups model to represent 3 × 8. 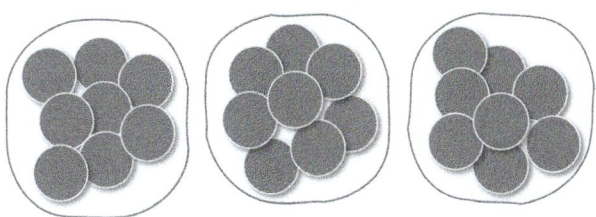	Partner A explains how the model represents the fact and records the equation on the activity page. "My model shows 3 groups of 8. That equals 24." 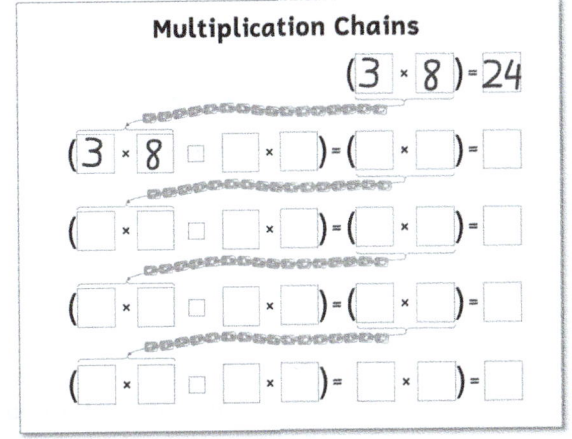
Partner B draws a 4 and places it over the 3, changing the expression to 4 × 8. Partner B changes the model by adding a fourth group of 8. 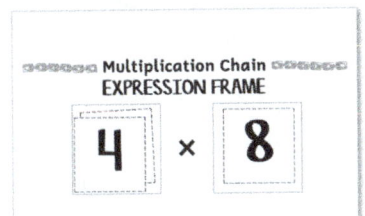 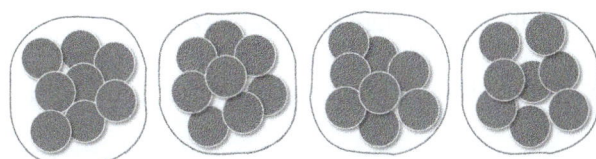	Partner B explains how the new fact is related to the original fact and records the change on the activity page. "There were 3 groups of 8. I added 1 group of 8. Now there are 4 groups of 8. That equals 32." 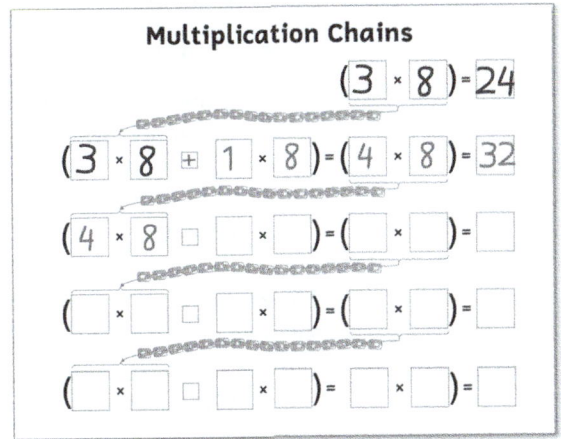
Partner A draws a 7 and places it over the 8, changing the expression to 4 × 7. Partner A changes the model by removing one counter from each of the four groups. 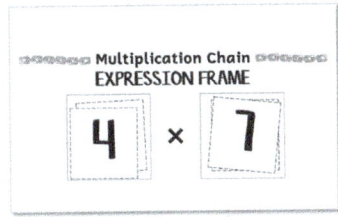 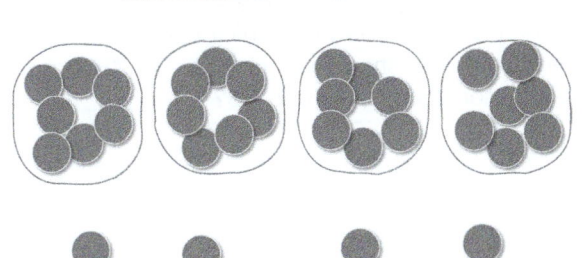	Partner A explains how the new fact is related to the original fact and records the change on the activity page. "There were 4 groups of 8 and I subtracted 4 groups of 1. Now there are 4 groups of 7. That equals 28." 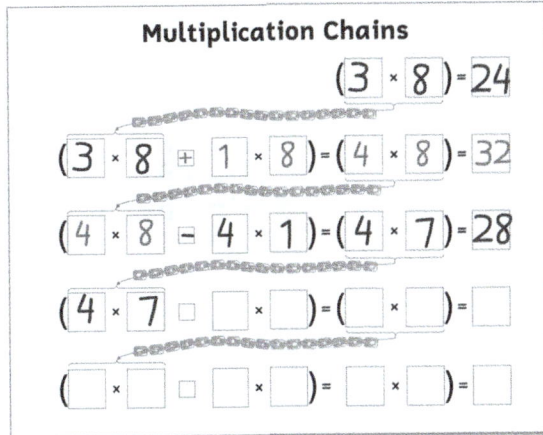

It may be helpful to work through an example together as a whole group before students begin working in their partner pairs. Consider limiting the range of factors by only using number cards 1–4 for students' first few rounds with this activity. As students build confidence with the game, they can expand the range of factors in their number set.

Engage in Math Discourse (Make the Mathematics Visible):

Observation and **Show Me**: As students work, encourage them to notice important relationships and push on early and partial understandings.

- Check that students are following the convention that the first factor represents the number of groups and the second factor represents number per group (e.g., 3 × 8 represents three groups of eight). *Note that this is a convention in the United States. Other countries have different conventions for how to interpret multiplication equations. Therefore, some students may have experiences in which an opposite convention was used (for example, they may have interpreted 3 × 8 as 3 taken 8 times).* As needed, remind students of the agreed-upon convention. Sharing a common convention allows students to engage in effective discussion and build meaning around this concept.

- Show me where you see the + (1 × 8) part of this equation in your model.

- How are these two products related? How much greater/less is this product? How does that connect to the part you added/took away?

- What might happen if you added 2 × 8 instead? How would that have changed the model? How would that change the product?

Interview: After the activity, bring the students together for a discussion. Suggested prompts:

- What did you notice about your chain of facts?

- What are some facts you know that you could chain to 6 × 7? (*Listen for students to suggest 5 × 7.*) Oh, I heard some of you suggest 5 × 7. That seems like a fact a lot of you know well! What is the product of 5 × 7? How could making a multiplication chain help you figure out 6 × 7 if you know 5 × 7?

> **Bridging Prompt (Prompt for Classroom Teacher to Use During New Lesson Content):**
>
> - Show the class an equal-groups model for 5 × 2. Call on the students who participated in the intervention group to share their thinking. Ask, "How could we use this model to help solve 5 × 3?"

Variation 2

Learning Target: Students will create chains of related multiplications by adding or subtracting rows or columns on a physical array model.

Variation Directions:

To begin this variation, organize students into pairs. Each pair will need a full set of multiplication chain number cards, a collection of counters or square-inch tiles, a multiplication chain expression frame, a multiplication chain activity page, and table or floor space to create their model.

Student pairs will generate chains of related facts as follows:

1. Partner A draws two multiplication chain number cards and places them in the expression frame.
2. Partner A builds an array model to represent the fact.
3. Partner A explains how their model represents the multiplication fact.
4. Partner B draws a new multiplication chain number card and places it over one of the two factors in the expression frame.
5. Partner B changes the model to make it match the new fact.
6. Partner B explains how the new fact is related to the original fact and records the addition/subtraction to get the new fact.
7. Next, Partner A takes a turn by drawing a new multiplication chain number card to change the expression (and model) again.

Continue taking turns as time permits.

Figure 5.16.2 Multiplication Chains Variation 2 example

Partner A draws 4 and 2 and places the number cards in the expression frame to make the expression 2 × 4.	Multiplication Chain EXPRESSION FRAME 2 × 4

(Continued)

(Continued)

Partner A builds an array model to represent 2 × 4. 	Partner A explains how the model represents the fact and records the equation on the activity page. "My model shows 2 rows of 4. That equals 8." 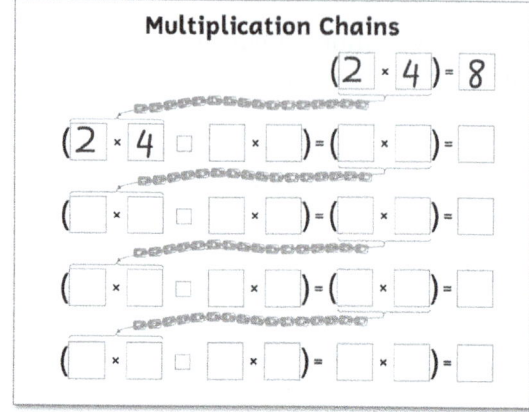
Partner B draws a 7 and places it over the 4, changing the expression to 2 × 7. Partner B changes the model by adding three more inch tiles to each row in the array. 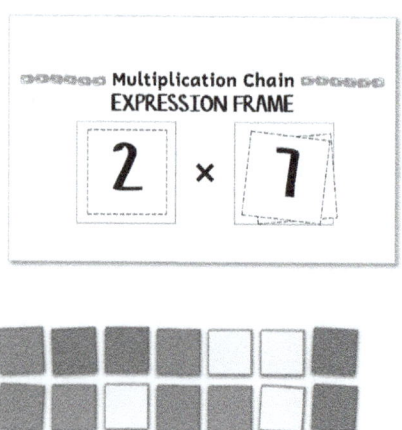	Partner B explains how the new fact is related to the original fact and records the change on the activity page. "There were 2 rows of 4. I added 3 to each row. Now, there are 2 rows of 7. That equals 14." 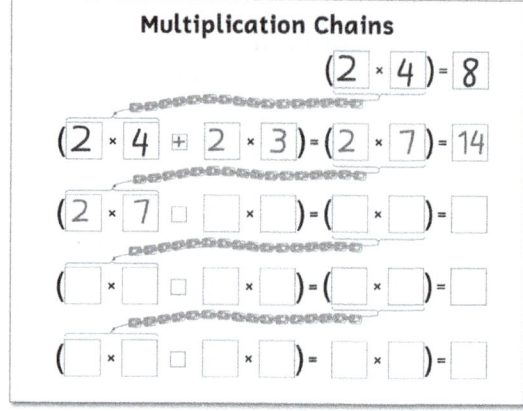
Partner A draws another 3 and places it over the 2, changing the expression to 3 × 7. Partner A changes the model by adding one more row of 7 to the array. 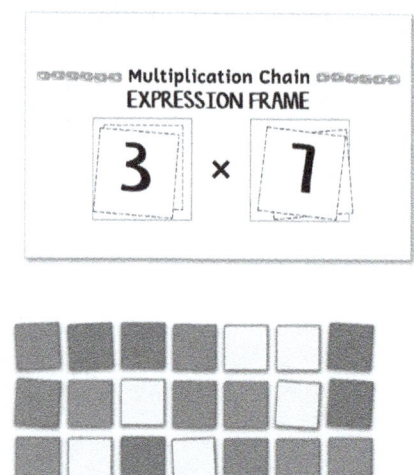	Partner A explains how the new fact is related to the original fact and records the change on the activity page. "There were 2 rows of 7 and I added 1 more row of 7. Now there are 3 rows of 7. That equals 21." 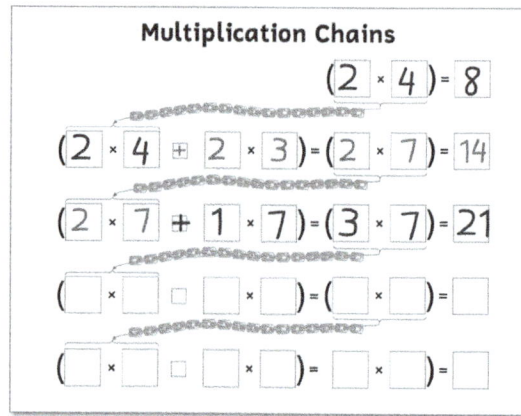

Engage in Math Discourse (Make the Mathematics Visible):

Observation and **Show Me**: As students work, encourage them to notice important relationships and push on early and partial understandings.

- Check that students are following the convention that the first factor represents the number of rows and the second factor represents number in each row (e.g., 2 × 4 represents two rows of four). *Note that this is a convention in the United States. Other countries have different conventions for how to interpret multiplication equations. Therefore, some students may have experiences in which an opposite convention was used (for example, they may have interpreted 2 × 4 as two columns of four).* As needed, remind students of the agreed-upon convention. Sharing a common convention allows students to engage in effective discussion and build meaning around this concept.
- Show me where you see the + (2 × 3) part of this equation in your model.
- How are these two products related? How much greater/less is this product? How does that connect to the part you added/took away?
- What might happen if you added (2 × 1) instead? How would that have changed the model? How would that change the product?

Interview: After the activity, bring the students together for a discussion. Suggested prompts:

- What did you notice about your chain of facts?
- What are some facts you know that you could chain to 7 × 9? *(Listen for students to suggest 7 × 10.)* Oh, I heard some of you suggest 7 × 10. That seems like a fact a lot of you know well! What is the product of 7 × 10? How could making a multiplication chain help you figure out 7 × 9 if you know 7 × 10?

Bridging Prompt (Prompt for Classroom Teacher to Use During New Lesson Content):

- Show the class an array for 6 × 5. Call on the students who participated in the intervention group to share their thinking. Ask, "How could we use this array to help solve 6 × 4?"

Variation 3

Learning Target: Students will create chains of related multiplications by adding or subtracting areas on a grid-paper area model.

Variation Directions:

To begin this variation, organize students into pairs. Each pair will need a full set of multiplication chain number cards, grid paper (large-size grids are suggested; for students with motor skill issues, consider having some precut grid-paper area models available), a multiplication chain expression frame, a multiplication chain activity page, and table space to create their model.

Student pairs will generate chains of related facts as follows:

1. Partner A draws two multiplication chain number cards and places them in the expression frame.
2. Partner A cuts out a grid-paper area model to represent the multiplication fact.
3. Partner A explains how their model represents the multiplication fact.
4. Partner B draws a new multiplication chain number card and places it over one of the two factors in the expression frame.
5. Partner B changes the model to make it match the new fact. To change the model, Partner B will need to **either** cut off a given number of rows or columns **or** cut out an area to *add to* the existing area.
6. Partner B explains how the new fact is related to the original fact and records the addition/subtraction to get the new fact.
7. Next, Partner A takes a turn by drawing a new multiplication chain number card to change the expression (and model) again.

Continue taking turns as time permits.

Figure 5.16.3 Multiplication Chains Variation 3 example

Partner A draws 8 and 3 and places the number cards in the expression frame to make the expression 8 × 3.	**Multiplication Chain EXPRESSION FRAME** 8×3
Partner A cuts out a grid-paper area representing 8 × 3.	Partner A explains how the model represents the fact and records the equation on the activity page. "I cut out a rectangle that has a length of 8 and a width of 3. The area is 24." **Multiplication Chains** $(8 \times 3) = 24$

Partner B draws a 9 and places it over the 3, changing the expression to 8 × 9. Partner B changes the model by adding on an area of 8 × 6.

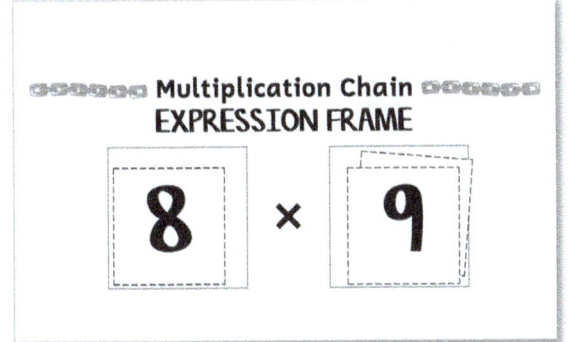

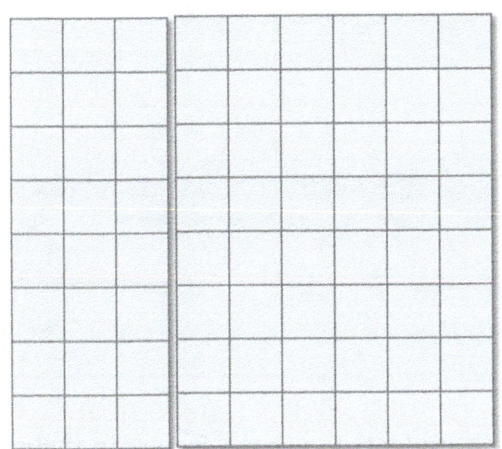

Partner B explains how the new fact is related to the original fact and records the change on the activity page.

"The rectangle had a length of 8 and a width of 3. I added on a rectangle that was 8 long and 6 wide. The new rectangle has a length of 8 and a width of 9. The total area is 72."

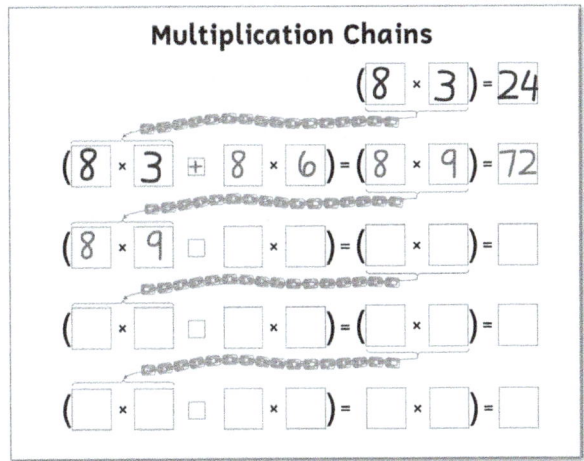

Partner A draws a 7 and places it over the 9, changing the expression to 8 × 7. Partner A changes the model by cutting off an area of 8 × 2.

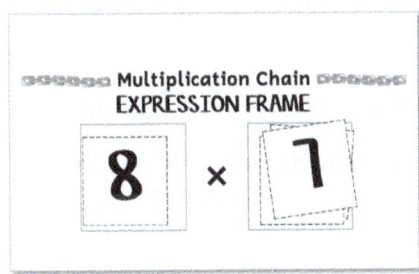

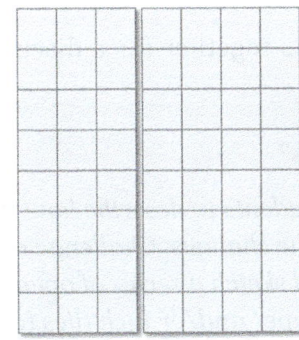 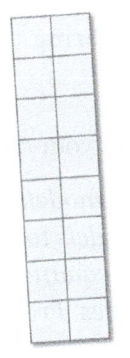

Partner A explains how the new fact is related to the original fact and records the change on the activity page.

"The rectangle had a length of 8 and a width of 9. I cut off a rectangle that was 8 long and 2 wide. The new rectangle has a length of 8 and a width of 7. The total area is 56."

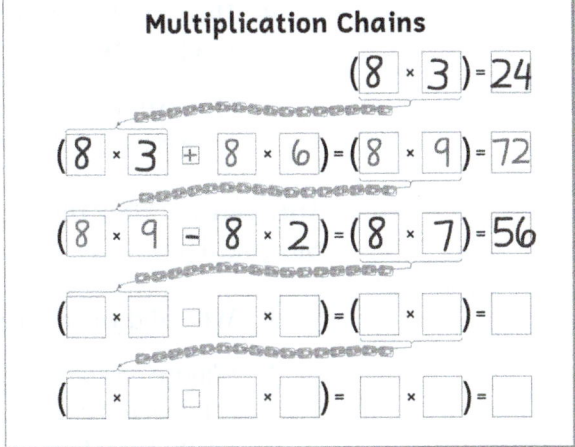

Engage in Math Discourse (Make the Mathematics Visible):

Observation and **Show Me**: As students work, encourage them to notice important relationships and push on early and partial understandings.

- Check that students are making sure to change models in ways that maintain a rectangular area. For example, watch for models that might look like this:

Figure 5.16.4 Student model

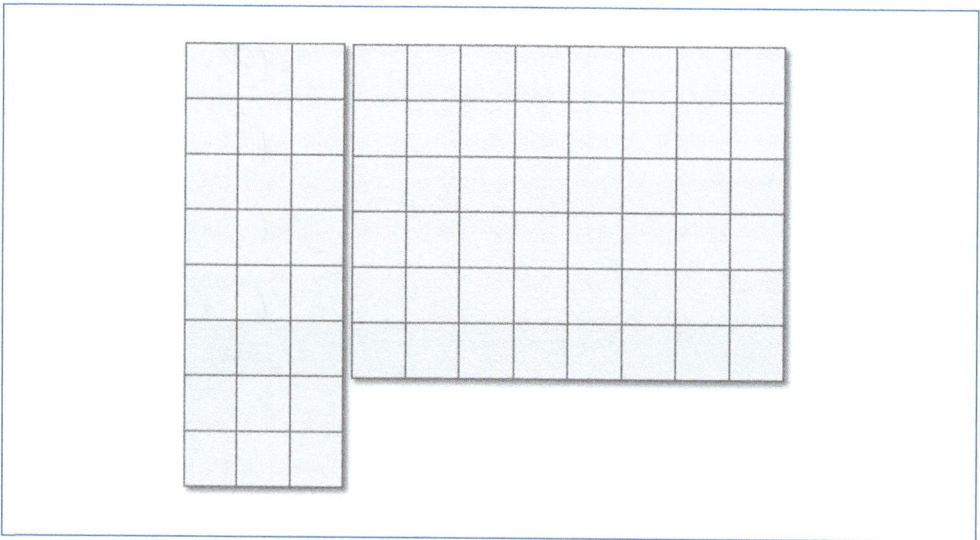

If you see models like this, ask students to use the factors in their expression to describe their model in terms of "rows of square units." Ask, "How many rows are there in your model? How many square units in each row?" When students notice that all the rows don't have the same number of square units, help them think about how to adjust their model to show the area represented by their expression.

- Show me where you see the − (8 × 2) part of this equation in your model.
- How are these two products related? How much greater/less is this product? How does that connect to the part you added/took away?
- How would it be different if you cut off (2 × 9) instead? How would that have changed the model? How would that change the product?

Interview: After the activity, bring the students together for a discussion. Suggested prompts:

- What did you notice about your chain of facts?
- *Choose one of the student models and ask students to describe how they could draw open area models to represent the changes. Challenge each group to go back to their equation chain and sketch a series of open area models to show the changes. Provide grid paper and/or inch tiles to scaffold as needed.*

CHAPTER 5. Intervention Tasks

Figure 5.16.5 Using open area models to represent a multiplication chain

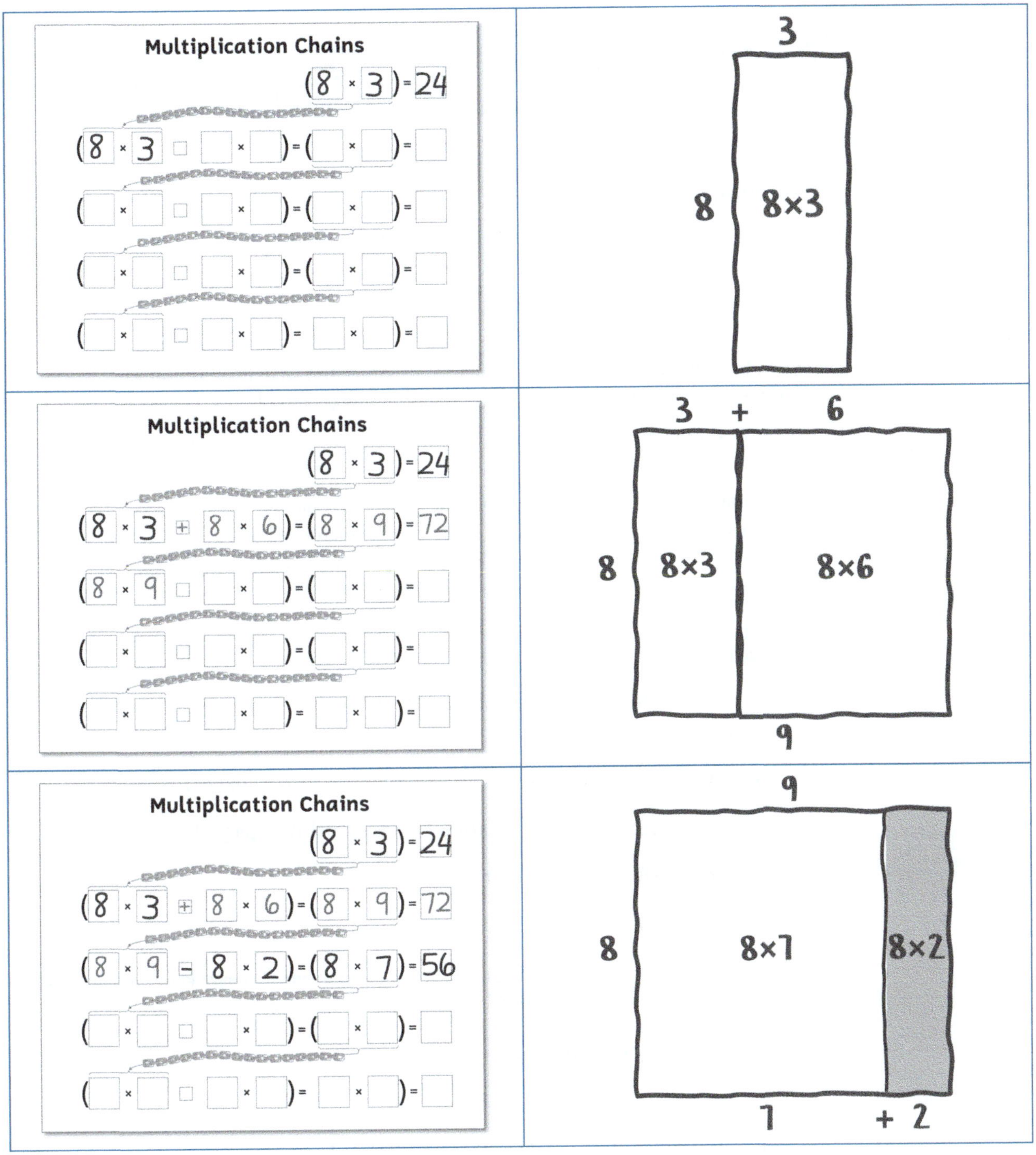

Multiplication Chains

$(8 \times 3) = 24$
$(8 \times 3 \,\square\, \square \times \square) = (\square \times \square) = \square$
$(\square \times \square \,\square\, \square \times \square) = (\square \times \square) = \square$
$(\square \times \square \,\square\, \square \times \square) = (\square \times \square) = \square$
$(\square \times \square \,\square\, \square \times \square) = \square \times \square = \square$

[Area model: rectangle, width 3, height 8, labeled 8×3]

Multiplication Chains

$(8 \times 3) = 24$
$(8 \times 3 \,+\, 8 \times 6) = (8 \times 9) = 72$
$(8 \times 9 \,\square\, \square \times \square) = (\square \times \square) = \square$
$(\square \times \square \,\square\, \square \times \square) = (\square \times \square) = \square$
$(\square \times \square \,\square\, \square \times \square) = \square \times \square = \square$

[Area model: rectangle split into 8×3 and 8×6, widths 3 + 6 = 9, height 8]

Multiplication Chains

$(8 \times 3) = 24$
$(8 \times 3 \,+\, 8 \times 6) = (8 \times 9) = 72$
$(8 \times 9 \,-\, 8 \times 2) = (8 \times 7) = 56$
$(\square \times \square \,\square\, \square \times \square) = (\square \times \square) = \square$
$(\square \times \square \,\square\, \square \times \square) = \square \times \square = \square$

[Area model: rectangle width 9, height 8, with 8×7 region and shaded 8×2 region subtracted; 7 + 2]

Bridging Prompt (Prompt for Classroom Teacher to Use During New Lesson Content):

- Show the class an area model sketch for 3 × 5. Call on the students who participated in the intervention group to share their thinking. Ask, "How could we use this model to help solve 3 × 7?"

Task 17: Speed Estimator

General Objective:
Students will use estimation to approximate products.

This activity might be used with:

Students who can (prerequisite knowledge):	Students who are Primed to (getting ready to learn):
• construct array models • fluently solve some facts using a strategy (e.g., skip counting, related facts, etc.)	• use and apply the distributive property to multiply • use related fact strategies to derive unknown facts • estimate and approximate

Materials:

- PRINTABLE: Estimation Game Board A (front side: array; back side: area)
- PRINTABLE: Array Estimation Game Board B (2 pages)
- PRINTABLE: Area Estimation Game Board B (2 pages)
- PRINTABLE: First Factor Cards
- PRINTABLE: Factor Cards

 Printables for this task are available for download at **https://companion.corwin.com/courses/ProactiveMathIntervention**.

Recommended Children's Literature:

- *Let's Estimate* – Adler, 2017
- *Millions, Billions, & Trillions* – Adler, 2013

Task Overview:

This is a partner game. Both students will use the same first factor (depending on the game board) and draw from a pile to get the second factor of a multiplication expression. They will write their expression on the game board next to whatever available expression they think is the *nearest estimate* of their product. Once both students have filled every spot on their game board, students will sketch the area model for their *actual* products on the game board and find the difference between each actual product and estimate. Students add up their differences (all differences are positive). Lowest total wins. *Note that the game may take more than one intervention session.*

Select (or create) a variation that focuses on the prior knowledge you are working to shore up for your student group.

Variation 1

Learning Target: Students will make connections between the concrete base ten manipulatives, physical arrays, and the area model as tools for estimating when multiplying a one-digit number by a two-digit number.

Variation Directions:

To begin this variation, organize students into pairs. Each student needs Estimation Game Board A (array side) and the pair needs 1 set of factor cards. Students will use tens and ones pieces from physical base ten materials to build an array for their multiplication expression atop the game board.

Students choose matching game boards and lay out a first factor card to match the game board. Then, they take turns as follows:

1. Partner A draws a factor card. This is the second factor in their multiplication expression.

2. Partner A looks at their game board for an available array they think is the nearest available estimate of their product. They write the factor they drew in the "Actual" expression next to that estimate.

3. Next, Partner B draws a factor card. This is the second factor in their multiplication expression.

4. Partner B looks at their game board for an available array they think is the nearest available estimate of their product. They write the factor they drew in the "Actual" expression next to that estimate.

5. Play continues until both players have filled all available spots on their game boards.

Once both students have filled every spot on their game board, students take out base ten materials and use the tens and ones pieces to build an array representing each actual product on top of the arrays for the estimates and find the difference between each actual product and estimate. Students add up their differences (all differences are positive). Lowest total wins.

Note that students should not get the place value blocks out until they are done choosing estimates. Also note that students are not finding actual products to determine differences for the game—they should be looking at the "extra" or "missing" amounts to see the differences. The goal of this activity is that students would **notice how these products are related to the estimates, NOT that they would calculate and then compute differences.*

Figure 5.17.1 Speed Estimator Variation 1 example

In this example, the student pair gets a game board with 3 for the first factor.	
Students lay out the first factor card:	**3 ×**
Partner A draws a factor card: **17**	Partner A chooses to record their product of 3 × 17 next to the estimate 3 × 20 on the game board: 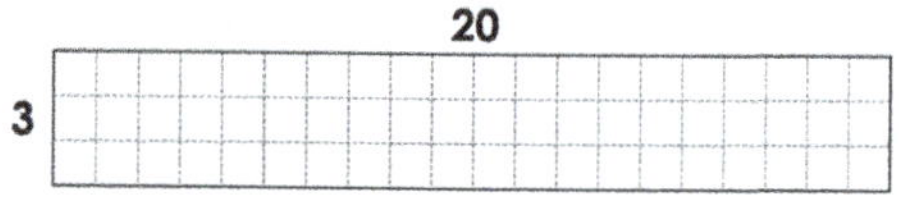 20 3 [grid] Estimate: 3 × 20 Actual: 3 × 17 Difference: _____
This makes the expression: **3 × 17**	
Next, Partner A uses base ten tens and ones to build an array representing their actual product on top of the array for the estimate and finds the difference between the two products. 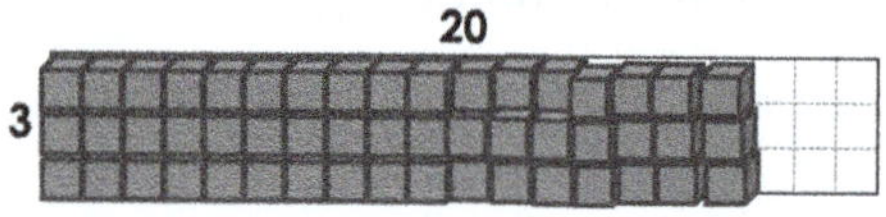 20 3 [array] Estimate: 3 × 20 Actual: 3 × 17 Difference: __9__	

Engage in Math Discourse (Make the Mathematics Visible):

Observation: As students work, encourage them to notice important relationships and push on early and partial understandings.

- Listen for language from students that suggests they are beginning to notice the Distributive Property at work. For example, a student who chooses 3 × 10 as an estimate for 3 × 11 might say something like, "I know the difference will be 3 because it's only one more group!"

- Watch for students who may only be looking for estimates with a factor that is *greater than* (or less than) their actual factor. Ask them to explain their thinking.

Interview: After the activity, bring the intervention group together for a discussion. Suggested prompts:

- What strategies did you use to pick the closest estimate?
- Did you have to count each square (or each unit cube) one at a time to figure out the difference each time? How did you see it? *(Elicit the idea of using multiplication.)*

Show Me: Choose one or two examples from student work and ask, "Is there a way we could use the estimate to figure out the actual product?

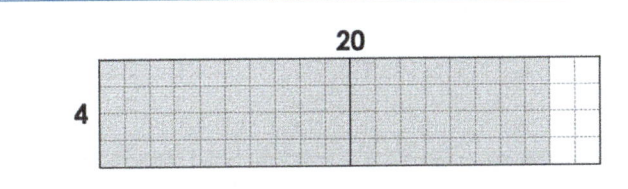

Bridging Prompt (Prompt for Classroom Teacher to Use During New Lesson Content):

Show students one of the arrays from Game Board A that has been shaded in to represent an area less than the full array. For example:

Figure 5.17.2 Game Board A shaded to show 4 × 18

Ask, "What is the total area that is shaded? How did you figure it out?" Call on the students who participated in the intervention group to share their thinking.

Variation 2

Learning Target: Students will use area models and the distributive property of multiplication over addition to estimate products when multiplying a one-digit number by a two-digit number.

Variation Directions:

In this variation, students will sketch an area model for their product on top of the estimated area model on the game board. Slide the game boards into sheet protectors (or laminate them) and have students use dry-erase markers.

To begin this variation, show the whole group the array side of a game board next to the area side of the same game board and ask them to compare corresponding images on the game boards. For example:

Figure 5.17.3 5 × 20 array and 5 × 20 open area model

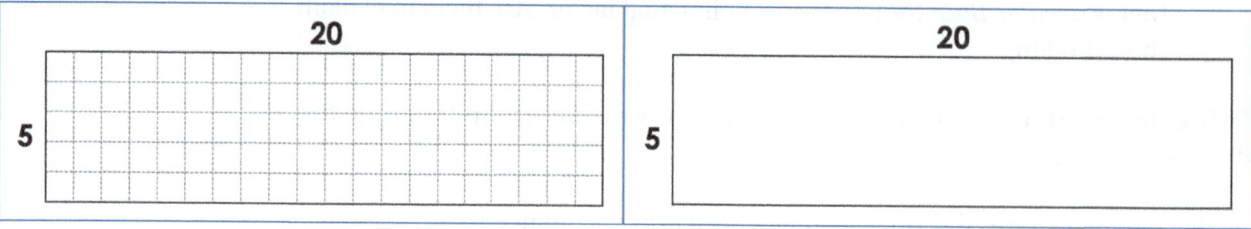

Ask students to discuss what the labels on each model represent. What is the same and what is different about the two models? Elicit from students that the array model (on the left) shows each individual square unit, while the area model (on the right) shows the side lengths but doesn't show each square unit. Discuss how students can *visualize* five rows of 20 square units in the open area model.

Take away the array game board and ask students to think about the area model. Ask them how an area of 5 × 17 would compare to this 5 × 20 area. How might we show it on this model? Elicit that it would be slightly shorter than this rectangle that is 20 units long. Have students show how they might sketch that area atop this image. Discuss what the parts of the image represent. Repeat for 5 × 22.

Figure 5.17.4 Representing 5 × 17 and 5 × 22 on a 5 × 20 open area model

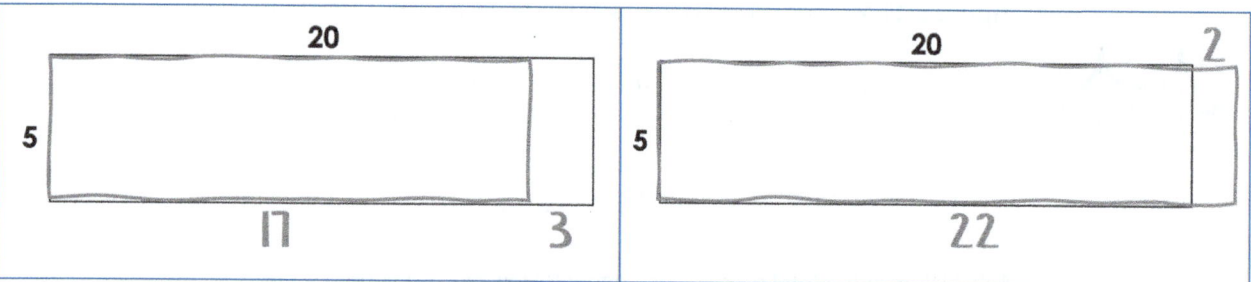

Next, organize students into pairs. Each student needs Estimation Game Board A (area side) and a dry-erase marker, and the pair needs 1 set of factor cards. Students choose matching game boards and lay out a first factor card to match the game board. Then, they take turns as follows:

1. Partner A draws a factor card. This is the second factor in their multiplication expression.

2. Partner A looks at their game board for an available area they think is the nearest available estimate of their product. They write the factor they drew in the "Actual" expression next to that estimate.

3. Next, Partner B draws a factor card. This is the second factor in their multiplication expression.

4. Partner B looks at their game board for an available area they think is the nearest available estimate of their product. They write the factor they drew in the "Actual" expression next to that estimate.

5. Play continues until both players have filled all available spots on their game boards.

Once both students have filled every spot on their game board, students will sketch area models for their actual products atop the estimates on the game board and find the difference between each actual product and estimate. Students add up their differences (all differences are positive). Lowest total wins.

*Note that students are not finding actual products to determine differences for the game—they should be finding the products for the "extra" or "missing" areas to see the differences. The goal of this activity is that students would **notice how these products are related to the estimates**, NOT that they would calculate and then compute differences.*

Figure 5.17.5 Speed Estimator Variation 2 example

In this example, the student pair gets a game board with 5 for the first factor.	
Students lay out the first factor card:	 5 ×
Partner A draws a factor card: 31 This makes the expression: 5 × 31	Partner A chooses an area model they think is *closest* to the product of 5 × 31 and records their expression on the game board: [area model: 30 by 5] Estimate: 5 × 30 Actual: 5 × 31 Difference: _____
Next, Partner A sketches an area model that represents the actual product right on top of the estimated area model and uses the model to find the difference between the products. Encourage students to label any new side lengths. Products are related to the estimates. [area model: 30 by 5, extended] Estimate: 5 × 30 Actual: 5 × 31 Difference: 5	

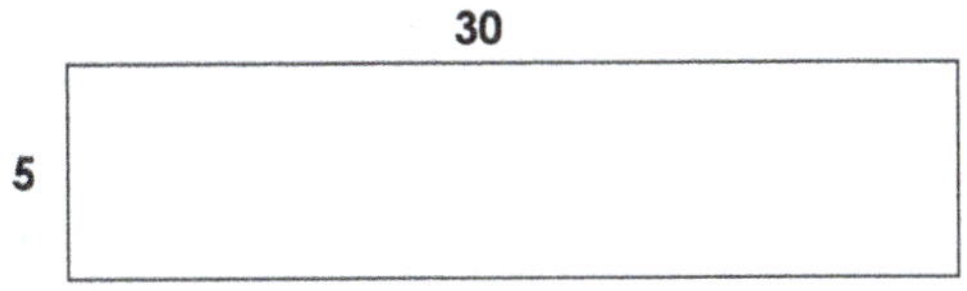

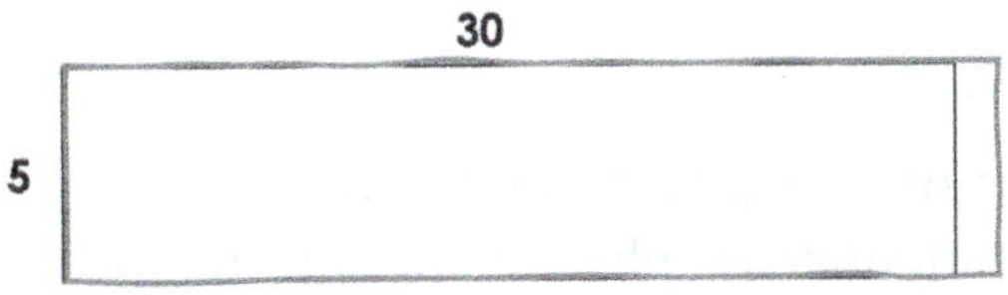

Engage in Math Discourse (Make the Mathematics Visible):

Observation: As students work, encourage them to notice important relationships and push on early and partial understandings.

- Be comfortable with some imprecision from students. Their sketches do not need to be to scale!

- Ask students to explain to you what the different parts of their sketches represent. Where do they see the difference in the models?

Interview: After the activity, bring the intervention group together for a discussion. Suggested prompts:

- What strategies did you use to pick the closest estimate?

- Did you have to count each square (or each unit cube) one at a time to figure out the difference each time? How did you see it? *(Elicit the idea of using multiplication.)*

Show Me: Choose one or two examples from student work and ask, "Is there a way we could use the estimate to figure out the actual product?"

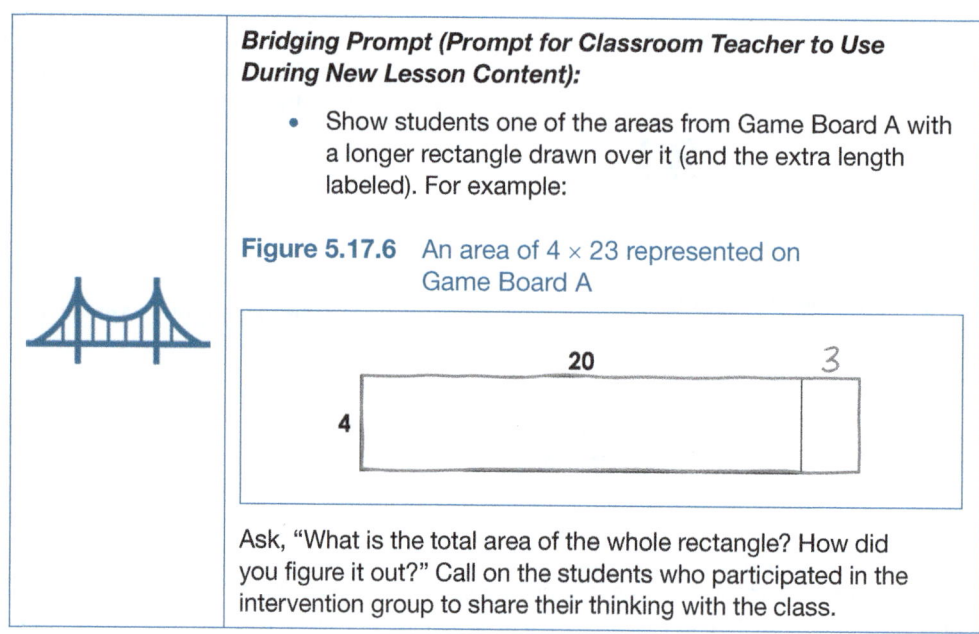

Bridging Prompt (Prompt for Classroom Teacher to Use During New Lesson Content):

- Show students one of the areas from Game Board A with a longer rectangle drawn over it (and the extra length labeled). For example:

Figure 5.17.6 An area of 4 × 23 represented on Game Board A

Ask, "What is the total area of the whole rectangle? How did you figure it out?" Call on the students who participated in the intervention group to share their thinking with the class.

Variation 3

Learning Target: Students will make connections between the concrete base ten manipulatives, physical arrays, and the area models as tools for estimating when multiplying a two-digit number by a two-digit number.

To begin this variation, organize students into pairs. Each student needs Array Estimation Game Board B, and the pair needs 1 set of factor cards (7, 8, and 9 may be removed if desired). Students will use hundreds, tens, and ones pieces from physical base ten materials to build an array for their multiplication expression atop the game board.

Students will take turns as follows:

1. Partner A draws *two* factor cards. These two cards are used to create a multiplication expression.

2. Partner A looks at their game board for an available array they think is the nearest available estimate of their product. They write their "Actual" expression next to that estimate.

3. Next, Partner B draws *two* factor cards. These two cards are used to create a multiplication expression.

4. Partner A looks at their game board for an available array they think is the nearest available estimate of their product. They write their "Actual" expression next to that estimate.

5. Play continues until both players have filled all available spots on their game boards.

Once both students have filled every spot on their game board, students use base ten hundreds, tens, and ones to build an array representing each actual product on top of the arrays for the estimates and find the difference between each actual product and estimate. Students add up their differences (all differences are positive). Lowest total wins.

*Note that students should not get the place value blocks out until they are done placing estimates. Also note that students are not finding actual products to determine differences for the game—they should be looking at the "extra" or "missing" amounts to see the differences. The goal of this activity is that students would **notice how these products are related to the estimates**, NOT that they would calculate and then compute differences.*

Figure 5.17.7   Speed Estimator Variation 3 example

Partner A draws the factor cards 17 and 23 and makes the expression: 	Partner A chooses to record their product of 17 × 23 next to the estimate 20 × 20 on the game board: 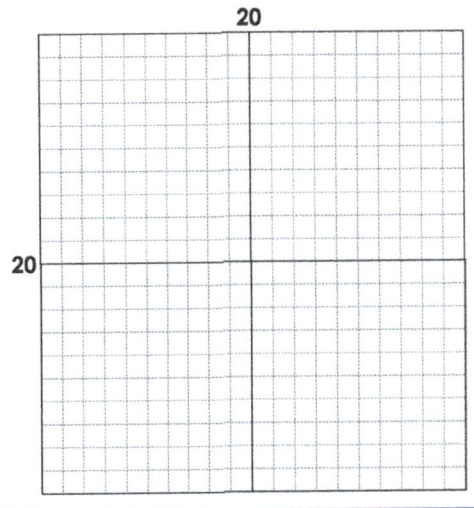

(Continued)

(Continued)

Next, Partner A uses base ten hundreds, tens, and ones pieces to build an array representing their actual product on top of the array for the estimate.

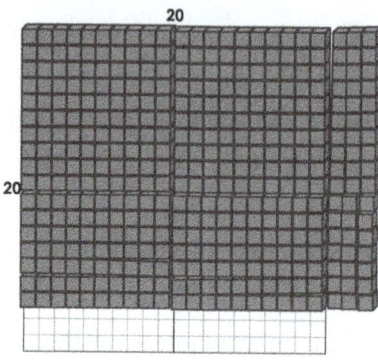

Estimate: 20 × 20   Actual: 17 × 23   Difference: _____

Partner A sees that there are 51 "extra" square units in the rows (17 rows of three "extra" square units) and 60 unfilled square units in the columns (three rows of 20 unfilled square units).

To determine the difference between the actual product and the estimate, Partner A rearranges the base ten pieces to cover as many of the empty square units as possible.

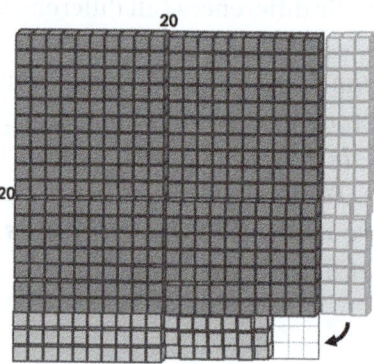

Estimate: 20 × 20   Actual: 17 × 23   Difference: 9

Engage in Math Discourse (Make the Mathematics Visible):

Observation: As students work, encourage them to notice important relationships and push on early and partial understandings.

- Listen for language from students that suggests they are beginning to notice the distributive property at work. For example, a student who chooses 20 × 20 as an estimate for 16 × 19 might say something like, "I know 16 × 19 is closer to 20 × 20 than 10 × 10 because 16 and 19 are both closer to 20 than they are to 10!"

- Watch for students who may only be looking for estimates with a factor that is *greater than* (or less than) their actual factor. Ask them to explain their thinking.

- Watch for students who may be struggling to determine the differences between their estimates and expressions. Encourage the use of rearrangement to make sense of the relationships.

Interview: After the activity, bring the intervention group together for a discussion. Suggested prompts:

- What strategies did you use to pick the closest estimate?
- Did you have to count each square (or each unit cube) one at a time to figure out the difference each time? How did you see it? *(Elicit the idea of using multiplication.)*
- What do you notice about the relationship between the number of "extra" cubes and the number of unfilled spaces? How does that relate to the difference you found?

Show Me: Choose one or two examples from student work and ask, "Is there a way we could use the estimate to figure out the actual product?"

Bridging Prompt (Prompt for Classroom Teacher to Use During New Lesson Content):

- Show students one of the arrays from Game Board B that has been shaded in to represent an area less than the full array. For example:

Figure 5.17.8 Game Board B shaded to show 20 × 18

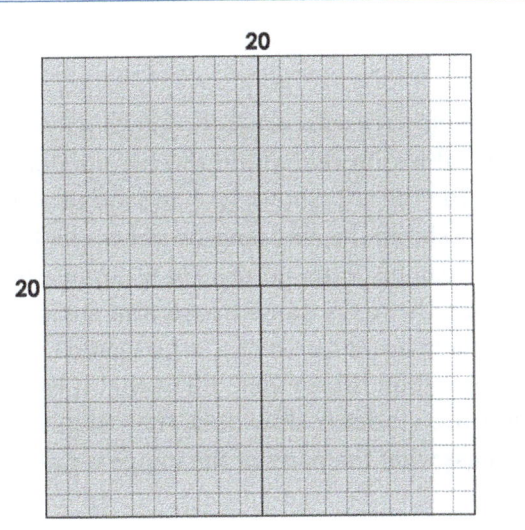

Ask, "What is the total area that is shaded? How did you figure it out?" Call on the students who participated in the intervention group to share their thinking.

Variation 4

Learning Target: Students will use area models and the distributive property of multiplication over Addition to estimate products when multiplying a two-digit number by a two-digit number.

Variation Directions:

In this variation, students will sketch an area model for their product on top of the estimated area model on the game board. Slide the game boards into sheet protectors (or laminate them) and have students use dry-erase markers.

To begin this variation, show the whole group an array game board and an area model game board and ask them to compare corresponding images on the game boards. For example:

Figure 5.17.9 20 × 30 array and 20 × 30 open area model

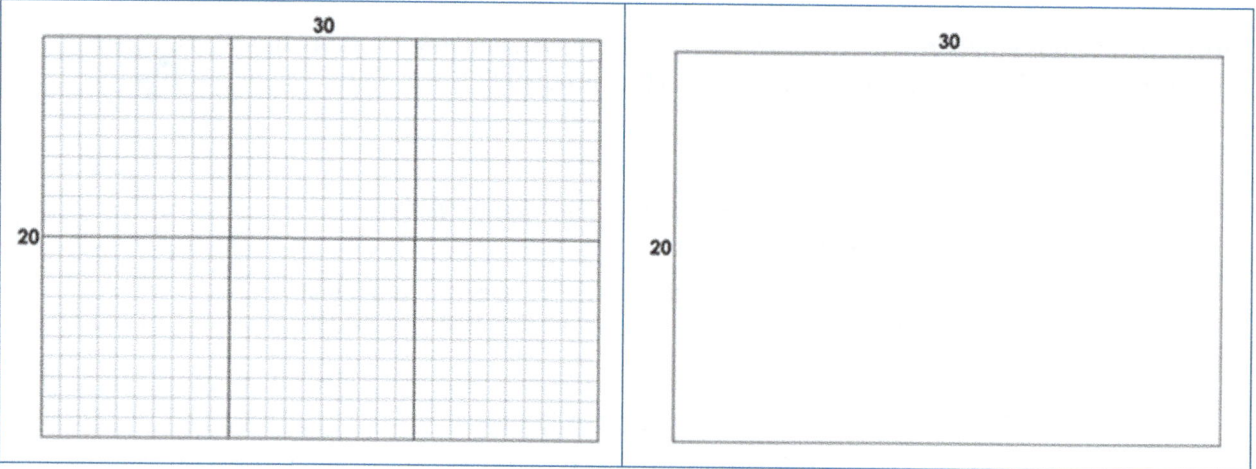

Ask students to discuss what the labels on each model represent. What is the same and what is different about the two models? Elicit from students that the array model (on the left) shows each individual square unit, while the area model (on the right) shows the side lengths but doesn't show each square unit. Discuss how students can *visualize* 20 rows of 30 square units in the open area model.

Take away the array game board and ask students to think about the area model. Ask them how an area of 17 × 30 would compare to this 20 × 30 area. How might we show it on this model? Elicit that it would be slightly shorter than this rectangle. Have students show how they might sketch that array on top of this image. Discuss what the parts of the image represent. Repeat for 22 × 30.

Figure 5.17.10 Representing 17 × 30 and 22 × 30 on a 20 ×30 open area model

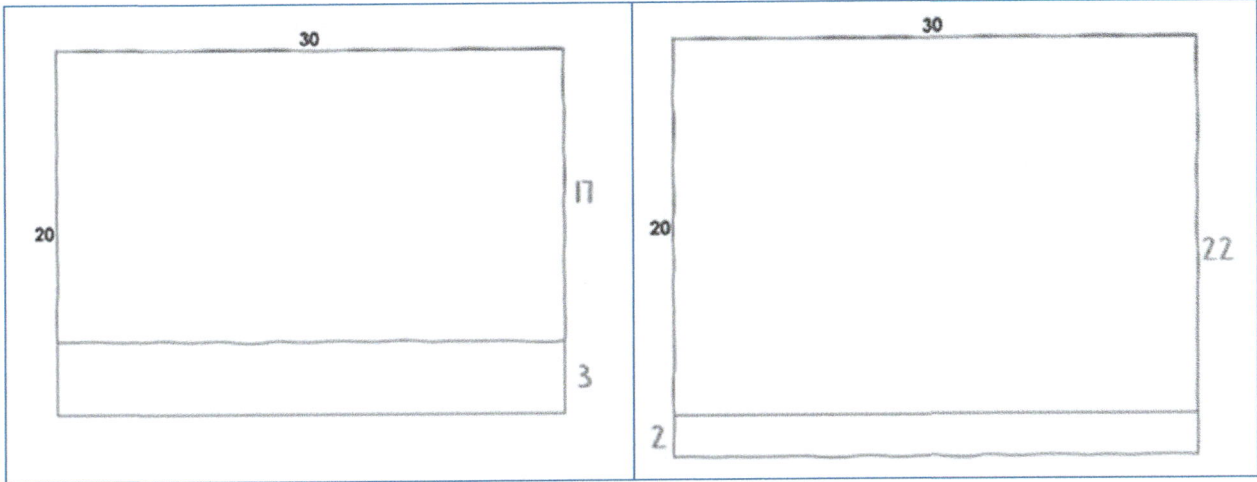

Next, organize students into pairs. Each student needs Area Estimation Game Board B and a dry-erase marker, and the pair needs 1 set of factor cards. Students will take turns as follows:

1. Partner A draws *two* factor cards. These two cards are used to create a multiplication expression.

2. Partner A looks at their game board for an available area model they think is the nearest available estimate of their product. They write their "Actual" expression next to that estimate.

3. Next, Partner B draws *two* factor cards. These two cards are used to create a multiplication expression.

4. Partner A looks at their game board for an available array they think is the nearest available estimate of their product. They write their "Actual" expression next to that estimate.

5. Play continues until both players have filled all available spots on their game boards.

Once both students have filled every spot on their game board, students will sketch area models for their actual products atop the estimates on the game board and find the difference between each actual product and estimate. Students add up their differences (all differences are positive). Lowest total wins.

*Note that students are not finding actual products to determine differences for the game—they should be finding the products for the "extra" or "missing" areas to see the differences between the estimate and the actual problem. The goal of this activity is that students would **notice how these products are related to the estimates**, NOT that they would calculate and then compute differences.

Figure 5.17.11 Speed Estimator Variation 4 example

Partner A draws the factor cards 11 and 14 and makes the expression: 	Partner A chooses to record their product of 11 × 14 next to the estimate 10 × 10 on the game board: 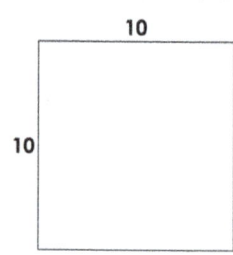 Estimate: <u>10 × 10</u> Actual: <u>11 × 14</u> Difference: _____

Next, Partner A sketches an area model that represents the actual product right on top of the estimated area model and uses the model to find the difference between the products. Encourage students to label any new side lengths.

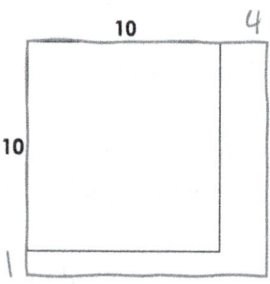

Estimate: <u>10 × 10</u>   Actual: <u>11 × 14</u>   Difference: _____

Encourage students to mark up the model as needed to identify the difference in the areas. In this example, Partner A further decomposes the area model to find that there are 40 + 10 + 4 "extra" square units in the area model.

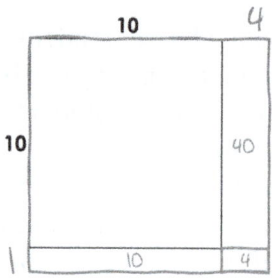

Estimate: <u>10 × 10</u>   Actual: <u>11 × 14</u>   Difference: <u>54</u>

Engage in Math Discourse (Make the Mathematics Visible):

Observation: As students work, encourage them to notice important relationships and push on early and partial understandings.

- Be comfortable with some imprecision from students. Tell students that their math sketches do not need to be to scale!
- Ask students to explain to you what each part of their math sketches represent. Where do they see the difference in the models?

Interview: After the activity, bring the intervention group together for a discussion. Suggested prompts:

- What strategies did you use to pick the closest estimate?
- Did you have to count each square (or each unit cube) one at a time to figure out the difference each time? How did you see it? *(Elicit the idea of using multiplication.)*

Show Me: Choose one or two examples from student work and ask, "Is there a way we could use the estimate to figure out the actual product?"

Bridging Prompt (Prompt for Classroom Teacher to Use During New Lesson Content):

- Show the class one of the areas from Game Board B with a longer rectangle drawn over it (and the extra length labeled). For example:

Figure 5.17.12   An area of 10 × 32 represented on Game Board B

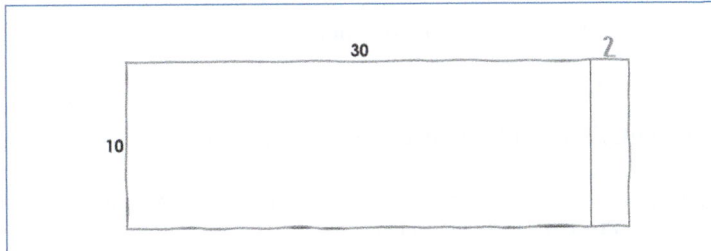

Ask, "What is the total area of the whole rectangle? How did you figure it out?" Call on the students who participated in the intervention group to share their thinking.

Task 18: Magic Pot Patterns

General Objective:

Students will explore number patterns and relationships that follow a given rule and use a story and hands-on activities to understand algebraic thinking.

This activity might be used with:

Students who are able to (prerequisite knowledge):	Students who are Primed to (getting ready to learn):
• recognize a pattern in addition and multiplication tables	• explore number patterns to look for additive or multiplicative relationships between a series of numbers • use patterns to develop generalizations

Materials:

- Cards and markers
- Various small items (coin purse, hairpin, etc.)
- Sticky notes (to change the rule for the pot for different groups)
- PRINTABLE: Input–Output Table chart
- PRINTABLE: Function Machine
- PRINTABLE: Magic Pot
- DIGITAL RESOURCE (function machine): https://qrs.ly/pggjc83

 Printables for this task are available for download at **https://companion.corwin.com/courses/ProactiveMathIntervention**.

Recommended Children's Literature:

- *Two of Everything* – Hong, 1996
- *One Grain of Rice: A Mathematical Folktale* – Demi, 1997

Task Overview:

This task sets the stage for an exploration of patterns and relationships in mathematics, fostering a deeper understanding of algebraic concepts.

Two of Everything by Lily Toy Hong recounts a Chinese folktale about a farmer who discovers a magic pot that doubles everything placed inside it. This humorous story serves as an excellent introduction to function machines and input/output tables, helping teachers transition to the concept of the "doubling pot" and recording data in a structured way to identify patterns. A function is a rule that assigns one output to each input. Teachers can then change the rule for the magic pot and keep it a secret. Students provide input numbers, and the teacher records the corresponding output numbers. The students then try to figure out the current rule of the magic pot after looking for patterns. This activity can be extended into a growing pattern by using the last output as the next input, applying the rule, and repeating the process.

Figure 5.18.1 Magic Doubling Pot connected to *Two of Everything* story

1. Read aloud: You don't need to read the entire book *Two of Everything* by Hong. Sharing an excerpt of retelling the folktale in a few sentences will provide a funny context for doubling and help students connect to multiplicative thinking. Briefly discuss the story context and its key elements, focusing on the magic pot's doubling power.

2. Discussion: Talk about the concept of doubling and how it relates to addition and multiplication.

In each variation of this task, students will move from connecting to the magic pot that serves as a function machine to other related function machines to explore number patterns and relationships.

Select (or create) a variation that focuses on the prior knowledge you are working to shore up for your student group.

Variation 1

Learning Target: Students will connect to the story context and explore the doubling pattern using the magic pot as a function machine.

Variation Directions:

1. Story reenactment: Have students summarize the story by reenacting some of the doubling events using the magic pot and small items (e.g., coin purse, hairpin, etc.).

2. **Record keeping:** As students retell the story, create a table showing the input (items put into the pot) and the output (doubled items).

Figure 5.18.2   Sample input–output table

Input	Output
1 coin	2 coins
1 hairpin	2 hairpins

3. **Pattern recognition:** Give students sticky notes. Ask them to write a number and predict what the output might be if they put that number into the magic pot. Guide students to find the pattern and generate a rule for the doubling process.

Figure 5.18.3   Sample table

Input	Output	Rule
3		
6		
7		
5		
4		

4. **New Rule:** Introduce a new rule for the magic pot/function machine.

Figure 5.18.4   Magic Pot function machine

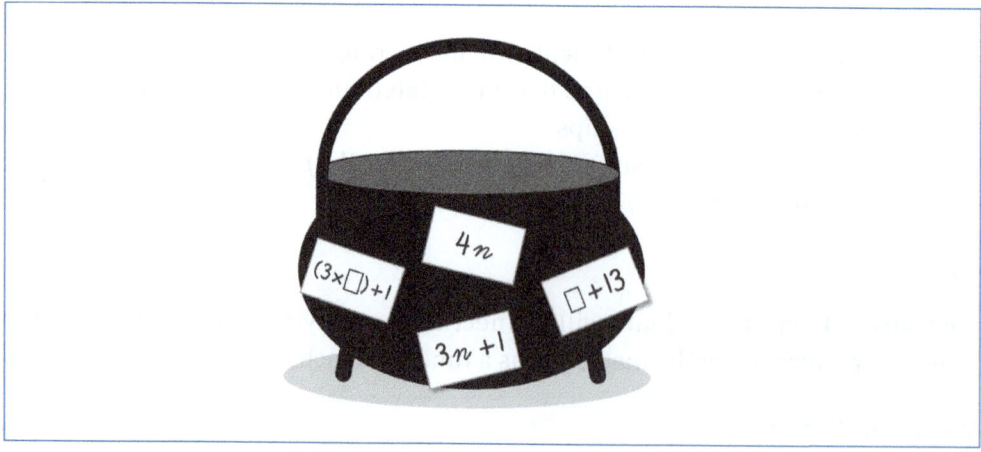

Source: magic pot image from istock.com/Daria Khivrenko

Ask students to write a number on a sticky note and drop it into the pot.

5. **Output generation:** Change the number according to this new rule (e.g., multiply 3 and add 1 to the input number). Record the new input–output pairs.

Figure 5.18.5 Sample input–output table

Input	Output	Rule
2	7	
3	10	
4		

Using the magic pot or the function machine, show a few inputs and outputs and ask students what they notice first when looking at the table. Then, ask what the rule might be and if they can predict what is next if they input a 4.

6. **Pattern analysis:** Have students analyze the new input–output table to determine the rule. Encourage them to make and test their conjectures.

Here are additional examples. Consider starting with:

- Simple rules: Start with basic addition or subtraction rules.
- Complex rules: Introduce more complex rules as students' understanding deepens and moves toward functional thinking. Use order of operations, telling them that what is in the parentheses goes first because grouping symbols are attended to first, such as with (number + 5) × 2 or $2(n + 5)$.
- Look at another example and ask students what they notice first when looking at the table. Then, ask what the rule might be and if they can predict what the next output will be if we continue the pattern.

This technology connection can help students continue this exploration: https://polypad.amplify.com/p#functions.

Engage in Math Discourse (Make the Mathematics Visible):

Observation: As students work, encourage them to notice important relationships and push on early and partial understandings.

- As students are engaged in the activity, check to see if students' rules work for all cases.

- Ask what students are noticing about the output.
- Encourage students to figure out the rule. Ask, "Does the rule you suggested work for the next input? If not, what could be another rule that would work for all the outputs?"

Interview: After the activity, bring the intervention group together for a discussion. Suggested prompts:

- What do you notice when we "add 3" and the starting number is 1? What if we start at the number 2 and add 3? Students should generate terms in the resulting sequence and observe that the terms appear to alternate between odd and even numbers. Have them explain informally why the numbers will continue to alternate in this way.
- Start at the number 5 and add 5. What do you notice about the pattern (5, 10, 15, 20, 25, 30)?

	Bridging Prompt (Prompt for Classroom Teacher to Use During New Lesson Content):
	• Have students who participated in the intervention group show the class a magic pot that has a hidden rule. Have them ask students in the class to offer a number to go into the pot. The students from the intervention group will provide the output following an agreed-upon rule. Record in a table the input and output. Challenge the class to identify the rule once they notice a pattern that repeats.
	• What is an input–output machine?
	• What happens in between, when you input a number and the machine gives you an output?
	• How might testing out rules help you find the hidden rule or function in an input–output machine?

Variation 2

Learning Target: Given the rule and the output, ask students to determine what the input was. This allows students to develop their algebraic thinking of "doing and undoing" an operation.

Variation Directions:

Begin this variation by showing students a function machine with the given rule and the output. Ask them how they might work backward to figure out what the input was. Tell students to notice that we are using a dot to indicate multiplication to reduce confusing "x" with the × symbol.

Figure 5.18.6 Function machines

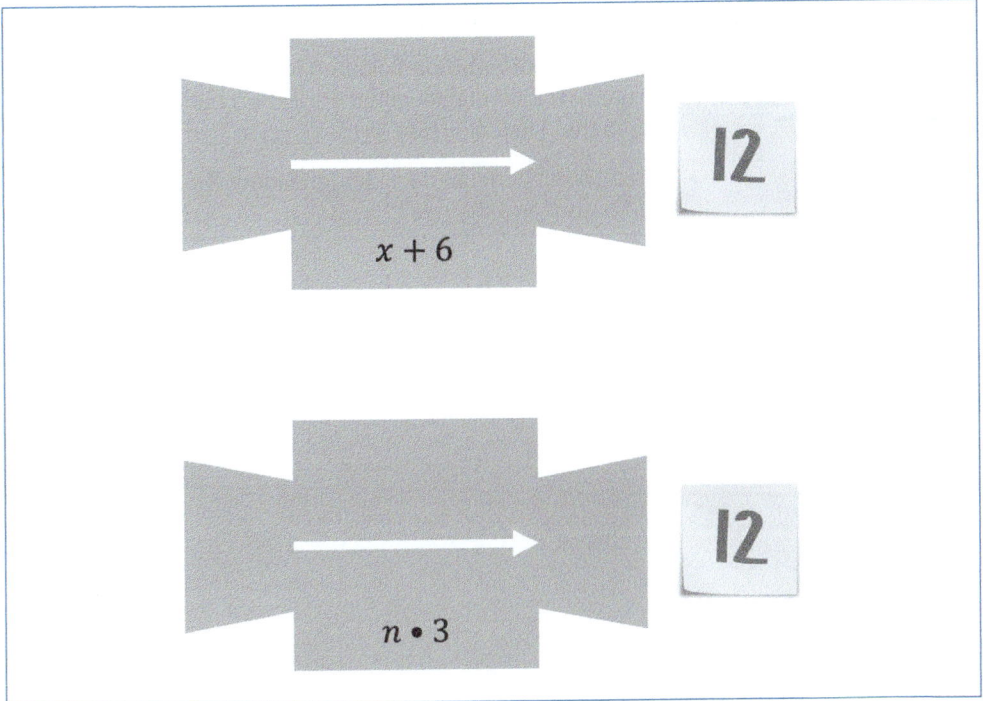

Engage in Math Discourse (Make the Mathematics Visible):

Observation: As students work, encourage them to notice important relationships and push on early and partial understandings.

- Encourage students to look back at the previous input and output table so that they can figure out a method. Students should see that there is an inverse relationship when going from the input to the output and going from the output to the input.

- For example:
  - in the first case, if the rule is to add 6 from input to output, we must then do the opposite (inverse operation), which is subtract 6 from the output to get to the input number.
  - In the second case, if the rule is to multiply 3 from input to output, we must then do the inverse operation, which is divide 3 from the output to get to the input number.

Interview: After the activity, bring the intervention group together for a discussion. Suggested prompts:

- What thinking was involved when you were "undoing" the rule to find the input? Some students might see that they were using the inverse operation of subtracting 6 to undo the addition to get the input of 6 as in the first example.

- Discuss how in the second example, the inverse operation to multiplication is division.

	**Bridging Prompt (Prompt for Classroom Teacher to Use During New Lesson Content):** • Have students who participated in the intervention group share a function table with the rule and output revealed. Have them ask students in the class to find the input. • Ask: "What can you do to find the input if you know only the output and the rule?"

Fraction Frenzy

TASK 19

General Objective:

Students will understand ways to decompose fractions greater than 1 in different ways and will write equivalent expressions to represent decompositions.

This activity might be used with:

Students who can (prerequisite knowledge):	Students who are Primed to (getting ready to learn):
• partition wholes into unit fractions • recognize equivalence in physical models (e.g., notice when two strips are the same length) • add whole numbers within 20	• understand patterns in writing 1 whole as a fraction • decompose a set of unit fractions into addend parts and compose the set into a sum • name fractions greater than 1 • recognize and understand how to write fractions greater than 1 as mixed numbers and recognize that they represent equivalent values (e.g., the fraction greater than 1, $\frac{5}{2}$, can be written as the mixed number $2\frac{1}{2}$, and $\frac{5}{2} = 2\frac{1}{2}$)

Materials:

- Physical fraction manipulatives
- PRINTABLE: Unit Fraction Strips (tabloid sized)

    or

    PRINTABLE: Unit Fraction strips (letter sized)
- PRINTABLE: Unit Fraction Circles

 Printables for this task are available for download at **https://companion.corwin.com/courses/ProactiveMathIntervention**.

Recommended Children's Literature:

- *Enchanted Numbers: Cinderella's Magical Journey* – Rowland, 2024 (adapt to use "fraction greater than 1" instead of the outdated language "improper fraction")

Task Overview:

This activity can be done individually, with partners, or in a small group. Create collections of unit fraction pieces (either physical fraction tiles or cut-apart unit fraction strips) that include as many "wholes" as possible (e.g., a collection of 30 one-fifth strips would create six wholes). Be sure the collection includes several "1 whole" strips as well.

In each variation of this task, students will be asked to grab handfuls of unit fractions (all the same unit), organize them into groups of 1 whole, and write equivalent expressions to represent the collection.

Variation 1

Select (or create) a variation that focuses on the prior knowledge you are working to shore up for your student group.

Learning Target: Recognize when a collection of unit fractions of the same size represents "1 whole."

Variation Directions:

Provide students with a "1 whole" strip for reference. Have each student grab a handful of unit fraction strips (each student should use only one unit size at a time). Ask students to determine how many $\frac{1}{x}$ strips will equal 1 whole. As students determine equivalencies, have them record their discoveries on a whole-group chart.

If needed, unit fraction wedges from fraction circles can be provided to students as a "just-in-time" scaffold. Fraction circles are a helpful scaffold in that they make the "whole" more readily apparent to students—it is easy to see when "1 whole" is complete, particularly for students with visual-spatial needs or challenges with fine motor skills.

Figure 5.19.1 Fraction Frenzy Variation 1 example

A student grabs a handful of one-sixths strips:	They find that six sixths equal 1 whole.
[illustration of scattered $\frac{1}{6}$ strips]	[illustration of a "1" whole bar above six $\frac{1}{6}$ strips]

The student adds their discovery to a whole-group chart:

Unit Fraction	How many in 1 whole?	1 =	Math sketch
$\frac{1}{6}$	6	$1 = \frac{6}{6}$ $1 = \frac{1}{6}+\frac{1}{6}+\frac{1}{6}+\frac{1}{6}+\frac{1}{6}+\frac{1}{6}$	

*Recording these relationships helps to make the mathematics visible because it allows students to see the patterns they are generating and make generalizations.

Engage in Math Discourse (Make the Mathematics Visible):

Observation: As students are finding relationships, encourage them to add information to the whole-group chart as they discover it. Students need not complete all the columns at the same time. As they work, encourage them to notice important relationships and push on early and partial understandings.

- Do you have a prediction about how many of these unit fractions it might take to equal 1 whole? Why do you think that? What patterns do you notice?

- Last time, you had the unit fraction $\frac{1}{x}$. This time, you have the unit fraction $\frac{1}{y}$. Do you think it will take *more* or *fewer* $\frac{1}{y}$s than $\frac{1}{x}$s to equal 1 whole? Why?

Interview: After the activity, bring the intervention group together for a discussion. Suggested prompts:

- What are you noticing about the relationship between unit fractions and 1 whole?

- What information does a unit fraction give you? If I told you the unit fraction was $\frac{1}{50}$, what would that make you picture in your mind? How would that unit fraction piece compare to the unit fractions we have on our chart? What about the unit fraction $\frac{1}{20}$?

Bridging Prompt (Prompt for Classroom Teacher to Use During New Lesson Content):

- Display the chart created in the intervention. Ask students who participated in the intervention group to explain the chart to the class. Ask the class to create a new entry on the chart and discuss it.

Variation 2

Learning Target: Recognize and write expressions to represent fractions greater than 1.

Variation Directions:

Provide students with a "1 whole" strip for reference. Have each student grab a handful of unit fraction strips (each student should use only one unit at a time). Ask students first to estimate how many wholes they think they have (without counting). Then, have students count the actual amount of unit fractions in their handful. Each student should name their amount using the sentence frame, "I have (how many) (size of unit fraction). That is (less than/equal to/greater than) 1 whole." Ask students to adjust their estimates at this point if desired/needed.

Next, students should arrange their fraction strips into groups to make as many wholes as they can. Then, ask them to name their amount using the sentence frame, "I have (how many) wholes plus (number of unit fractions)." As students share their sentence frames, help them record corresponding expressions on a whole-group chart.

Figure 5.19.2 Fraction Frenzy Variation 2 example

A student grabs a handful of one-sixths strips:	The student looks at the handful and estimates:
	"I think I have about 5 wholes here."

The student counts the strips in the handful:	The student says:
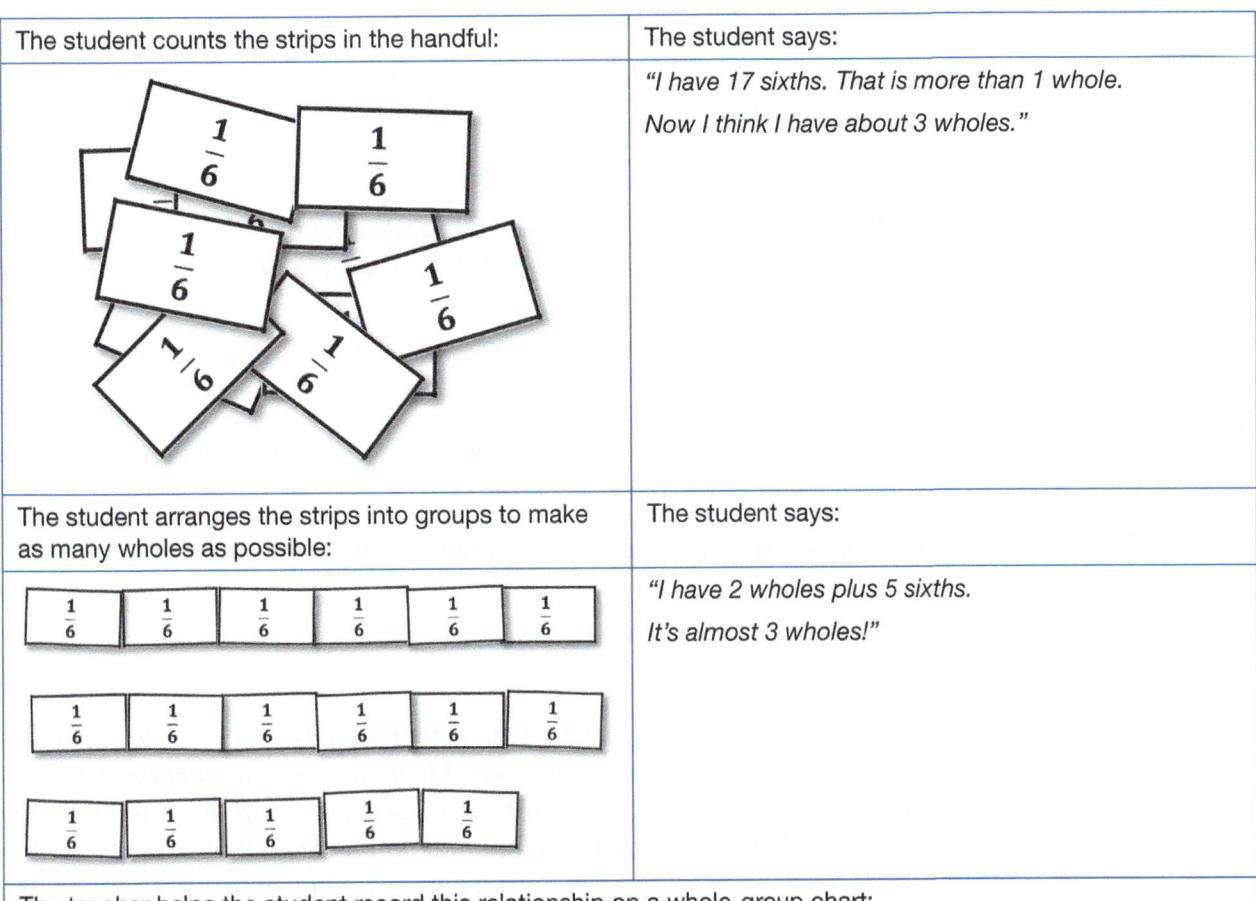	"I have 17 sixths. That is more than 1 whole. Now I think I have about 3 wholes."
The student arranges the strips into groups to make as many wholes as possible: $\frac{1}{6}$ $\frac{1}{6}$ $\frac{1}{6}$ $\frac{1}{6}$ $\frac{1}{6}$ $\frac{1}{6}$ $\frac{1}{6}$ $\frac{1}{6}$ $\frac{1}{6}$ $\frac{1}{6}$ $\frac{1}{6}$ $\frac{1}{6}$ $\frac{1}{6}$ $\frac{1}{6}$ $\frac{1}{6}$ $\frac{1}{6}$ $\frac{1}{6}$	The student says: "I have 2 wholes plus 5 sixths. It's almost 3 wholes!"

The teacher helps the student record this relationship on a whole-group chart:

Fraction	Mixed Number	Equivalent Expressions
$\frac{17}{6}$	$2\frac{5}{6}$	

*Recording these relationships helps to make the mathematics visible because it allows students to see the patterns they are generating and make generalizations.

Engage in Math Discourse (Make the Mathematics Visible):

Observation: As students are working, encourage them to notice important relationships and push on fragile understandings.

- Do you think your fraction will be greater than 1 whole? How do you know?
- Now that you have counted your unit fractions, do you have a prediction for how many wholes you can make? How did you make your prediction?
- Now that you have counted your unit fractions, do you think there will be any "extra" unit fraction pieces? Or will you be able to group them all into wholes? How do you know?

Interview and **Show Me:** After the activity, bring the intervention group together for a discussion. Suggested prompts:

- What are you noticing about what happens when you make groups of as many unit fractions into wholes as possible?
- What do you notice about all the fractions we found that were greater than 1?

Encourage students to analyze their physical models and consider different ways they could count the fraction strips. Elicit various equivalent expressions for each fraction/mixed number pair from students and add these to the whole-group chart. Challenge students to find and add their own equivalent expressions, being certain to relate each expression to the student's model.

Figure 5.19.3 Sample whole-group chart

Fraction	Mixed Number	Equivalent Expressions
$\dfrac{17}{6}$	$2\dfrac{5}{6}$	$2 + \dfrac{5}{6} = \dfrac{6}{6} + \dfrac{6}{6} + \dfrac{5}{6} = 1 + 1 + \dfrac{5}{6} = \dfrac{12}{6} + \dfrac{5}{6}$

CHAPTER 5. Intervention Tasks

$\frac{17}{6} = \frac{6}{6} + \frac{6}{6} + \frac{5}{6}$	(fraction tiles showing 6/6, 6/6, 5/6)
$\frac{17}{6} = 2 + \frac{5}{6}$	(fraction tiles showing two groups of 6/6 circled as 2, and 5/6)

Bridging Prompt (Prompt for Classroom Teacher to Use During New Lesson Content):

- Display the chart created in the intervention. Have students who participated in the intervention group model one of the expressions from the chart and explain their thinking. Identify another expression on the chart and have the class model it with fraction tiles.

Variation 3

Learning Target: Increase comfort with nontraditional decompositions of mixed numbers to support regrouping in addition and subtraction of fractions.

Variation Directions:

Provide students with a "1 whole" strip for reference. Have each student grab a handful of unit fraction strips (each student should have only one unit at a time, and each student should grab a large handful so they definitely have more than two wholes). Ask students to identify a fraction greater than 1 and a mixed number to represent that amount. Students should arrange their unit fraction tiles to prove the relationship between the fraction greater than 1 and the mixed number.

Ask each student to add their values to a different row on a whole-group chart labeled as shown in Figure 5.19.4 (for now, leave the center blank):

Figure 5.19.4 Sample whole-group chart

Fraction Greater than 1		Mixed Number
All wholes are ungrouped		As many wholes as possible are grouped

Next, remind students that they have been practicing writing different expressions to represent the same value. Note that students have a strength in writing fractions greater than 1 (where all the wholes are ungrouped) and mixed numbers (where they have grouped as many wholes as possible), but sometimes, it can be useful to think about a number in a nontraditional format (for example, when we need to regroup for addition and subtraction of whole numbers, that includes grouping and ungrouping amounts, e.g., grouping ten ones together to make a 10 or ungrouping a 10 to create ten ones). Sometimes, with fraction situations, we might want to *group some* of the wholes and leave some of the wholes *ungrouped*. Use a student example as a model for discussion:

Figure 5.19.5 Constructing a whole-group chart

Teacher says: "Student A grabbed 27 sixths. To make a mixed number, they grouped as many wholes as possible. Student A, how many wholes were you able to group?" (4)

Teacher asks the intervention group: "What if Student A only grouped 1 whole? How many ungrouped sixths would there be?" (21)

Teacher writes $1\frac{21}{6}$ in the center column and asks how students might read this number.

Teacher adds a header to the center column and challenges students to find other ways they might regroup $\frac{27}{6}$ so that *some wholes are grouped and some wholes are ungrouped*.

Challenge students to find different equivalent expressions to add to other rows on the chart.

Fraction Greater than 1 All wholes are ungrouped	Some wholes are grouped and some wholes are ungrouped	Mixed Number As many wholes as possible are grouped
$\frac{27}{6}$	$1\frac{21}{6}, 3\frac{9}{6}, 2\frac{15}{6}$	$4\frac{3}{6}$
$\frac{19}{3}$		$6\frac{1}{3}$
$\frac{14}{5}$		$2\frac{4}{5}$

*Recording these relationships helps to make the mathematics visible because it allows students to see the patterns they are generating and make generalizations.

Engage in Math Discourse (Make the Mathematics Visible):

Observation: As students are working, encourage them to notice important relationships and push on early and partial understandings.

- How do you know that this fraction is a fraction greater than one? What should we look for?
- How do you know that you have grouped as many wholes as possible to make this mixed number?
- Your mixed number is $1\frac{4}{5}$. Will you be able to find any expressions to put in the middle column? What makes you say that?

Interview: After the activity, bring the intervention group together for a discussion. Suggested prompts:

- What are you noticing about the expressions we are finding for that middle column?
- Can we call those numbers we wrote in this middle column *mixed numbers*? (No. A mixed number has to have as many wholes grouped as possible. This means the fraction part of a mixed number will be *less than 1*.)

Bridging Prompt (Prompt for Classroom Teacher to Use During New Lesson Content):

- Display the chart created by the intervention group. Display a value that could fit in the middle column (but isn't already there) for one of the rows in the chart. Challenge students to use fraction tiles to prove whether it does or does not represent an equivalent value.

TASK 20: Cake Time

General Objective:

Compare fractions by reasoning about their size.

This activity might be used with:

Students who are able to (prerequisite knowledge):	Students who are Primed to (getting ready to learn):
• partition basic shapes into two, three, or four equal shares and describe the shares using the terms *halves*, *thirds*, *half of*, *a third of*, etc., and describe the whole as two halves, three thirds, four fourths • recognize that equal shares of identical wholes need not have the same shape • understand a fraction $\frac{1}{b}$ as the quantity formed by 1 part when a whole is partitioned into b equal parts • understand a fraction $\frac{a}{b}$ as the quantity formed by parts of size $\frac{1}{b}$	• compare two fractions by reasoning about their size • recognize that comparisons are valid only when the two fractions refer to the same whole

Materials:

- Pattern block manipulatives (with the orange squares and tan rhombuses removed)
- Blank paper
- Crayons (yellow, red, blue, and green)
- PRINTABLE: Pattern Block Cutouts
- PRINTABLE: Hexagons Recording Sheet
- PRINTABLE: Comparing Fractions Activity Page
- PRINTABLE: Fraction Compare Cards
- DIGITAL RESOURCE (pattern blocks): https://qrs.ly/pfgjc84

 Printables for this task are available for download at **https://companion.corwin.com/courses/ProactiveMathIntervention**.

Recommended Children's Literature:

- *Give Me Half* – Murphy, 1996
- *The Wishing Club: A Story About Fractions* – Napoli, 2007

Task Overview:

Third-grade students are expected to become proficient with comparing two fractions with the same numerator or the same denominator by reasoning about their size. In this task, students will partition rectangles into halves, thirds, and fourths, describe each portion as a fraction, and compare the fractional pieces. Students will use pattern blocks to describe the hexagon whole as fractions $\left(\frac{2}{2}, \frac{3}{3}, \frac{6}{6}\right)$ and reason about the size of each unit fraction part $\left(\frac{1}{2}, \frac{1}{3}, \frac{1}{6}\right)$.

Select (or create) a variation that focuses on the prior knowledge you are working to shore up for your student group.

Variation 1

Learning Target: Students will use concrete objects to compare fractions by reasoning about their size.

Variation Directions:

If desired, a literature book about fractions such as *Give Me Half* by Stuart J. Murphy, 1996, can be used to introduce the activity. Another approach is to talk about an event with a cake that will be cut into same-sized pieces.

Conduct this task in a whole group. Organize the students into pairs and distribute sets of pattern block shapes. Explain that the yellow hexagon represents one whole cake. Tell students that their task is to find all the ways to use the other pattern block shapes to make the hexagon cake, using pieces of the same size.

Say, "If the yellow hexagon is a whole cake, what fractional part does the red trapezoid represent?" Allow time for students to cover the hexagon with trapezoids. Encourage them to either trace the trapezoids onto the hexagon shape or paste pattern block cutouts onto the shape. Once complete, record the fractional part of each trapezoid on a whole-group chart.

Figure 5.20.1 Whole-group chart

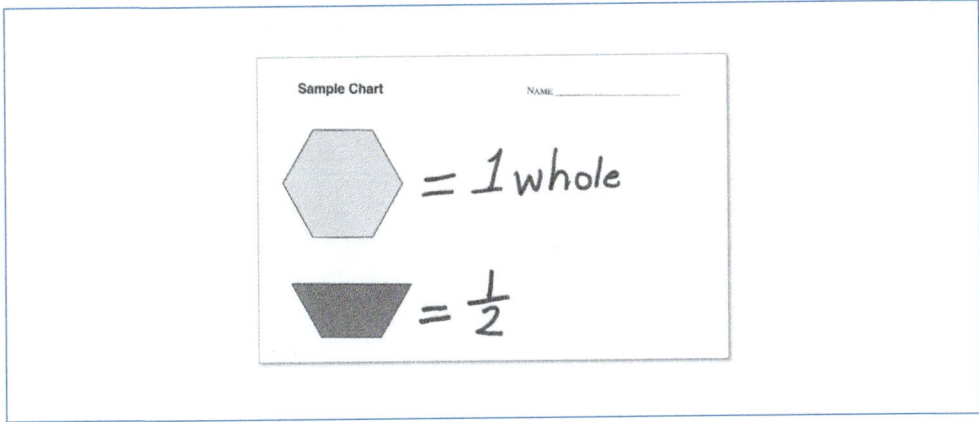

As students determine the fractional value of each pattern block, emphasize math language. For example, "The trapezoid represents $\frac{1}{2}$ of the hexagon whole. How many trapezoid pieces does it take to equal a whole hexagon? So $\frac{1}{2}$ represents 1 of 2 equal pieces and $\frac{2}{2}$ represents 2 of 2 equal pieces." Use this opportunity to discuss the meaning of the numerator (how many parts you have or are shaded) and the denominator (how many equal parts in the whole) and terms such as *equivalent fractions*.

Next, have students repeat the process by covering new hexagons with rhombuses and then triangles. Each time, ask students to describe how many equal parts make up the hexagon (e.g., two trapezoids, three rhombuses, or six triangles). After sharing their findings with the group, label each trapezoid with the fraction $\frac{1}{2}$, each rhombus with $\frac{1}{3}$, and each triangle with $\frac{1}{6}$ on the whole-group chart.

Figure 5.20.2 Whole-group chart

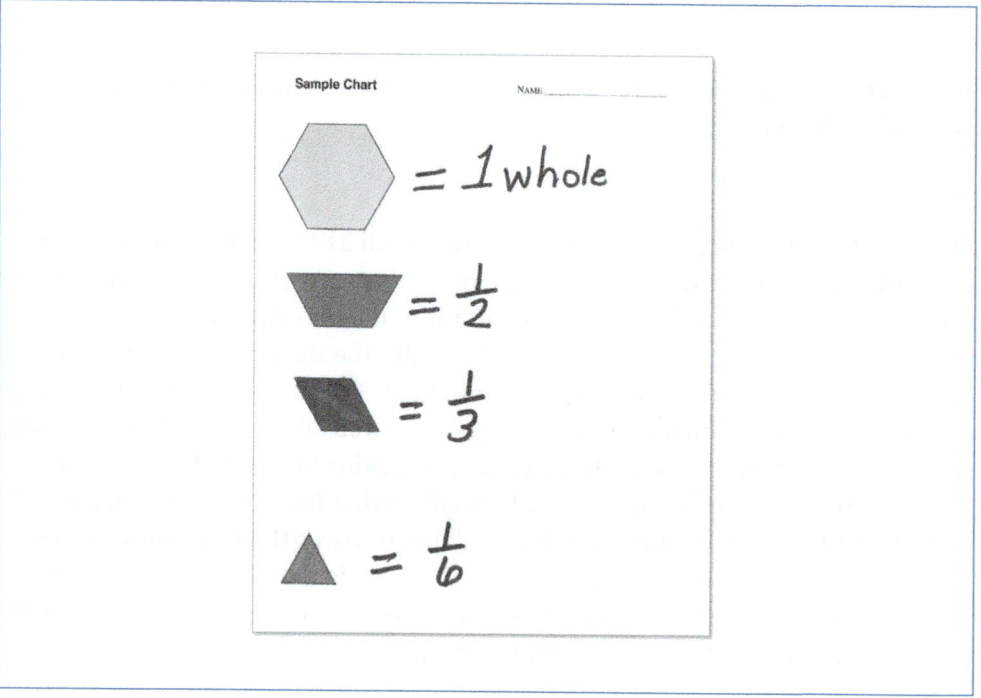

Review the chart and the four ways that they made the hexagon cake with the pattern blocks. Elicit from students that they identified three ways to make the hexagon cake with pattern blocks of the same size. For example:

Figure 5.20.3 Representing relationship with the pattern blocks

| The rhombus represents $\frac{1}{3}$ of the hexagon whole because it takes three rhombuses to cover the hexagon. | |

The triangle represents $\frac{1}{6}$ of the hexagon whole because it takes six triangles to cover the hexagon.	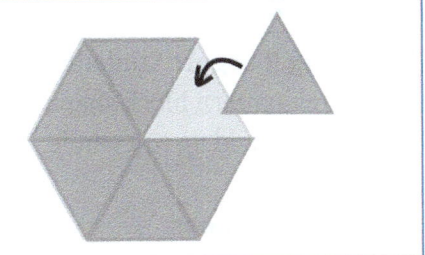

Engage in Math Discourse (Make the Mathematics Visible):

Observation: As students work, encourage them to notice important relationships and push on early and partial understandings.

- What do you notice about the size of the rhombus compared to the trapezoid?
- Explain how you decided the fractional value of each piece.
- Do you think the rhombus or the triangle will need more pieces to completely cover the hexagon? Why?
- What happens if you combine two triangles? What part of the hexagon does that make?

Interview and **Show Me:** After the activity, bring the intervention group together for a discussion. Suggested prompts:

- What are the different ways you discovered to cover the hexagon?
- How does the size of the pattern block affect the fraction it represents?
- Explain why the trapezoid represents $\frac{1}{2}$ of the hexagon.
- What is similar about the fractions for the rhombus and the triangle? What is different?

	Bridging Prompt (Prompt for Classroom Teacher to Use During New Lesson Content):
	• Show the class a yellow hexagon and tell them it represents a whole pizza.
	• Invite two students who participated in the intervention group to show how they could represent sharing the pizza equally using the trapezoid pattern blocks.
	• Ask the class, "If you had to share the pizza with your friends using smaller slices, how could you divide it? What other-sized slices could you use to divide the pizza into equal parts?"

Variation 2

Learning Target: Students will explore equivalency and fractional parts by combining pattern block shapes to create a whole hexagon to show how different fractions can add up to 1 whole.

Variation Directions:

Organize the students into pairs and distribute sets of pattern block shapes and pattern block cutouts along with a Hexagons Recording Sheet. Explain that this task involves finding as many different combinations as possible to create a whole hexagon cake *using a mix of different pattern block shapes*. Encourage students to experiment by combining two or more types of shapes to make the hexagon. Once students find a combination, instruct them to trace the shapes or paste the cutouts onto their recording sheet and label each shape with its fractional value. For example, if they use one trapezoid $\left(\frac{1}{2}\right)$, one rhombus $\left(\frac{1}{3}\right)$, and one triangle $\left(\frac{1}{6}\right)$, to cover the whole they should label each accordingly to show how the pieces add up to 1 whole.

Figure 5.20.4 Combining pattern blocks to equal 1 whole

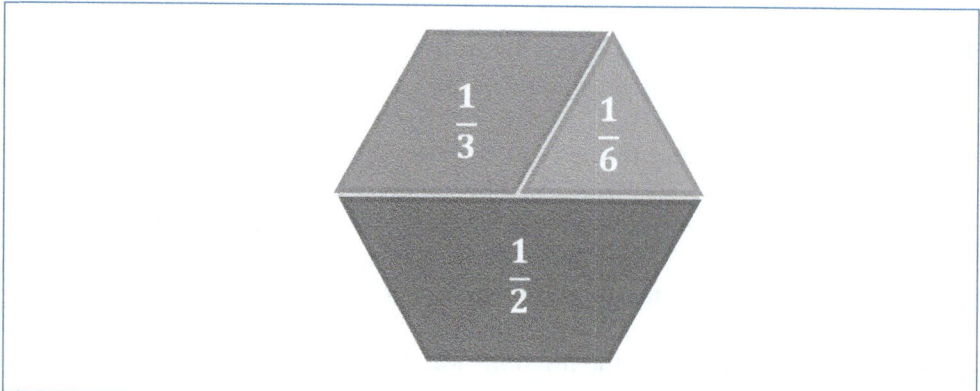

Bring the students together and facilitate a whole-group discussion. Using chart paper, record students' responses as they share their findings. Encourage students to share any unique combinations they discovered using mixed shapes. Record these combinations on the chart, emphasizing how the fractional parts combine to equal 1 whole. For example, write, "One trapezoid $\left(\frac{1}{2}\right)$, one rhombus $\left(\frac{1}{3}\right)$, and one triangle $\left(\frac{1}{6}\right)$ together equal 1 whole hexagon."

Figure 5.20.5 Whole-group chart

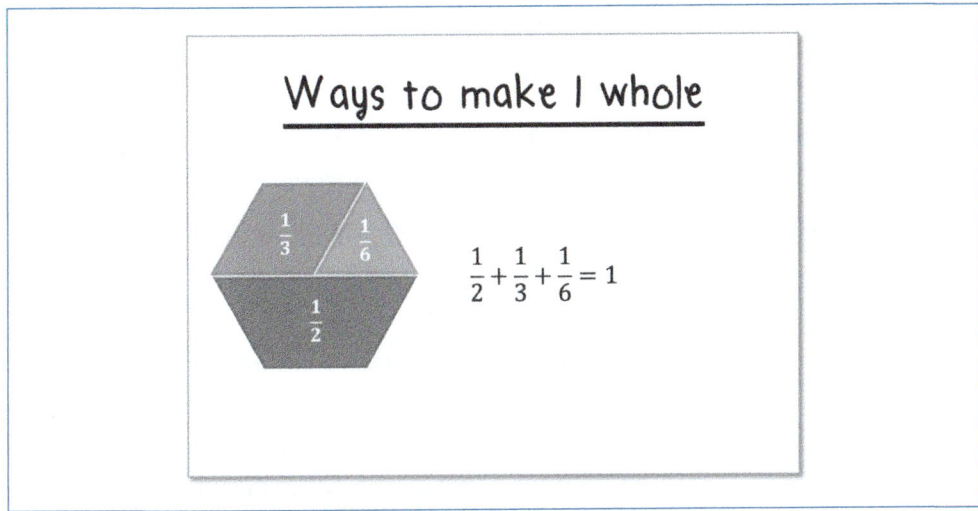

Bring the students' attention to the chart with all the recorded combinations and explain that each of the combinations represents 1 whole hexagon cake. Emphasize the concept of equivalency by pointing out the different combinations of shapes that result in the same total value. For example:

Two trapezoids equal 1 whole hexagon, so $\frac{1}{2}+\frac{1}{2}=1$.

Three rhombuses equal 1 whole hexagon, so $\frac{1}{3}+\frac{1}{3}+\frac{1}{3}=1$.

Six triangles equal 1 whole hexagon, so $\frac{1}{6}+\frac{1}{6}+\frac{1}{6}+\frac{1}{6}+\frac{1}{6}+\frac{1}{6}=1$.

One trapezoid and three triangles equal 1 whole hexagon, so $\frac{1}{2}+\frac{1}{6}+\frac{1}{6}+\frac{1}{6}=1$.

One trapezoid, one rhombus, and two triangles together equal 1 whole hexagon, so $\frac{1}{2}+\frac{1}{3}+\frac{2}{6}=1$.

Engage in Math Discourse (Make the Mathematics Visible):

Observation: As students work, encourage them to notice important relationships and push on early and partial understandings.

- What equivalencies are you finding?
- What do you notice about the number of pieces needed to make 1 whole when using triangles versus rhombuses?
- Why are two trapezoids equivalent to one hexagon? What fraction does each trapezoid represent?
- What are three triangles equivalent to? What fraction does each triangle represent?

 **Interview** and **Show Me:** After the activity, bring the intervention group together for a discussion. Suggested prompts:

- What equivalencies do you find?
- What other combinations of shapes can you find that add up to 1 whole hexagon? How do you know they're equal to 1 whole?
- How many of this shape (point to a triangle, rhombus, trapezoid) did you use to cover the hexagon? What does that tell us about its fraction?

Bridging Prompt (Prompt for Classroom Teacher to Use During New Lesson Content):

- Share the following scenario with the class:
  - Imagine you are bakers working together to create the perfect batch of cookies, and each cookie is shaped like a hexagon. But you've run out of full hexagon molds! You need to combine smaller molds (trapezoid, rhombus, triangle) to make whole cookie shapes.
  - If these molds represent fractional parts of a whole, what other combinations can we use to make 1 whole cookie? Explain why these combinations are equal to 1 whole.
- Invite students who participated in the intervention group to show one of the ways they found to make 1 whole.
- Challenge the class to find as many additional ways as they can.

Variation 3

Learning Target: Students will compare fractional parts to determine if they are equal, greater than, or less than using pattern blocks, visual representations, and math language to explain their reasoning.

Variation Directions:

Pull the students together for a whole-group discussion. Tell the students, "You are planning a birthday party, and you want to share a hexagon-shaped cake fairly with your guests. Each guest should get an equal-sized piece of the cake. But what happens if some pieces are bigger or smaller than others?" Distribute pattern blocks, blank paper, pencils, and/or crayons to students and pose the questions:

- If one guest gets $\frac{1}{2}$ of the cake and another guest gets $\frac{1}{3}$, will they each get the same amount?
- If one guest gets $\frac{1}{2}$ of the cake and another gets $\frac{3}{6}$, will they each get the same amount?

Encourage students to use pattern blocks to explore and explain their thinking. As students share their ideas, record their comparisons using =, <, and > on a whole-group chart. Discuss how equivalent fractions like $\frac{1}{2}$ and $\frac{3}{6}$ represent the same portion of the whole but use different numbers of parts.

$$\frac{1}{2} > \frac{1}{3}$$

$$\frac{1}{2} = \frac{3}{6}$$

Engage in Math Discourse (Make the Mathematics Visible):

Observation: As students work, encourage them to notice important relationships and push on early and partial understandings.

- How are you using the pattern blocks to compare the fractions?
- Explain your answer using math language (one-half, one-fourth, and one-sixth, numerator, denominator).
- What does the numerator represent? What does the denominator represent?

Interview: After the activity, bring the intervention group together for a discussion. Suggested prompts:

- How did you decide whether the fractions were greater than, less than, or equal?
- Which combinations are equivalent?

	Bridging Prompt (Prompt for Classroom Teacher to Use During New Lesson Content):
	• Show the class the yellow hexagon, the red trapezoid, and the green triangle. Have students who participated in the intervention group explain to the class how to identify the unit fraction for the red trapezoid and the green triangle if the yellow hexagon represents 1 whole.
	• Ask how many green triangles it takes to cover the red trapezoid completely. What fractions are represented? What can we say about the two fractions?

Variation 4

Learning Target: Students will use mathematical symbols (>, =, <) and descriptive language to compare fractions that are not equivalent and justify their reasoning using math sketches, pattern blocks, or other visual representations.

Variation Directions:

Bring the students together for a whole-group discussion. Begin by holding up a deck of fraction cards $\left(e.g., \frac{1}{2}, \frac{2}{2}, \frac{1}{3}, \frac{2}{3}, \frac{3}{3}, \frac{1}{6}, \frac{2}{6}, \frac{3}{6}, \frac{4}{6}, \frac{5}{6}, \frac{6}{6}, \frac{1}{4}, \frac{2}{4}, \frac{3}{4}, \frac{4}{4}\right)$.

Say: "Today, we'll use fraction cards to compare fractions. Your goal is to draw two cards, compare the fractions, and use pattern blocks or math sketches to show your answer. Then, you'll record your comparison using symbols and words." Play the game with a student. Draw one fraction card and have the student, as your partner, draw the other fraction card. For example,

Say: "I drew $\frac{1}{3}$ and my partner drew $\frac{4}{6}$. Now, my partner and I will use pattern blocks to model the fractions we drew." Place a blue rhombus (representing $\frac{1}{3}$) on a yellow hexagon. Ask the student to represent $\frac{4}{6}$ using four triangles on the yellow hexagon. Show the fraction manipulatives under a document camera or display so all the students can see. Ask the students to compare the fractions they see and determine which fraction is larger.

Represent the fraction comparison for the students:

$$\frac{1}{3} < \frac{4}{6}$$

Play one more time and have the students determine the correct symbol to compare the fractions.

Next, organize students in pairs to play the game. Provide each pair with a set of fraction cards, pattern blocks, and pattern block cutouts.

Directions:

1. Partner A draws one card and Partner B draws a card.

2. Each partner should then use the pattern blocks to represent the fraction card they drew.

3. Students use pattern blocks or math sketches to justify their answers and record the comparison with symbols and words.

4. Together, the partners explain their fraction representation and decide which fraction is larger.

5. The partner with the fraction with the greater value gets a point.

Engage in Math Discourse (Make the Mathematics Visible):

Observation: As students work, encourage them to notice important relationships and push on early and partial understandings.

- How are you using the fraction blocks to compare the fractions?
- Explain your answer using math language (one-half, two-fourths, and three-sixths, numerator, denominator)?

Interview and **Show Me:** After the activity, bring the intervention group together for a discussion. Suggested prompts:

- How did you use pattern blocks to compare the fractions and decide which fraction was greater?
- What does the numerator represent? What does the denominator represent?
- What happens to the size of the fraction as the numerator increases but the denominator stays the same?
- What happens to the size of the fraction if the denominator increases while the numerator stays the same?

Bridging Prompt (Prompt for Classroom Teacher to Use During New Lesson Content):

- Show the class two fraction cards, such as $\frac{1}{3}$ and $\frac{2}{3}$. Have students who participated in the intervention group use blue rhombuses to represent them on a yellow hexagon.
- Ask, "What do you notice about these two fractions? How are they similar? How are they different? How does changing the numerator affect the fraction?"
  - Have students who participated in the intervention represent the class thinking symbolically: $\frac{1}{3} < \frac{2}{3}$.
- Repeat with a different set of fractions, this time with a common numerator. For example, $\frac{2}{3}$ and $\frac{2}{6}$.
- Ask: What do you notice about the denominators? How does changing the denominator affect the size of the pieces?"

TASK 21

Line 'Em Up

General Objective:

Students will locate fractions on a number line.

This activity might be used with:

Students who can (prerequisite knowledge):	Students who are Primed to (getting ready to learn):
• put whole numbers in order • place whole numbers in order on a number line • build unit fractions and nonunit fractions using fraction bar models • read and write fractions	• compare fractions with common numerators • compare fractions with common denominators • order fractions with common numerators • order fractions with common denominators • compare fractions to common benchmarks (e.g., 0, $\frac{1}{2}$, 1, $1\frac{1}{2}$, 2)

Materials:

- Physical fraction manipulatives
- String or rope
- Clothespins
- PRINTABLE: Unit Fraction Strips (tabloid sized)

    or

    PRINTABLE: Unit Fraction Strips (letter sized)
- PRINTABLE: Unit Fraction Circles
- PRINTABLE: Line 'Em Up Common Numerator Cards
- PRINTABLE: Line 'Em Up Common Denominator Cards
- PRINTABLE: Line 'Em Up Benchmark Cards
- PRINTABLE: Compare 'Em Cards Set A
- PRINTABLE: Compare 'Em Cards Set B

- PRINTABLE: Compare 'Em Cards Set C
- PRINTABLE: Compare 'Em Cards Set D
- PRINTABLE: Compare 'Em Cards Set E

 Printables for this task are available for download at **https://companion.corwin.com/courses/ProactiveMathIntervention**.

Recommended Children's Literature:

- *One . . . Two . . . Three . . . Sassafras* – Murphy, 2002 (putting simple whole numbers in order)

Task Overview:

In this activity, students will use physical length models (fraction strips) to compare and order fractions. Students will build physical number lines (using clothespins and string/rope) as they make comparisons.

Select (or create) a variation that focuses on the prior knowledge you are working to shore up for your student group.

Variation 1

Learning Target: Students will compare fractions that have common denominators.

Variation Directions:

This activity can be done as a whole group or in pairs/triads. Each group should have a set of Line 'Em Up Common Denominator cards, a string/rope number line, a collection of clothespins, and a set of fraction strips.

Group members will take turns as follows:

1. The group draws a starting fraction card and clips it to the number line somewhere around the middle of the line.
2. Student A draws a fraction card.
   a. Student A decides if their fraction is *greater than* or *less than* the starting fraction on the number line. The student should build both fractions using fraction strips to check their thinking.
   b. When student A decides, they will clip their fraction to the number line on the appropriate side of the starting fraction and explain their reasoning to the group. *Group members should be encouraged to challenge the student to "prove it" using fraction strips.*
3. Student B draws a fraction card.
   a. Student B decides where their fraction should be placed on the number line. They will need to decide if it is *greater than both, less than both,* or *between* the fractions already positioned on the number line. The student should also build the fractions using fraction strips to check their thinking.

262 PART 2. PUTTING PRIMING INTO ACTION

b. When student B decides, they will clip their fraction to the number line in the appropriate place and explain their reasoning to the group. If needed, the student can adjust the placement of the fraction(s) already on the number line to make room. *Group members should be encouraged to challenge the student to "prove it" using fraction strips.*

4. Students continue to take turns placing fractions on the number line until all fractions in the set have been placed on the line.

Figure 5.21.1 Line 'Em Up Variation 1 example

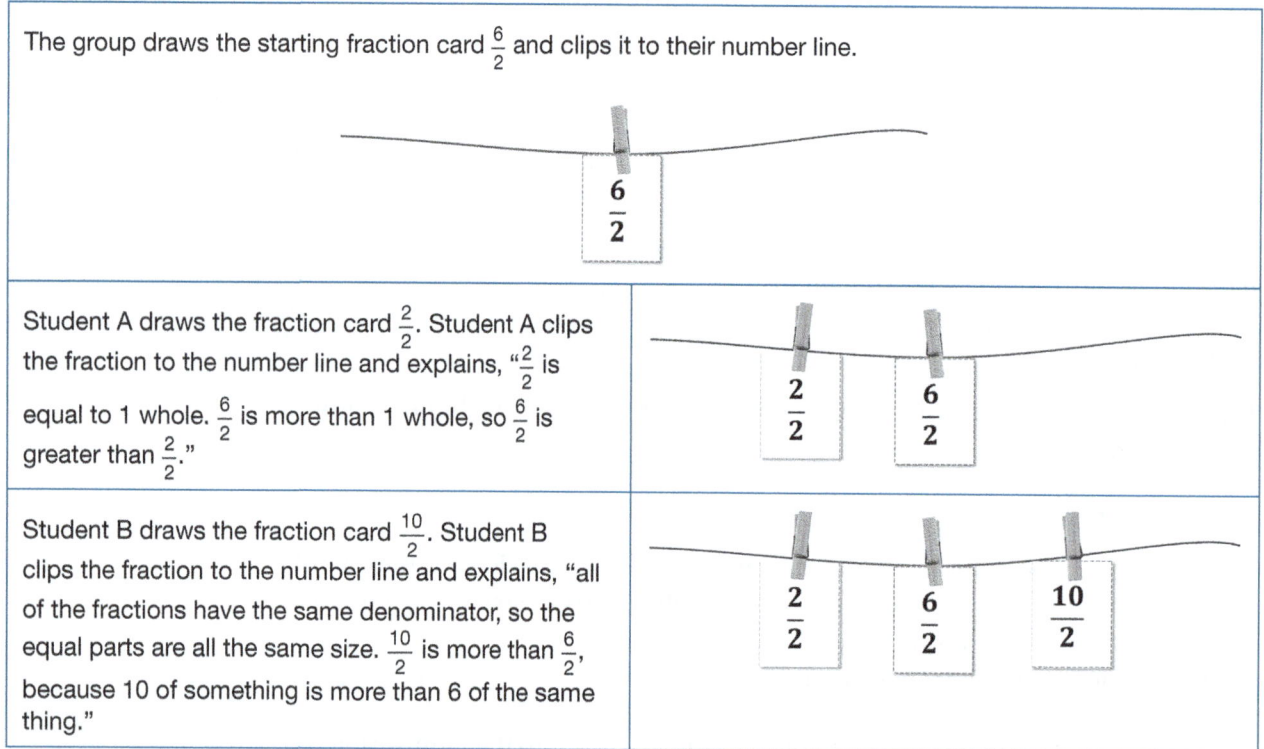

Engage in Math Discourse (Make the Mathematics Visible):

Observation: As students work, encourage them to notice important relationships and push on early and partial understandings.

- What are you noticing about the order of the fractions? Why do you think that is happening?

- What do you notice about the numerators of these fractions?

- What do you notice about the denominators of the fractions?

- *Choose two fractions on the students' number line to discuss:* What can you tell me about how the fraction strips for $\frac{7}{8}$ compare to the fraction strips for $\frac{3}{8}$?

CHAPTER 5. Intervention Tasks   263

Interview: After the activity, bring the intervention group together for a discussion. Suggested prompts:

- Look around the room at all of our number lines. What do you notice? What do you wonder?

- What conjecture might you make about comparing fractions that have the same denominator? Why did this happen? What do you know about the size of the unit fractions that could explain this?

Bridging Prompt (Prompt for Classroom Teacher to Use During New Lesson Content):
• Show the class a fraction clipped to a number line string. Show the class another fraction that has the same denominator. Ask: "How can we use what we know about the size of unit fractions to decide where this fraction should be placed on the number line?"
• Look for opportunities to invite students who participated in the intervention to share their reasoning.

Variation 2

Learning Target: Students will compare fractions that have common numerators.

Variation Directions:

This variation works the same way as Variation 1, except that students will be comparing fractions with common numerators in this variation. Each group should have a set of Line 'Em Up Common Numerator cards, a string/rope number line, a collection of clothespins, and a set of fraction strips.

Group members will take turns as follows:

1. The group draws a starting fraction card and clips it to the number line somewhere around the middle of the line.

2. Student A draws a fraction card.

    a. Student A decides if their fraction is *greater than* or *less than* the starting fraction on the number line. Student should build both fractions using fraction strips to check their thinking.

    b. When student A decides, they will clip their fraction to the number line on the appropriate side of the starting fraction and explain their reasoning to the group. *Group members should be encouraged to challenge the student to "prove it" using fraction strips.*

3. Student B draws a fraction card.
   a. Student B decides where their fraction should be placed on the number line. They will need to decide if it is *greater than both*, *less than both*, or *between* the fractions already on the number line. The student should also build the fractions using fraction strips to check their thinking.
   b. When student B decides, they will clip their fraction to the number line in the appropriate place and explain their reasoning to the group. If needed, the student can adjust the placement of the fraction(s) already on the number line to make room. *Group members should be encouraged to challenge the student to "prove it" using fraction strips.*
4. Students continue to take turns placing fraction cards on the number line until all fractions in the card set have been placed on the line.

Figure 5.21.2 Line 'Em Up Variation 2 example

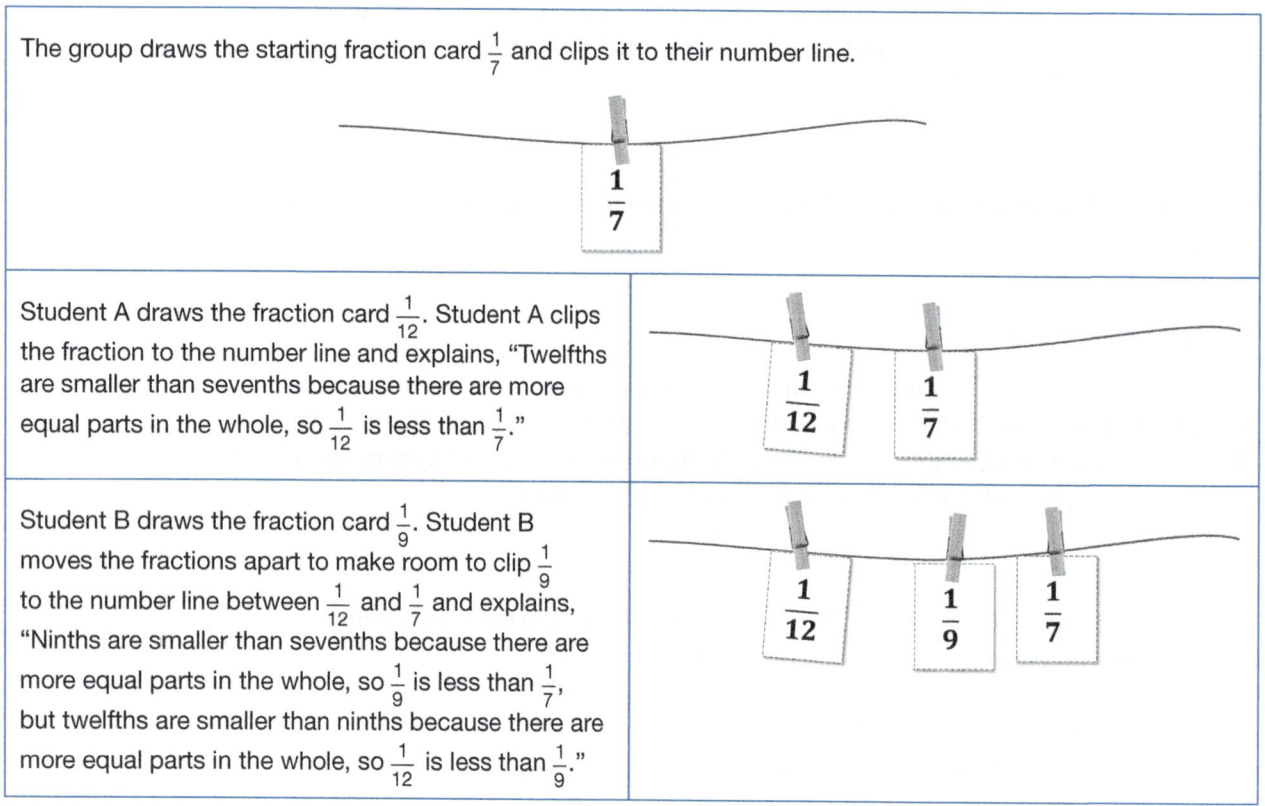

The group draws the starting fraction card $\frac{1}{7}$ and clips it to their number line.	
Student A draws the fraction card $\frac{1}{12}$. Student A clips the fraction to the number line and explains, "Twelfths are smaller than sevenths because there are more equal parts in the whole, so $\frac{1}{12}$ is less than $\frac{1}{7}$."	
Student B draws the fraction card $\frac{1}{9}$. Student B moves the fractions apart to make room to clip $\frac{1}{9}$ to the number line between $\frac{1}{12}$ and $\frac{1}{7}$ and explains, "Ninths are smaller than sevenths because there are more equal parts in the whole, so $\frac{1}{9}$ is less than $\frac{1}{7}$, but twelfths are smaller than ninths because there are more equal parts in the whole, so $\frac{1}{12}$ is less than $\frac{1}{9}$."	

Engage in Math Discourse (Make the Mathematics Visible):

Observation: As students work, encourage them to notice important relationships and push on early and partial understandings.

▸ What are you noticing about the order of the fractions? Why do you think that is happening?

- What do you notice about the numerators of these fractions?
- What do you notice about the denominators of the fractions?
- *Choose two fractions on the students' number line to discuss:* What can you tell me about how the fraction strips for $\frac{3}{5}$ compare to the fraction strips for $\frac{3}{8}$?

Interview: After the activity, bring the intervention group together for a discussion. Suggested prompts:

- Look around the room at all of our number lines. What do you notice? What do you wonder?
- What conjecture might you make about comparing fractions that have the same numerator?
- Why did this happen? What do you know about the size of the unit fractions that could explain this?

	Bridging Prompt (Prompt for Classroom Teacher to Use During New Lesson Content):
	• Show the class a fraction clipped to a number line string. Show the class another fraction that has the same numerator. Ask: "How can we use what we know about the size of unit fractions to decide where this fraction should be placed on the number line?"
	• Look for opportunities to invite students who participated in the intervention to share their reasoning.

Variation 3

Learning Target: Students will compare fractions to common benchmark numbers such as $0, \frac{1}{2}, 1, 1\frac{1}{2}, 2$, etc. by reasoning about relative size of the number.

Variation Directions:

This activity can be done as a whole group or in pairs/triads. Each group should have a set of Compare 'Em cards and a set of fraction strips. Students will need a string/rope number line and floor or table space but will not use clothespins for this variation.

Give each group one or more benchmark numbers to place on their number line. If more than one benchmark number is used, students should leave space between numbers.

Group members will take turns as follows:

1. Student A draws a fraction card.
   a. Student A decides if their fraction is *greater than* or *less than* the benchmark number(s). The student can build the fractions and the benchmark number(s) using fraction strips to check their thinking if desired.

b. When student A decides, they will place their fraction in the appropriate region of the number line (see example) and explain their reasoning to the group. *Group members may challenge the student to "prove it" using fraction strips.*

2. Student B draws a fraction card.

   a. Student B decides if their fraction is *greater than* or *less than* the benchmark number(s). The student can build the fractions and the benchmark number(s) using fraction strips to check their thinking.

   b. When student B decides, they will place their fraction in the appropriate region of the number line (see example) and explain their reasoning to the group. *Group members may challenge the student to "prove it" using fraction strips.*

3. Students continue to take turns placing fractions on either side of the benchmark(s) until all fractions in the set have been placed. *Note that students are not ordering the numbers in this variation. They are only deciding if each number is greater or less than the given benchmark(s).*

Figure 5.21.3 Line 'Em Up Variation 3 example

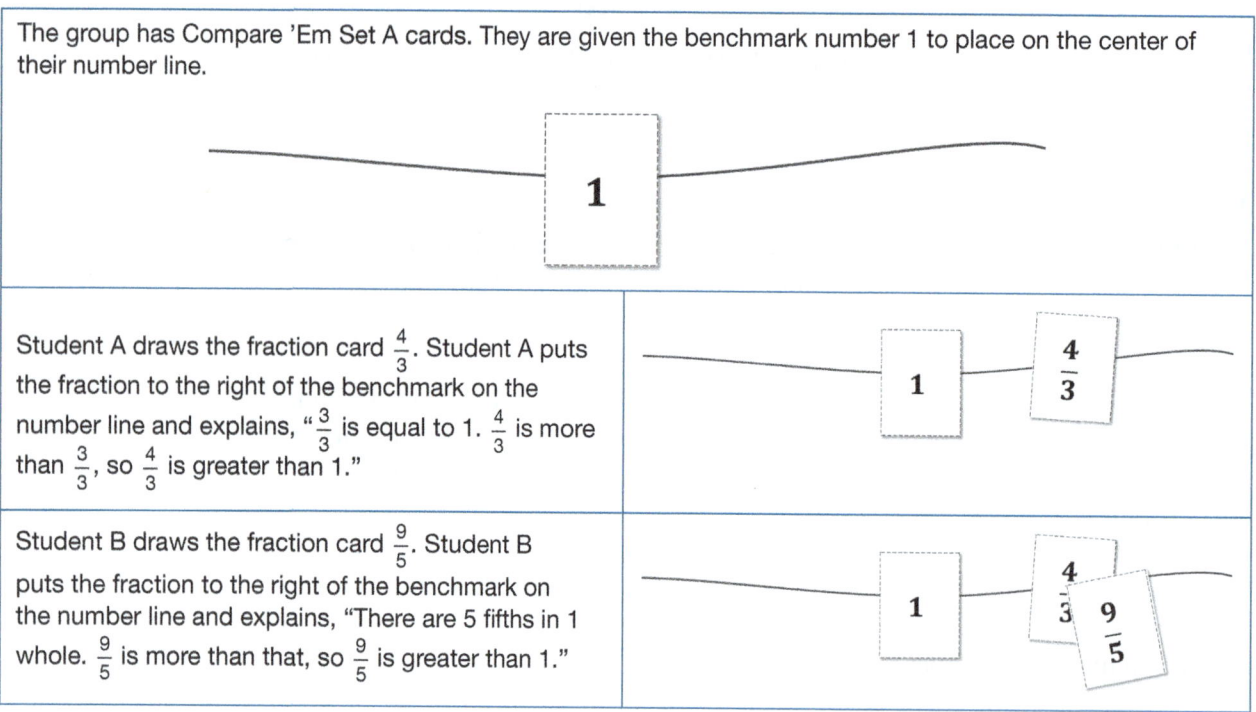

Engage in Math Discourse (Make the Mathematics Visible):

Observation: As students work, encourage them to notice important relationships and push on early and partial understandings.

- What are you noticing about the fractions that are less than this benchmark?
- What are you noticing about the fractions that are greater than this benchmark?
- Would there be any fractions you would place exactly **on** the benchmark? What would those fractions look like?

Interview: After the activity, bring the intervention group together for a discussion. Suggested prompts:

- Look around the room at all of our number lines. What do you notice?
- What conjecture might you make about fractions less than *(benchmark)*? About fractions greater than *(benchmark)*?

Bridging Prompt (Prompt for Classroom Teacher to Use During New Lesson Content):
• Show the class a number line string with cards for 0, $\frac{1}{2}$, and 1 clipped to the line. Ask, "How could we use these benchmarks to decide if $\frac{1}{8}$ or $\frac{9}{10}$ is greater?"
• Look for opportunities to lift up the thinking of students who participated in the intervention.

Variation 4

Learning Target: Students will compare fractions with common numerators or denominators **and** compare them to common benchmark numbers such as $0, \frac{1}{2}, 1, 1\frac{1}{2}, 2$, etc. by reasoning about relative size of the number.

Variation Directions:

This variation combines Variation 1 or Variation 2 with Variation 3. Follow the directions for Variation 1 or 2, except place benchmark numbers on the number line instead of making a math sketch of the fraction to start each round. Students will need to compare the fractions they draw to the benchmarks *as well as* to the other fractions in the set. Students should use the Common Numerator or Common Denominator Line 'Em Up cards, *not the Compare 'Em cards*, for this variation.

Figure 5.21.4 Line 'Em Up Variation 4 example

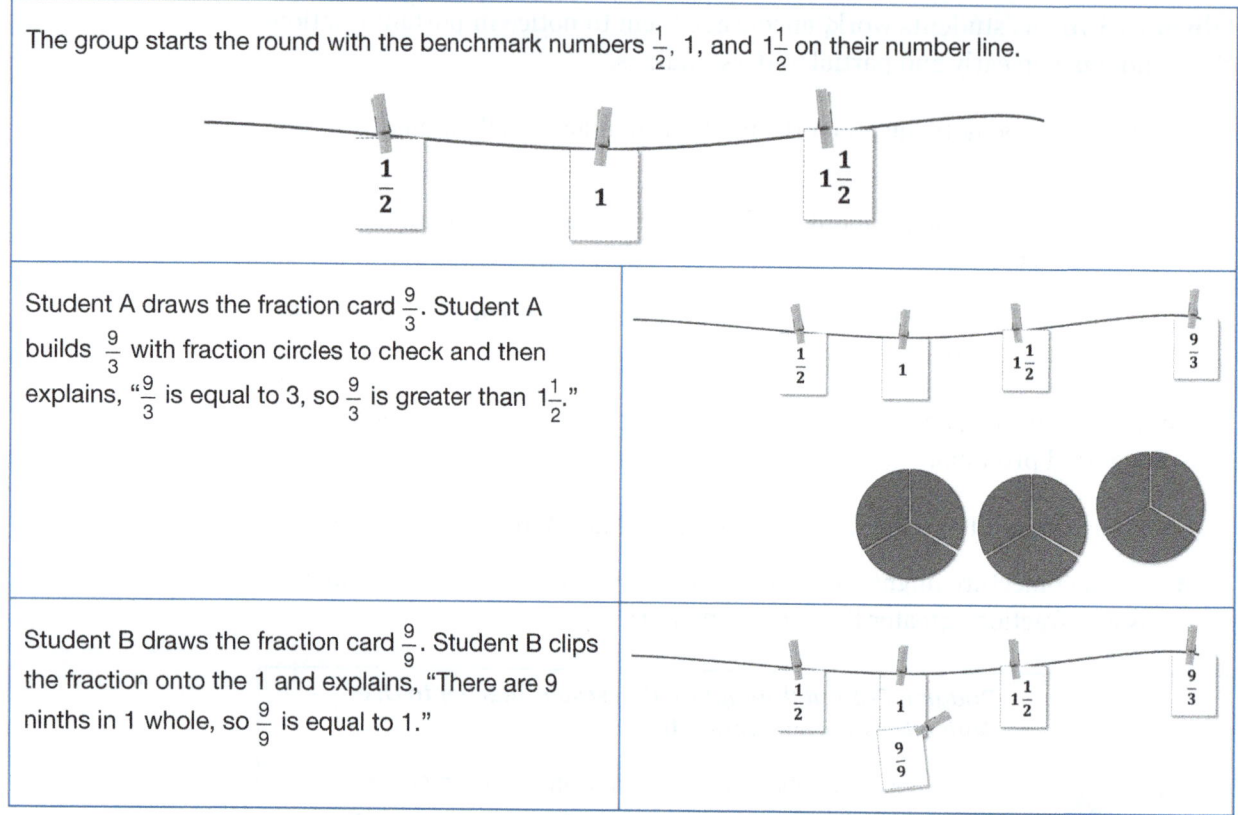

Engage in Math Discourse (Make the Mathematics Visible):

Observation: As students work, encourage them to notice important relationships and push on early and partial understandings.

- How are you deciding where to place your fractions? Are you looking at the numerator/denominator, or are you thinking about the benchmarks first?
- How do the benchmarks change the number line for you?

Interview: After the activity, bring the intervention group together for a discussion. Suggested prompts:

- Look around the room at all of our number lines. What do you notice? What do you wonder?
- What are you noticing about the relationships between the fractions and the benchmark numbers?

Bridging Prompt (Prompt for Classroom Teacher to Use During New Lesson Content):

- What strategy would you use to compare $\frac{2}{6}$ to $\frac{2}{4}$? Call on the students who participated in the intervention group to share their thinking.

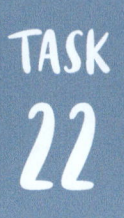

TASK 22

To Be or Not to Be: Equivalent Fractions

General Objective:

Students will explain their reasoning for generating equivalent fractions by using visual fraction models.

This activity might be used with:

Students who are able to (prerequisite knowledge):	Students who are Primed to (getting ready to learn):
• name and write fractions and mixed numbers represented by a model • represent unit fractions as $\frac{1}{2}, \frac{1}{4}, \frac{1}{8}$, as well as $\frac{1}{3}, \frac{1}{6}$ • understand that the denominator names the total number of parts in the whole or group and the numerator is the number of parts being indicated	• represent equivalent fractions. Students can create a model to represent an equivalent fraction. • find equivalent fractions by "renaming the fraction" (becomes useful when adding and subtracting fractions with unlike denominators)

Materials:

- Fraction number line
- Fraction squares
- Clothesline as a number line
- Square paper like origami paper
- PRINTABLE: 1-inch grid paper
- PRINTABLE: Brownie Tray Activity Sheet
- PRINTABLE: Unit Fraction Strips (tabloid sized)

    or

    PRINTABLE: Unit Fraction Strips (letter sized)
- PRINTABLE: Unit Fraction Circles
- DIGITAL RESOURCE (fraction models): https://qrs.ly/bigjc86
- DIGITAL RESOURCE (number line): https://qrs.ly/ipgjc88

- DIGITAL RESOURCE (fraction strips): https://qrs.ly/2wgjc8c
- DIGITAL RESOURCE (fraction circles): https://qrs.ly/lxgjc8d

 Printables for this task are available for download at **https://companion.corwin.com/courses/ProactiveMathIntervention**.

Recommended Children's Literature:

- *The Lion's Share: A Tale of Halving Cake and Eating It Too* – McElligott, 2009

Task Overview:

Students will engage in a relevant context to explore fraction representations of area and length models to find equivalent fractions. Students will explore ways to rename fractions or find equivalent fractions using visuals. "Renaming a fraction" means to write an equivalent fraction with different numbers in the numerator and denominator, essentially changing how the fraction looks while still representing the same value and the same position on a number line; this is initially done using models and patterns and, eventually, by multiplying both the numerator and denominator of the fraction by the same number.

Area models are the most familiar because they tend to be easier for students to grasp, and length models (e.g., number line, measurement) and set models tend to be more difficult. Set models offer a particular challenge in representing equivalence. Students may need additional support bridging their understanding of area models to help support their understanding of measurement and set models.

Select (or create) a variation that focuses on the prior knowledge you are working to shore up for your student group.

Variation 1

Learning Target: Students understand two fractions as equivalent (equal) if they represent the same point on a number line.

Variation Directions:

Use the Walking Club context to help students make sense of finding equivalent fractions on a number line.

> **Walking Club (Number Line)**
>
> Four friends are in a walking club. Clara said she walked $\frac{1}{2}$ mile over the weekend, Maria walked $\frac{3}{6}$ mile. Eunbee walked $\frac{4}{8}$ mile, and Alex walked $\frac{2}{4}$ mile. Clara said, "We all walked the same distance!" Is what Clara said true?
>
> Explain your thinking using models, math sketches, words, and symbols.
>
> **What are other ways to rename these fractions as equivalent fractions?**

Begin with a quick exercise. Give students a square piece of paper (like origami paper) and ask them to fold it in half. Shade in half. If some students are challenged by paper folding, you can use pre-folded fraction strips for this activity.

Engage in Math Discourse (Make the Mathematics Visible):

Figure 5.22.1 Folding halves

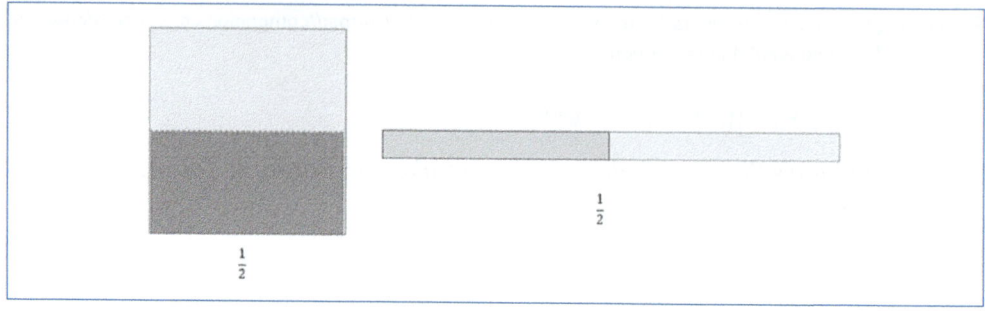

Give another square piece of paper (that is the same size) and fold it in half and then half again and shade $\frac{2}{4}$. Students can create a length model to represent a fraction using strips of paper. Ask students to draw a line and mark the $\frac{1}{2}$ way mark. Then they can fold it in half again and ask them to mark $\frac{1}{4}$ to see if they know the magnitude of fractions on a number line.

Figure 5.22.2 Folding fourths

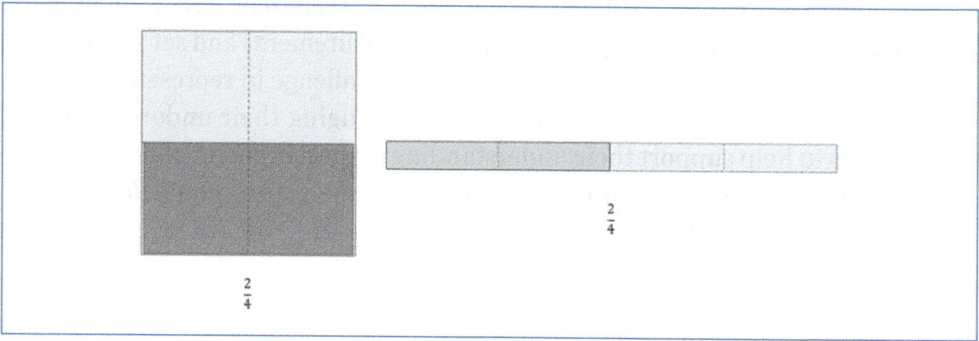

Emphasize the students making the connection that when they mark $\frac{1}{2}$ and $\frac{2}{4}$ on the number line, the two fractions are in the same position and hence equivalent.

Observation: As students work, encourage them to notice important relationships and push on early and partial understandings.

- Think about how they folded the paper. If they need help seeing the different fractions, have them fold a fraction strip (strips of paper the same size).

- Count the units, and remind students that the denominator names the total number of equal parts in the whole or group (if a set model), and the numerator is the number of parts being indicated or counted.

- Describe their math sketches. Students might use tape diagrams or a number line and mark the fractions.

Interview and **Show Me:** After the activity, bring the intervention group together for a discussion. Suggested prompts:

- What are the parts of your whole? How many parts are there?
- How do you know fractions are equivalent when using number lines?
- Can you draw a math sketch or create another model to prove that the fractions are equivalent?

Bridging Prompt (Prompt for Classroom Teacher to Use During New Lesson Content):

- Show the class a number line with one point labeled with two equivalent fractions [for example, $\frac{1}{2}$ and $\frac{2}{4}$]. Suggest that this is where two students are in a race. Ask the class, "What does it mean when two fractions are positioned in the same place on the number line? For example, $\frac{1}{2}$ and $\frac{2}{4}$?"
- Look for opportunities to have students who participated in the intervention group share their reasoning.

Variation 2

Learning Target: Students understand two fractions as equivalent (equal) if they are the same size using the area model.

Variation Directions:

Use the Baking Club context to help students make sense of finding equivalent fractions with an area model.

Figure 5.22.3   Baking Club context

Baking Club (Area model)

Sammy is baking brownies for the school bake sale. He cuts the brownie trays in different ways. He wants to make sure he sells half of each tray. Shade in half of the tray and name each fraction.

What do you notice about the different names for halves?

How can you prove that the shaded trays are all halves?

Brownie Tray Activity Sheet   A

The purpose of the brownie tray task is to have students use what they know about area models to help them to explore equivalent fractions. When using the area model of fractions, students might notice that the area is the same size and find different names for halves.

Figure 5.22.4 Brownie tray shading

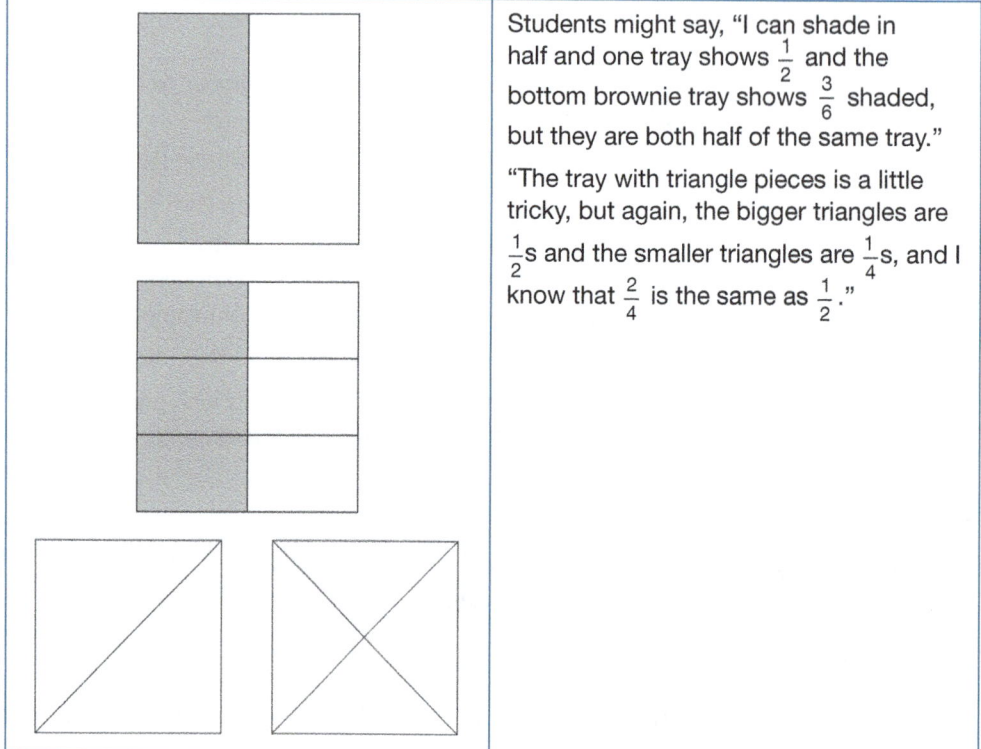

Engage in Math Discourse (Make the Mathematics Visible):

Observation: As students work, circulate between student pairs to observe students as they work and provide feedback. Use questioning as a tool to gain information about student understanding as they complete the tasks. Encourage students to notice important relationships and push on early and partial understandings.

- Encourage students to fully explain their thinking using models, math sketches, symbols, and words. Use the number line to find as many different equivalent fractions as they can for $\frac{1}{3}$, $\frac{2}{3}$, and other fractions.

- A geoboard or graph paper could be used to show these different partitions, or folding a piece of paper could provide students with the kinesthetic experience. Students should have access to concrete and/or virtual manipulatives to use with this task (e.g., fraction squares, grid paper).

- What strategy (or strategies) did you use to determine which fraction models are equivalent?

CHAPTER 5. Intervention Tasks

- Which fraction models are the easiest for you to use to identify equivalent fractions? What made it easy?

- Which fraction models are the hardest for you to use to identify equivalent fractions? What made it difficult?

- What relationships do you see?

- Can you create another equivalent model for $\frac{1}{2}$ that is different from the ones shown?

Interview and **Show Me:** After the activity, bring the intervention group together for a discussion. Suggested prompts:

- What did you learn about equivalent fractions from this activity?

- What methods did you use to show your thinking?

- How did you use math sketches or math tools to show if the fraction pairs were equivalent? Not equivalent?

- What equivalent fractions can you find that also represent parts of brownie trays?

- How do you know that your fractions are equivalent?

- Can you create a math sketch or another model to prove that the fractions are equivalent?

Bridging Prompt (Prompt for Classroom Teacher to Use During New Lesson Content):

- Present the following task to the class:

- Amanda created new brownie trays like these. Are the shaded portions a half of each tray?

Figure 5.22.5 Brownie trays

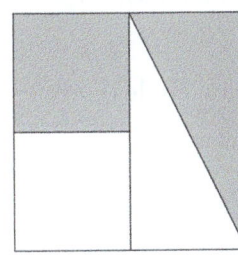 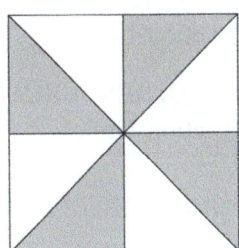

- Ask, "How did you use math sketches or math tools to show if the fraction pairs were equivalent? Not equivalent?"

- Encourage students who participated in the intervention group to share their thinking with the class.

TASK 23
Fraction Equivalence Through Quilts and Fringes

General Objective:

Students can explain why a fraction is equivalent using a visual model and the identity property for multiplication.

This activity might be used with:

Students who are able to (prerequisite knowledge):	Students who are Primed to (getting ready to learn):
• represent a part of a whole as a fraction and can name fractions	• name equivalent fraction using a visual model and can see how they are equivalent

Materials:

- Fraction bars/circles
- PRINTABLE: Unit Fraction Strips (tabloid sized)

    or

    PRINTABLE: Unit Fraction Strips (letter sized)
- PRINTABLE: Unit Fraction Circles
- DIGITAL RESOURCE (fraction bars): https://polypad.amplify.com/p#fraction-bars
- DIGITAL RESOURCE (fraction circles): https://polypad.amplify.com/p#fraction-circles

Recommended Children's Literature:

- *The All-Together Quilt* – Rockwell, 2020
- *Sweet Clara and the Freedom Quilt* – Hopkinson, 1995
- *Sam Johnson and the Blue Ribbon Quilt* – Campbell-Ernst, 1992

- *The Seasons Sewn: A Year in Patchwork* – Paul, 2000
- *The Kindness Quilt* – Wallace, 2006

Task Overview:

In this activity, students will explain why fractions are equivalent using visual models and the identity property for multiplication. "Renaming a fraction" means to write an equivalent fraction with different values in the numerator and denominator, essentially changing how the fraction looks while still representing the same value. Students will have hands-on experiences of folding, cutting, naming, and renaming fractions that allow them to see the pattern when finding equivalent fraction. Eventually, through patterns, they will discover that by multiplying both the numerator and denominator of the fraction by the same number, they can have an endless number of equivalent fractions. They should already understand fractions as parts of a whole and be able to identify equivalent fractions using models. The activity will involve creating visual representations of equivalent fractions and applying the identity property to demonstrate equivalency. By the end, students will articulate their understanding of equivalent fractions through visual and mathematical reasoning.

Select (or create) a variation that focuses on the prior knowledge you are working to shore up for your student group.

Variation 1

Learning Target: Students will investigate equivalent fractions by folding paper to visualize and rename fractions, enhancing their understanding through discussion and explanation of their observations using models, symbols, and words.

Variation Directions:

In this variation, students will take a square piece of paper and fold it and see if they can find equivalent fractions. We can show that some fractions are equal by using models, like math sketches or cut pieces of paper. Even if we change the number of parts, like cutting or folding a shape into more pieces, the amount it represents can still be the same, which means the two fractions are equivalent. This helps us see how we can find different fractions that represent the same value.

Paper folding should happen before using ready-made manipulatives to give more meaning to the math tools. Allow students to make their own fraction tool kit (i.e., paper fraction squares or circles) and keep them in a baggie to use as they move to fraction operations.

Connect this Fraction Folds activities to quilt designs. Quilts are often made of smaller pieces that fill a square or patch. *This is a great opportunity to link to some of the children's literature on this topic and show images or read excerpts from the books.*

Figure 5.23.1 Quilt design

Source: iStock.com/NNehring

Using the paper strips, have the students take one and fold it in half. Ask them how they know it is one half [keeping the edges even]. Then, ask students to shade in one of the halves. Next, they fold the halves in half. Notice what happens to the shaded fraction. How can you rename the fraction $\frac{1}{2}$? Take another piece of paper and shade one-third. Then keep folding. What are other names for the fraction $\frac{1}{3}$?

Figure 5.23.2 Fraction folding

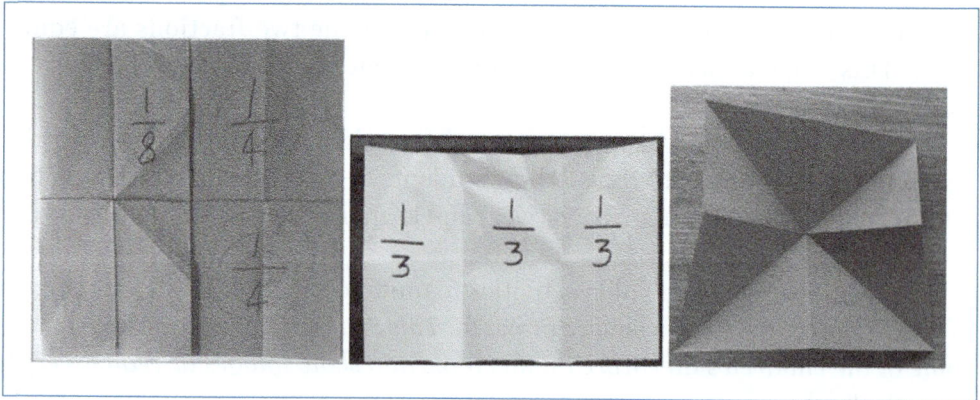

Engage in Math Discourse (Make the Mathematics Visible):

Observation: As students work, encourage them to notice important relationships and push on early and partial understandings.

- As students are folding, ask them to fully explain their thinking as they rename the unit fractions.
- When you fold the paper in smaller rectangles, what do you notice about the numerators and the denominators of the original fraction compared to the fraction after folding?

Interview and **Show Me:** After the activity, bring the intervention group together for a discussion. Suggested prompts:

- Explain how fractions can be renamed and still be equivalent. What do you see happening to the shaded area as you continue to fold?
- Think in terms of the patterns and how the numerator and denominators are renamed. How can we use math sketches or fold pieces of paper to show that two fractions are equivalent, even if we change the number of parts they are divided into?
- At the end of the task, have students name as many different equivalent fractions on a poster as possible. Ask them to notice and share an observation of what is happening or if they see a pattern.

	Bridging Prompt (Prompt for Classroom Teacher to Use During New Lesson Content):
	• Show the class some of the paper strips from the intervention group that represent different equivalencies for $\frac{1}{2}$. Ask, "How many different fraction names does $\frac{1}{2}$ have? How can we rename $\frac{1}{2}$?" • Ask students who participated in the intervention group to show how their paper folding can help answer the question.

Variation 3

Learning Target: Students will explore and identify equivalent fractions by creating a flip book with layered colored paper, demonstrating the identity property of multiplication through visual representations and discussions about fraction renaming.

Variation Directions:

Students will take three pieces of colored paper and layer them to fold into a flip book. Have students layer the three pieces as shown in Figure 5.23.3, leaving about a half-inch overlap for each piece at the end. Then, have students fold the pages over (on the dotted line shown) to create three more layers.

Figure 5.23.3 Directions for creating a Fraction Fringe book

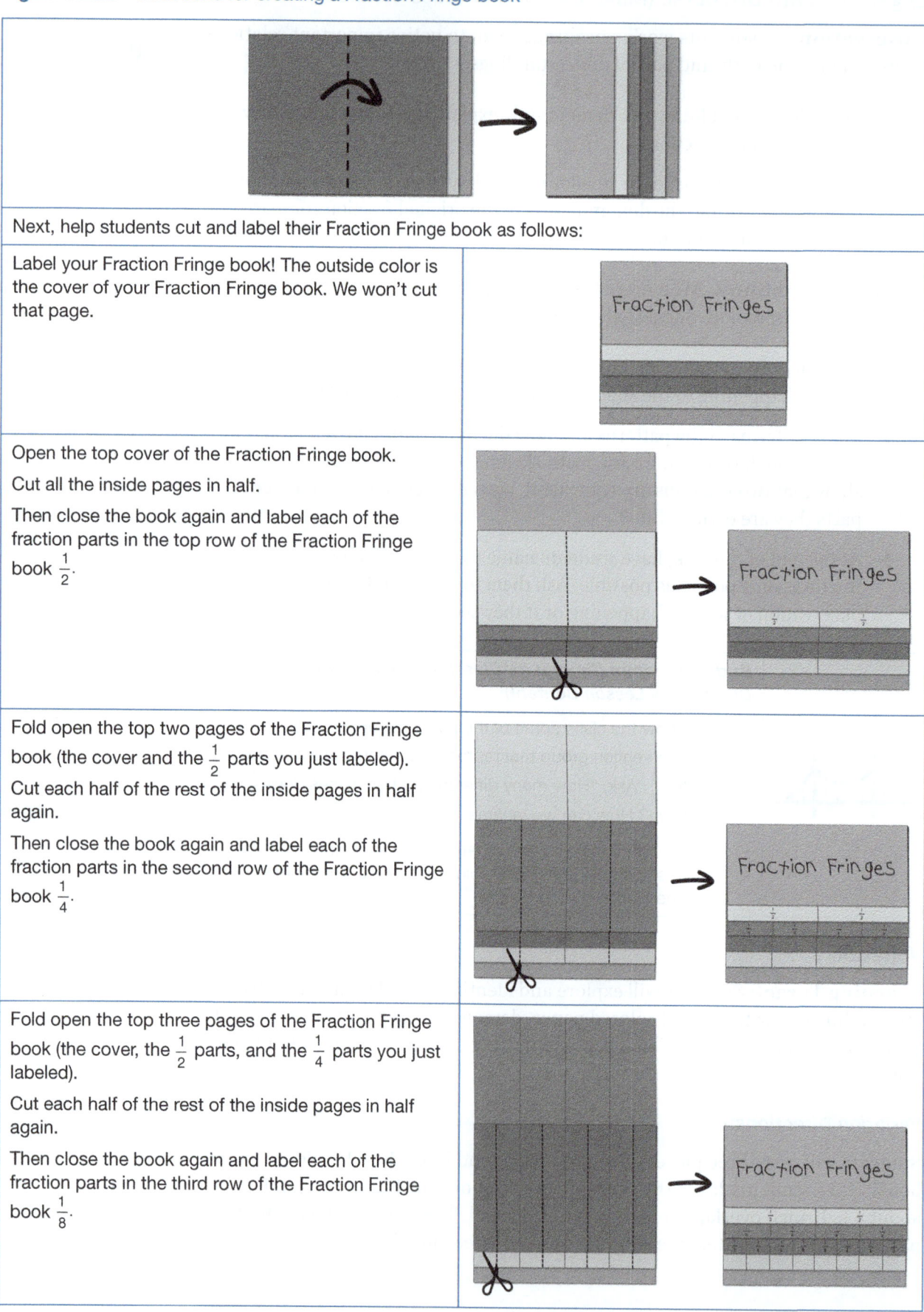

Fold open the top four pages of the Fraction Fringe book (the cover, the $\frac{1}{2}$ parts, the $\frac{1}{4}$ parts, and the $\frac{1}{8}$ parts you just labeled).

Cut each half of the last inside page in half again.

Then close the book again and label each of the fraction parts in the last row of the Fraction Fringe book $\frac{1}{16}$.

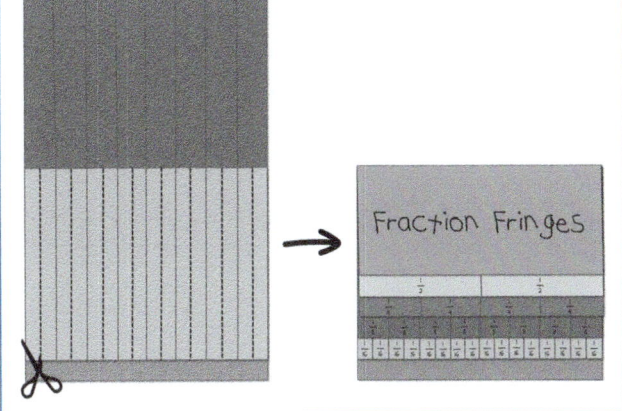

Tell a story: Identity property of multiplication states that when a number is multiplied by 1, the product is always the original number.

- Ask students, "Do you have a nickname? Just like you might have different names for the same person, fractions can be renamed with multiple fraction names. Renaming or changing a fraction to an equivalent fraction doesn't change its size or value."

- Ask students, "What happens when we multiply any number by 1?" (The answer will always be itself, and the number keeps its identity.) Well, that is true for fractions too!

- Let's take a look. Using the Fraction Fringes in your flip book, find all the names of fractions that are equal to $\frac{1}{2}$. What pattern do you notice? Record your thinking. Do you see a relationship or a pattern in the numerators and the denominators?

Figure 5.23.4 Fraction relationships

$$\frac{1}{2} = \frac{2}{4}$$
$$\frac{1}{2} = \frac{4}{8}$$
$$\frac{1}{2} = \frac{8}{16}$$

- There are many ways we can write a fraction that is equal to 1 such as $\frac{2}{2}$, $\frac{3}{3}$, and $\frac{10}{10}$. If we multiply a fraction by 1, we can create an endless number of equivalent fractions.

Figure 5.23.5 Multiplying by 1

$$\frac{1}{2} \times \boxed{\frac{2}{2}} = \frac{2}{4}$$

$$\frac{1}{2} \times \boxed{\frac{4}{4}} = \frac{4}{8}$$

$$\frac{1}{2} \times \boxed{\frac{8}{8}} = \frac{8}{16}$$

Engage in Math Discourse (Make the Mathematics Visible):

Observation: As students work, encourage them to notice important relationships and push on early and partial understandings.

- As students are creating their Fraction Fringes, ask them to count using the unit fraction for one of the Fraction Fringes. For example, they should count $\frac{1}{16}, \frac{2}{16}, \frac{3}{16}, \frac{4}{16}$, and then ask what are all the other fringes that are the equivalent to $\frac{4}{16}$, which is when they will notice that $\frac{4}{16}$ is the equivalent to $\frac{1}{4}$.

$$\frac{1}{4} \times \frac{4}{4} = \frac{4}{16}$$

Interview and **Show Me:** After the activity, bring the intervention group together for a discussion. Suggested prompts:

- How did you find other equivalent fractions using your Fraction Fringes? Explain.

- How can we use math sketches or models to show that two fractions are equal, even if they are divided into a different number of parts?

- How does folding paper help you see the equivalent fractions?

- What would you say to a friend who says $\frac{3}{3}$ is greater than $\frac{2}{2}$?

- If someone said $\frac{1}{2}$ is the same as $\frac{500}{1000}$, would you agree with them? Why or why not?

Bridging Prompt (Prompt for Classroom Teacher to Use During New Lesson Content):

- Show the class a set of Fraction Fringes created by students in the intervention group. Ask, "How can we use math sketches or models to show that two fractions are equal, even if they are divided into a different number of parts?"

- Encourage students who participated in the intervention group to share their ideas with the class.

TASK 24

Sport Stats

General Objective:

Students will learn how to connect fractions to decimals.

This activity might be used with:

Students who are able to (prerequisite knowledge):	Students who are Primed to (getting ready to learn):
• name equivalent fractional amounts using concrete and semiconcrete models for fractions less than 1 • name fractions with denominators of 2, 4, 5, 10, 20, 25, and 50 as equivalent fractions with a denominator of 100 and record in decimal form	• represent and identify equivalencies among fractions and decimals, with and without models

Materials:

- Index cards
- PRINTABLE: Decimal Grids
- PRINTABLE: Hundredths Wheel
- PRINTABLE: Fraction–Decimal Game Cards

 Printables for this task are available for download at **https://companion.corwin.com/courses/ProactiveMathIntervention**.

Recommended Children's Literature:

- *Who Was Babe Ruth?* – Holub, Who, & Hammond, 2012
- *Piece = Part = Portion: Fractions = Decimals = Percents* – Gifford & Thaler, 2011
- *The Grizzly Gazette* – Murphy, 2002
- *Sports Illustrated Kids Stats! The Greatest Number in Sports* – Editors of Sports Illustrated, 2013

Task Overview:

In this task, students begin to identify equivalence among fractions and decimals, starting with common fractions such as halves, thirds, fourths, and eighths as decimal fractions. For example, using a decimal grid and shading $\frac{1}{2}$ and $\frac{50}{100} = 0.5$. A double number line, decimal grids, and hundredths wheel are useful models to connect decimals and fractions as one moves beyond common fractions to continue the development of fraction–decimal equivalence.

Select (or create) a variation that focuses on the prior knowledge you are working to shore up for your student group.

Variation 1

Learning Target: Students will relate common fractions and decimal fractions to equivalent decimal representations.

Variation Directions:

Introduce the task to students:
   Before we get started, let's connect fractions with decimals using money.

- What if I say I have $\frac{3}{4}$ of a dollar—how much money would I have?
- What if I say I have $\frac{9}{10}$ of a dollar—how much money would I have?
- What strategies did you use in order to write each fraction as a decimal?

A **decimal fraction** is a fraction in which the denominator is a power of 10 (such as 10, 100, 1,000, etc.) and is expressed in decimal notation. Decimal fractions are often used to represent numbers that are not whole, allowing for a more precise representation of values.

$$\frac{9}{10} = 0.9$$

$$\frac{25}{100} = 0.25$$

$$\frac{4}{1000} = 0.004$$

Students will determine their friend's batting average using fractions.
   Explain that we can use fractions or decimals to compare batting skills. A batting average in baseball is a statistic that shows how good a player is at hitting the ball. For example, if you had 5 hits in 10 at-bats, your batting average would be 0.500 (which means you hit the ball one half the time).
   If you had 3 hits in 10 at bats, your batting average would be 0.300 (which means you hit the ball 3 out of 10 times at bat). Batting averages are always reported in thousandths because it provides a very precise measure when you compare different batters. You can use division to figure out the batting average. It is calculated by dividing the number of hits a player gets by the number of times they get to bat. For our batting averages, we can show two digits after the decimal point (e.g., 0.30).

Figure 5.24.1 Batting averages

Anna	Ben	Carlos	David	Emma	Frankie
$\frac{4}{10}$	$\frac{2}{5}$	$\frac{3}{10}$	$\frac{3}{20}$	$\frac{5}{25}$	$\frac{15}{50}$

See if the visual images of a decimal grid or hundredths wheel can help you connect fractions to decimals.

On the grid (see Figure 5.24.2), ask students, "How can we show $\frac{1}{10}$ by shading a portion of the grid?" We want students to align the shading with the use of concrete base ten materials, so explain that the shading should look like those materials. Reinforce by using the concrete versions first, pointing out the shift from whole-number values to decimal values, where the former hundred piece is now going to represent one whole, the ten piece represents one tenth, and the one piece represents one hundredth. Transitioning to the grid, they can align the "former ten piece" (now one-tenth) by setting it on the grid and shading it vertically. Ask them, "What does each column on the grid equal?" [one tenth]. Then they can shade the other numbers.

Figure 5.24.2 Blank grid and a grid shaded to show the representation for one-tenth

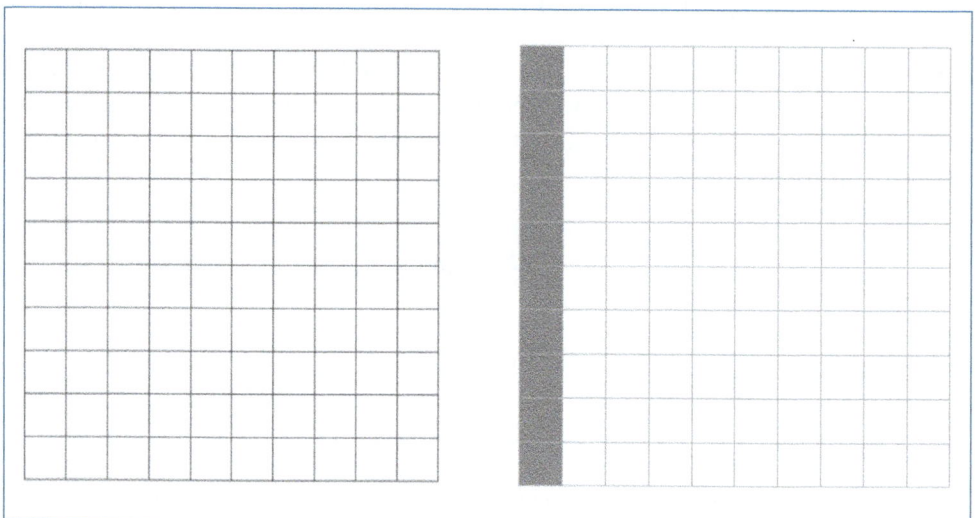

Another helpful representation for decimal values is the hundredths wheel. Use the printable version and cut two of the same-sized versions—one marked as shown and the other blank—out of two pieces of cardstock. Then, cut the radius for each (as shown) and overlap them. They can model fractions and angles too (see Task 29: Shape Shifting). Ask students to look at the marks around the edge. What do they notice? What do they wonder? Now, using the overlapping solid to cover up some of the hundredths wheel, ask, "How can you show $\frac{1}{10}$ of the wheel?" See Figure 5.24.3 for a visual of the hundredths wheel and a representation showing one-tenth shaded.

Figure 5.24.3 Hundredths wheel and hundredths wheel shaded to show one-tenth

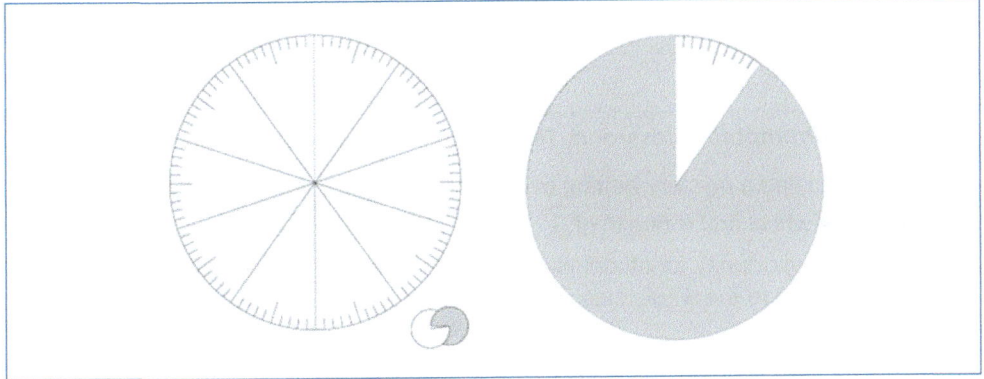

Source: Van de Walle et al., 2023.

Engage in Math Discourse (Make the Mathematics Visible):

Observation: As students work, encourage them to notice important relationships and push on early and partial understandings.

- How does the student recognize, identify, and name equivalent fractions and decimals using concrete, semiconcrete, or abstract models (the fraction or decimal written symbolically)?
- How does the student recognize, identify, and name equivalent fractions and decimals without using concrete or semiconcrete models?
- How does the student demonstrate an understanding that fractional models can be expressed in multiple equivalent decimal forms (e.g., 0.8, 0.80, etc.) and vice versa?
- How does the student recognize that there are different ways to name an equivalent fraction or decimal to represent a model or quantity, including using the simplest form?

Interview and **Show Me:** After the activity, bring the intervention group together for a discussion. Suggested prompts:

- What patterns did you notice when converting fractions to decimals?
- How would you explain why $\frac{25}{100}$ is the same as 0.25?
- What strategy did you use to figure out the decimal for $\frac{4}{10}$?
- How did using a decimal grid or hundredths wheel help you understand fraction–decimal equivalence?
- What other fractions can you think of that are equivalent to 0.50? How would you show that on a grid or hundredths wheel?
- How would you explain the process of finding a batting average to a friend?
- Can you think of other real-life examples in which fractions and decimals represent the same idea?

- If you wanted to find the decimal form of $\frac{7}{10}$, what steps would you take?
- How would you convert a fraction like $\frac{3}{4}$ to a decimal without a model?

Bridging Prompt (Prompt for Classroom Teacher to Use During New Lesson Content):

- One student had a mystery batting average. Show a model of a grid (or show a hundredths wheel and rotate it) of $\frac{2}{10}$. Ask who can tell us the amount this shows. Provide an opportunity for students who participated in the intervention group to explain the grid to the whole class and say how they know the batting average is two-tenths.
- What strategies can you use to find a decimal equivalent for common fractions (e.g., $\frac{1}{2}, \frac{1}{4}, \frac{1}{8}, \frac{2}{4}, \frac{3}{4}, \frac{1}{5}, \frac{1}{10}$)?
- How does thinking about money help you connect fractions to decimals?

Variation 2

Learning Target: Students will match fractions with decimals.

Variation Directions:

Organize students in pairs for this task. Give each pair a set of 10 index cards and ask them to write 5 pairs of fractions and decimals each or use the Fraction–Decimal Game Cards and select 10 matching pairs. Mix them up and place them face up in a rectangular display. Play the matching game, in which player one selects two cards and if the fraction shown is equal to the decimal shown, they get to keep the pair. Then, the second player tries to find a match. If the cards selected are not a match, the game continues with the next player. When done, take out more matching card pairs and make another array.

Engage in Math Discourse (Make the Mathematics Visible):

Observation: As students work, encourage them to notice important relationships and push on early and partial understandings.

- How are you deciding if two cards are a matched set?
- What strategies did you use to find a matching decimal for the fractions (e.g., $\frac{1}{2}, \frac{1}{4}, \frac{1}{8}, \frac{2}{4}, \frac{3}{4}$ or $\frac{1}{5}, \frac{1}{10}$)? Which strategy is the most efficient for you and why?
- How can one connect fractions and decimals with money, like one quarter, two quarters, three quarters?
- What strategies did you use to find a matching fraction equivalent for the decimals? Which strategy is the most efficient for you and why?
- How does a decimal grid or hundredths wheel help you find a fraction or decimal equivalent?

- How can finding a decimal fraction with the denominators of 10, 100, or 1,000 help you change fractions to decimals (e.g., $\frac{1}{4} = \frac{25}{100} = 0.25$ and $\frac{1}{8} = \frac{125}{1000} = 0.125$)?
- How might division help you find a decimal equivalent for a fraction?

Interview: After the activity, bring the intervention group together for a discussion. Suggested prompts:

- What are some patterns you were noticing as you played the game?

	***Bridging Prompt (Prompt for Classroom Teacher to Use During New Lesson Content):*** • Show four fraction and decimal cards face up (created by students in the intervention group for the matching game) on the document camera or other projection device. Ask the class to make a match with a fraction that is equal to a decimal among the four cards (two pairs should be there). • Ask, "How does thinking about decimal fractions help you change a fraction to a decimal?" • Provide an opportunity for students who participated in the intervention group to share their reasoning using a decimal grid or a hundredths wheel.

Task 25: Fraction Zap

General Objective:
Students will add/subtract fractions using a variety of strategies.

This activity might be used with:

Students who are able to (prerequisite knowledge):	Students who are Primed to (getting ready to learn):
• name and write fractions and mixed numbers represented by a model • represent fractions and mixed numbers with models and symbols • represent equivalent fractions	• solve practical problems that involve addition and subtraction with fractions having like denominators of 12 or less • solve practical problems that involve addition and subtraction with fractions having unlike denominators of 12 or less

Materials:
- Fraction circles/bars
- Fraction open line
- PRINTABLE: Fraction Zap!
- DIGITAL RESOURCE (fraction models): https://qrs.ly/bigjc86
- DIGITAL RESOURCE (fraction strips): https://qrs.ly/2wgjc8c
- DIGITAL RESOURCE (fraction circles): https://qrs.ly/lxgjc8d

 Printable for this task available for download at **https://companion.corwin.com/courses/ProactiveMathIntervention**.

Recommended Children's Literature:
- *Full House: An Invitation to Fractions* – Dodds, 2009

Task Overview:
Students use addition or subtraction strategies to determine how much was "zapped" and record their results. In the activity, students will discuss their strategies and reflect on how thinking in terms of addition can help solve subtraction problems with fractions.

Select (or create) a variation that focuses on the prior knowledge you are working to shore up for your student group.

Variation 1

Learning Target: Students will think about the missing addends to figure out what fraction was zapped. This will allow students to use an adding-up strategy (also called "think addition") or subtraction strategy to find the missing addend. The activity is played by figuring out the missing addend, which is a way to think about fraction subtraction. Fraction Zap will elicit strategies like adding up to the target number.

Variation Directions:

Monster loves key lime pies and key lime bars. When it sees these key lime snacks out of the oven, it uses zapping power to capture them and eat them. Find out how many key lime pies/bars Monster has zapped. The number in the circle shows how many pies/bars were baked before they were zapped.

In this variation, students use addition or subtraction strategies to determine how much was zapped.

Do this first example in the role-playing and then show this image on the white board after they have answers as you discuss their thinking.

Figure 5.25.1 Fraction Zap Variation 1 example

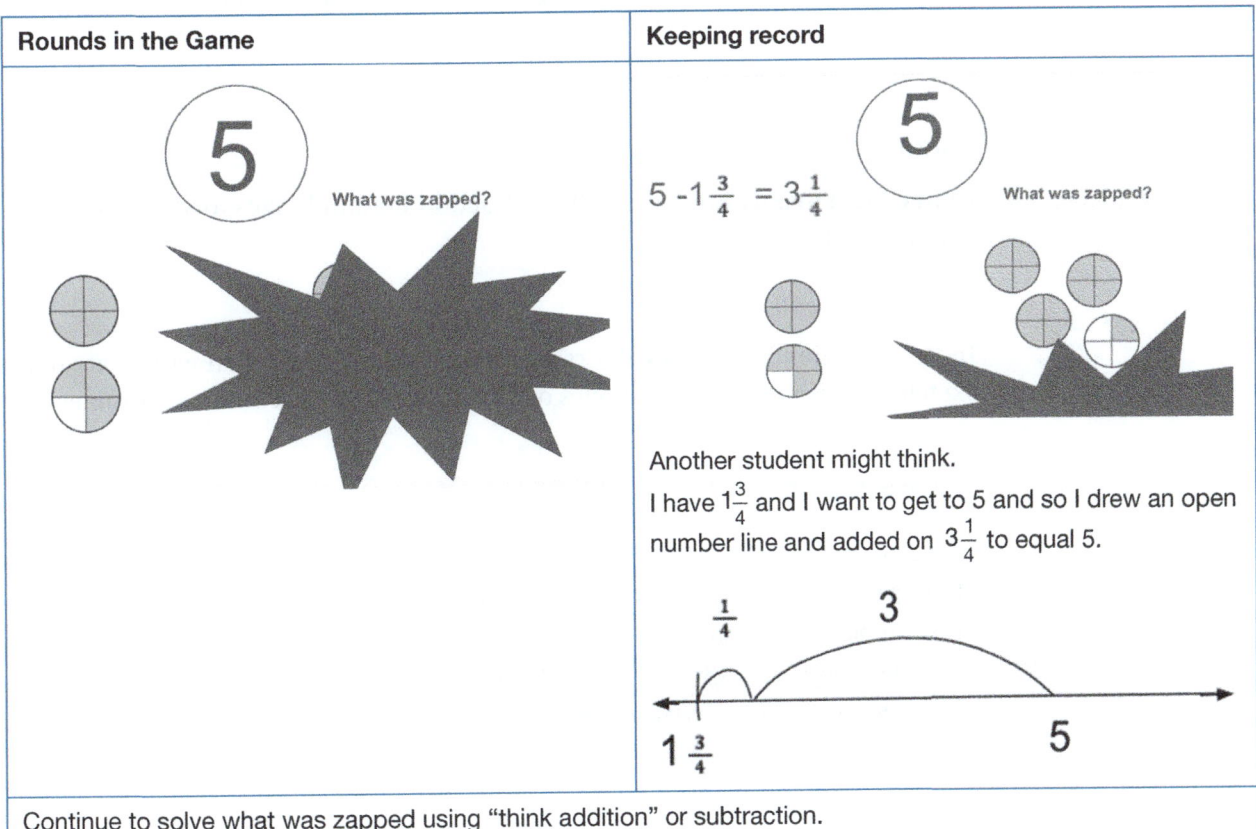

Engage in Math Discourse (Make the Mathematics Visible):

Observation: As students work, encourage them to notice important relationships and push on early and partial understandings.

- Listen for diverse strategies of adding up as "think addition" or subtracting to find the unknown. Ask them to explain their thinking using the fraction materials and the number line.

Interview: After the activity, bring the intervention group together for a discussion. Suggested prompts:

- What strategies did you use to solve for what was zapped?
- Why do you think other students suggest that "think addition" is easier than subtracting?

Bridging Prompt (Prompt for Classroom Teacher to Use During New Lesson Content):
• Show the class a fraction zap situation. Have students who participated in the intervention group explain the story context. Ask the class to find the "zapped" amount.
• Ask, "How can you use 'thinking addition' to solve other subtraction problems with fractions?"
• Look for opportunities for students who participated in the intervention group to share strategies using fraction manipulatives or a number line.

Variation 2

Learning Target: Students will use fraction manipulatives to solve some contextual problems.

Variation Directions:

In this variation, students will first use concrete materials then will create math sketches of fraction models (circles or bars) to solve these contextual problems.

Figure 5.25.2 Contextual problems

Fraction with like denominators
Example: You walked $\frac{3}{4}$ mile to the park and then walked another $\frac{3}{4}$ mile to the store. How far did you walk in total?
A student might put together $\frac{3}{4}$ and $\frac{3}{4}$ then combine the two sets to make a whole and notice that they would have walked $1\frac{2}{4}$ or $1\frac{1}{2}$ miles.

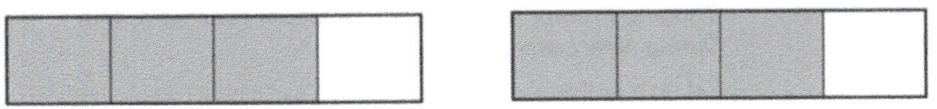

Try this by using fraction materials first then drawing a fraction model:

A recipe requires $1\frac{2}{3}$ cup of flour and $\frac{1}{3}$ cup of sugar. How much more flour than sugar does the recipe require?

Fraction Problems With Unlike Denominators

Example: You have $\frac{3}{4}$ of a box of pencils and you give away $\frac{1}{8}$ of the box to a friend. How much of the box of pencils do you have left?

(A student might draw $\frac{3}{4}$ of a box then split the box into eighths by splitting the box in half with the dotted line and take away $\frac{1}{8}$ to have $\frac{5}{8}$ left.)

Try this by using fraction materials first then creating a math sketch as a semiconcrete representation of the fraction model.

You drank $\frac{3}{10}$ liter of water in the morning and $\frac{1}{5}$ liter in the afternoon. How much water did you drink in total?

Engage in Math Discourse (Make the Mathematics Visible):

Observation, **Interview** and **Show Me**: As students work, encourage them to notice important relationships and push on early and partial understandings.

- Can you use a math sketch or a model to show how to make sense of the problem?
- How would you combine two fractions that have the same denominator? Can you explain your approach using the fraction models or the open number line?
- In looking at the context of this specific problem, what does the denominator tell us? What does the numerator tell us?

Interview: After the activity, bring the intervention group together for a discussion. Suggested prompts:

- How do the fraction tools and math sketches help us add and subtract fractions with like and unlike denominators?

Bridging Prompt (Prompt for Classroom Teacher to Use During New Lesson Content):

Set up a problem with models/visuals (walk on a hike) to help the students who participated in the intervention group get into familiar territory. Then, have them model some problems from this variation for the class using materials and the open number line.

- How do visual models help you solve the fraction problems in context?
- How does using the fraction tools and math sketches help you to add fractions with unlike denominators?
- Can using "thinking addition" work with all subtraction problems?

Task 26: Fraction Sums

General Objective:
Students will add fractions using a variety of strategies including making a whole.

This activity might be used with:

Students who are able to (prerequisite knowledge):	Students who are Primed to (getting ready to learn):
• name and write fractions and mixed numbers represented by a model • represent fractions and mixed numbers with models and symbols • represent equivalent fractions	• solve practical problems that involve addition and subtraction with fractions having like denominators of 12 or less • solve practical problems that involve addition and subtraction with fractions having unlike denominators of 12 or less

Materials:

Dice (regular or fraction) or a spinner with fractions

Fraction number line

Blank paper

Fraction circles

PRINTABLE: Unit Fraction Circles

PRINTABLE: Fraction Sum Games Pages

DIGITAL RESOURCE (fraction models): https://qrs.ly/bigjc86

DIGITAL RESOURCE (fraction strips): https://qrs.ly/2wgjc8c

DIGITAL RESOURCE (fraction circles): https://qrs.ly/lxgjc8d

 Printables for this task are available for download at **https://companion.corwin.com/courses/ ProactiveMathIntervention**.

Recommended Children's Literature:

The Wishing Club: A Story About Fractions – Napoli, 2007

▸ *Gator Pie* – Mathews, 1979

▸ *Math Chef* – D'Amico, 1997

Task Overview:

In "Make a Whole," students will work in pairs to practice adding fractions with like denominators. Each player rolls two dice to generate a numerator and denominator and create a fraction (or use a fraction die or fraction spinner), shading in portions of a circle as they add fractions to make a whole. *Note:* We always want to be cautious that students **don't** see a fraction as two different whole numbers, which is a common hiccup. A fraction is a number that we can count and perform operations on just as we do with whole numbers. We also want to avoid calling the numbers a "top number" and a "bottom number" because it reinforces students seeing fractions as two different numbers.

The goal of this game is to collect as many wholes as possible, encouraging strategic thinking around equivalent fractions and fraction renaming.

Select (or create) a variation that focuses on the prior knowledge you are working to shore up for your student group.

Variation 1

Learning Target: "Make a whole" uses measurement context to help students connect real-world fractions and real-life materials like measuring cups to add fractions.

Variation Directions:

Figure 5.26.1 Make a Whole warm-up

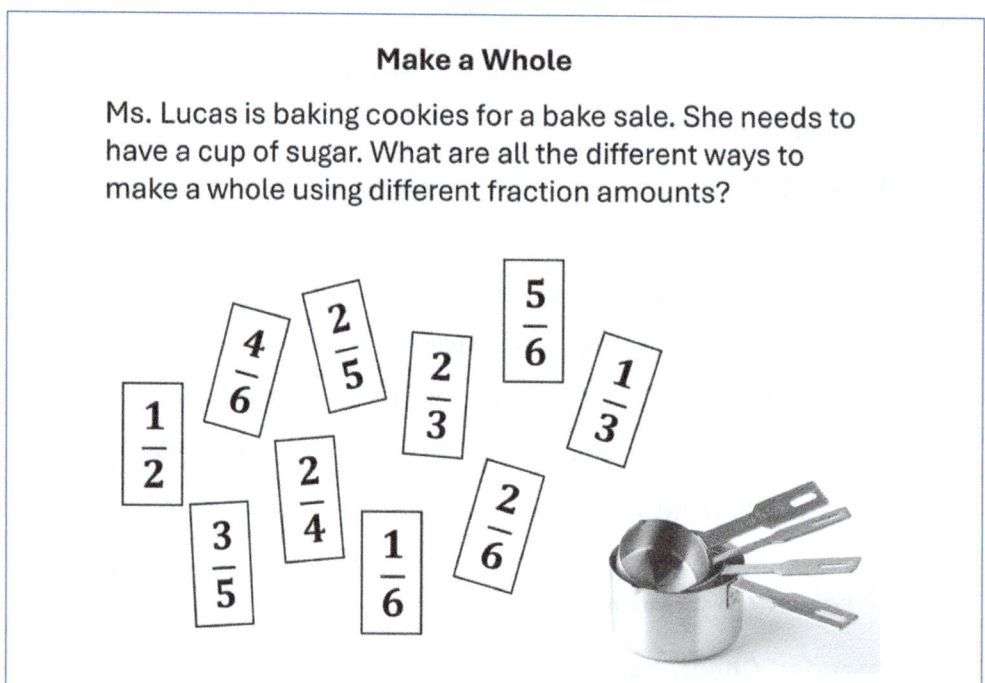

Source: measuring cups by iStock.com/Snappy_girl

"Make a Whole" is a warm-up that will support students' understanding of adding compatible fractions. Have a rich conversation around different ways to make a whole. Some students will find $\frac{2}{3}+\frac{1}{3}$ as an example of making a whole and $\frac{4}{6} + \frac{2}{6}$ making a whole, while others might match $\frac{2}{3} + \frac{2}{6}$, knowing that $\frac{2}{6}$ is equivalent to $\frac{1}{3}$. Use the context of measuring cups to take opportunities to discuss how to find equivalent fractions. For example, you might say, "I needed a $\frac{1}{2}$ cup of sugar but could not find a one-half cup. What other measuring cups could I use to get the same amount?" Students might say, "You can use a $\frac{1}{6}$ cup three times because $\frac{3}{6}$ is the same as $\frac{1}{2}$, or you can use a $\frac{1}{4}$ cup two times because $\frac{2}{4}$ is the same as $\frac{1}{2}$."

Engage in Math Discourse (Make the Mathematics Visible):

Observation and **Show Me:** As students work, encourage them to notice important relationships and push on early and partial understandings.

- Listen for language from students as they use combinations of the ingredient amounts to equal 1 whole. Ask them to explain their thinking as they make decisions of what fractions to combine.

- Students will notice that when they use fractions with the same denominator, it is easier to combine them. Once they focus on the denominator, ask them to show you how to build a fraction that equals 1 whole. How do they know what they need?

- Monitor to see if students are using equivalent fractions to make a whole. If not, challenge them by giving them a fraction and saying they cannot use fractions with the same denominator to reach 1 whole.

Interview and **Show Me**: After the activity, bring the intervention group together for a discussion. Suggested prompts:

- What strategies did you use as you added on a fraction to make a whole.

- Looking at the denominator, how do you know how many parts equals a whole? Is there a pattern you see for all fractions?

- I have another card that has $\frac{2}{5}$ on it. What can I use from the cards we have to equal a whole? What new card could I make?

Bridging Prompt (Prompt for Classroom Teacher to Use During New Lesson Content):
• Provide another set of recipe measurements. Have them on the document camera. Bring up two students who participated in the intervention group to select one card each that could when added together equal 1 whole.

Variation 2

Learning Target: Students will use combinations of fractions to add to 1 whole, using equivalent fractions when needed.

Variation Directions:

In this variation, students will roll two dice and make a fraction that they shade in and will add on until they can make a whole. This is a partner game. Both students try to collect as many wholes as possible. Students will roll two dice and make a fraction less than one to fill in the circles. For example, Player A rolls 1 and 4 and shades in $\frac{1}{4}$. Then, on his next turn, Player A rolls a 1 and 2 and shades $\frac{1}{2}$ or $\frac{2}{4}$, depending on for which fraction he is trying to complete a whole. Some students might say that with a roll of 1 and 2, they can make a fraction greater than a whole like $\frac{2}{1}$, which might make the game end too quickly. Thus, we say that for this game, we only can make fractions less than 1. This allows students to add fractions with like denominators and strategically think about renaming fractions to get 4 wholes before their partners do. Players must record their moves each round to keep track of the addition problems.

Figure 5.26.2 Fraction Sums game board

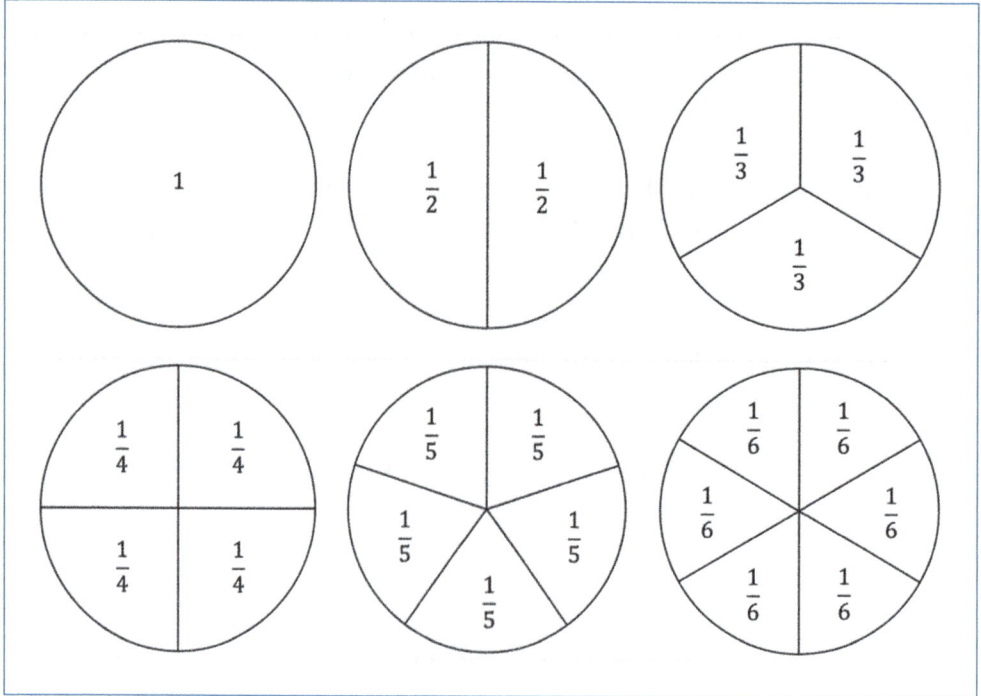

CHAPTER 5. Intervention Tasks

Figure 5.26.3 Fraction Sums Variation 2 example

Player A rolls 2 and 5 and shades in $\frac{2}{5}$.

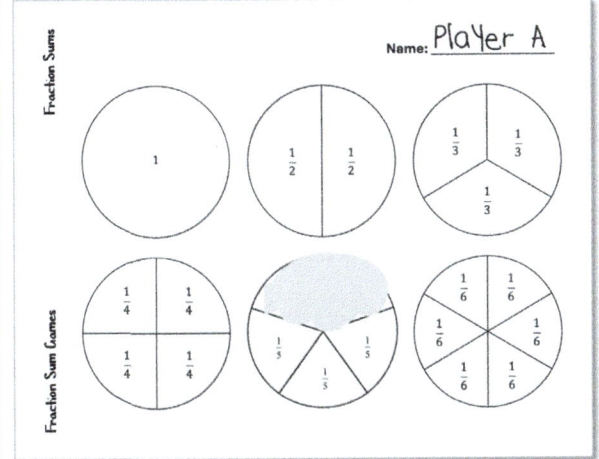

Player B rolls 2 and 3 and shades in $\frac{2}{3}$.

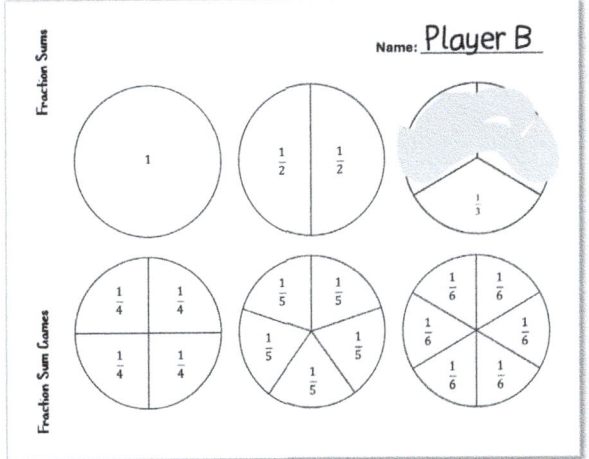

Player A then rolls 1 and 4 and shades in $\frac{1}{4}$.

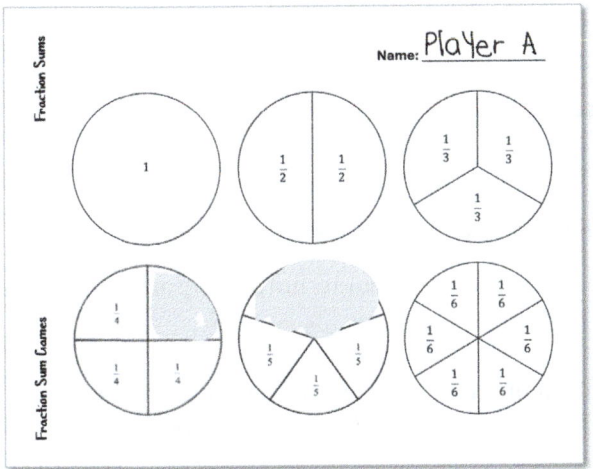

Player B rolls 3 and 3 and says, "That's 1 whole" and shades 1 whole to keep.

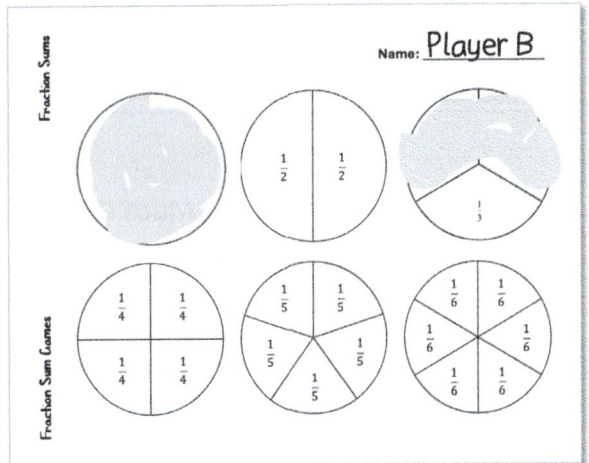

Player A then rolls 1 and 2 and shades $\frac{1}{2}$.

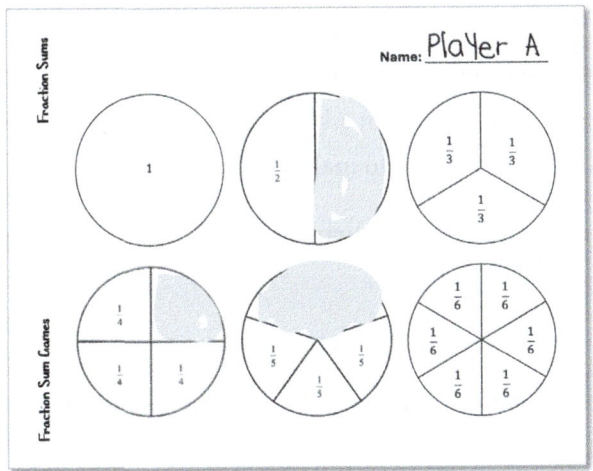

Player B then rolls 1 and 3 and shades $\frac{1}{3}$, which fills out his thirds to complete a whole.

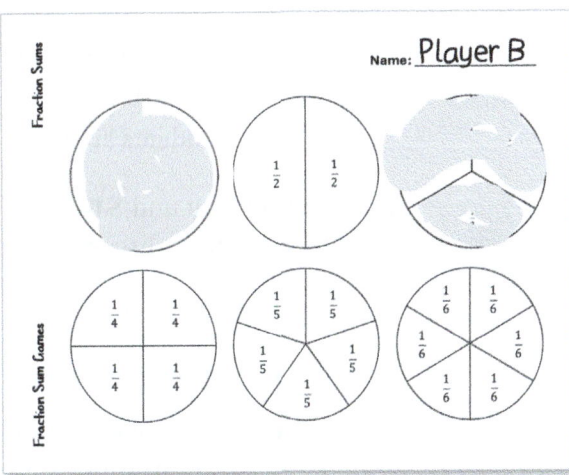

(Continued)

(Continued)

Player A rolls 2 and 4 and knows that $\frac{2}{4}$ is equivalent to $\frac{1}{2}$ and debates, "Should I add it to the fourths or to the halves because $\frac{1}{4}+\frac{2}{4}=\frac{3}{4}$ or take the halves by adding $\frac{1}{2}+\frac{2}{4}=1$ whole?" Player A decides to keep the whole.

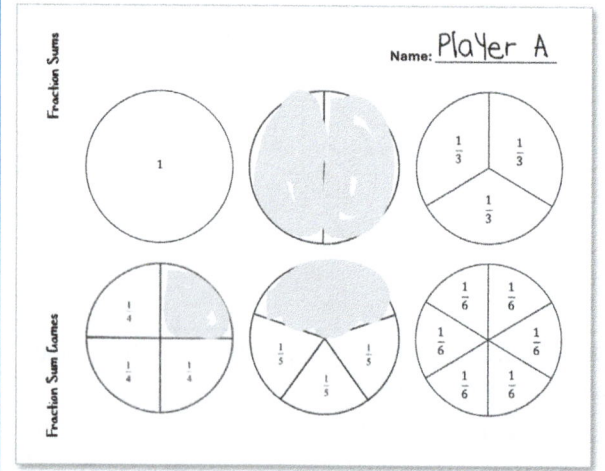

Player B rolls a 1 and 6 and shades in $\frac{1}{6}$.

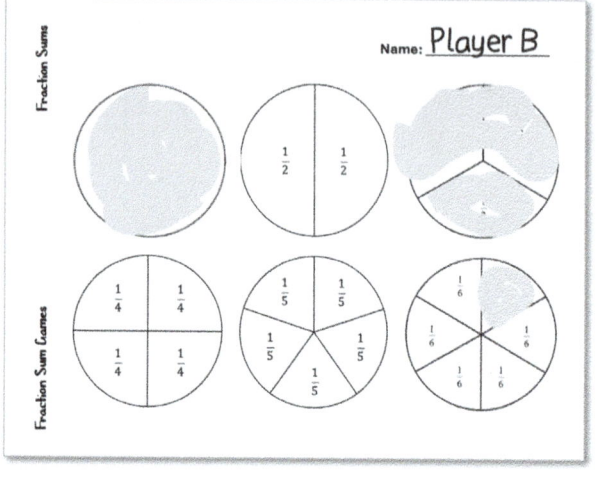

Play continues until one of the players has shaded 3 wholes.

Engage in Math Discourse (Make the Mathematics Visible):

Observation: As students work, encourage them to notice important relationships and push on early and partial understandings.

- Ask students if the two numbers they roll mean a fraction is really two whole numbers. Make sure they recognize that a fraction *is* a number and is not two whole numbers.

- Listen for language from students as they roll and make a fraction to add on to make a whole. Ask them to explain their thinking.

- Students will notice that when they shade $\frac{2}{4}$, they shade $\frac{1}{2}$ and might notice that rolling 1 and 2 (the fraction $\frac{1}{2}$), 2 and 4 (the fraction $\frac{2}{4}$), and 3 and 6 (the fraction $\frac{3}{6}$) would all give them a half shaded. Monitor to see if students are using equivalent fractions to make a whole.

Interview and **Show Me:** After the activity, bring the intervention group together for a discussion. Suggested prompts:

- What strategies did you use as you added on a fraction to make a whole?

- Did you find some different fractions that could be used to represent $\frac{1}{2}$?

- Choose one or two examples from student work and ask, "How do you add fractions with the same denominators?"

> ***Bridging Prompt (Prompt for Classroom Teacher to Use During New Lesson Content):***
>
> - Show the class a blank Fraction Sums game board. Have students who participated in the intervention group come up and roll two dice (or a single fraction die). Then, have them do a "think aloud" to explain how they would shade the game board. Give a student in the class the dice and roll for a fraction. Have them come up and talk about their decision for where to shade that fraction and if there is any way to equal 1 whole.
>
> If they cannot equal a whole with that roll, have students in the class write down another fraction they would like to roll to complete a whole.

Variation 3

Learning Target: Students will use addition of fractions to get to a target sum.

Variation Directions:

The goal of the game is to add fractions rolled on the dice to reach or come as close as possible to the target sum of 3 without going over. Organize students in pairs. Each pair will need two dice, one for the numerator and one for the denominator (preferably a 6-sided die for simplicity), and each player will need paper and a pencil to record their fraction sums.

1. On each turn, a player rolls both dice. The first die (numerator) is the number of parts of the fraction, and the second die (denominator) is the number of equal parts in the whole (e.g., if the first die rolls a 3 and the second die rolls a 4, the fraction is $\frac{3}{4}$).
2. The player then adds the rolled fraction to their running total.
3. Players continue taking turns, adding the fraction rolled to their total after each roll.
4. Each turn, players must decide whether to roll again or "pass." If they pass, they stop and keep their total.
5. If a player's total exceeds 3 at any point, they are "out" for the round.

Scoring:

- The player whose total is closest to 3, without going over, earns 1 point for the round.
- If a player hits exactly 3, they score 2 points for the round.
- If all players exceed 3, no points are earned for the round.

Play for a set number of rounds (e.g., 5 rounds). The player with the highest total score at the end of the game wins.

Other variations:

- Advanced fractions: Use a spinner or a die with more sides (like a 12-sided die) for larger fractions or fractions greater than 1.

- Target range: Instead of just aiming for a single number of wholes such as 3, players can aim for a specific range (e.g., between $2\frac{1}{2}$ and 3) and earn points for hitting that range.

- Target Sum 3 with dice is a fast-paced and engaging game that promotes practice in fraction addition while adding an element of strategy! Students love it. Save some partial running recording sheets to use for the Bridging Prompt, or create a fictitious one.

Engage in Math Discourse (Make the Mathematics Visible):

Observation: As students work, encourage them to notice important relationships and push on early and partial understandings.

- What is your current total, and how did you arrive at that?
- How do you decide whether to roll again or pass?
- How does the denominator affect your decision to keep your total or pass?
- If you roll a fraction close to 1, like a $\frac{4}{5}$, how does that impact your total? Do you think it's wise to keep it? Why or why not?
- If you're getting close to 3, should you take a chance and roll again? What will happen if you go over?

Interview and **Show Me:** After the activity, bring the intervention group together for a discussion. Suggested prompts:

- What strategies did you use to get close to the target sum of 3?
- Did you ever exceed 3? How did that affect your score for the round?
- If you were to play this game again, would you adjust your strategy? How?
- What did you learn about adding fractions through playing this game?
- How did you handle fractions with larger denominators? Was it harder to add them?

Bridging Prompt (Prompt for Classroom Teacher to Use During New Lesson Content):

Show the class a fictitious running record of rolls for a sample Target Sum 3 game or one that is unfinished from the intervention class. Have students who participated in the intervention explain the game to the class. After describing the rules, ask them what they need to roll to get a sum of 3. This would allow them to explain how they can find the residual part to make the whole.

- If you had a board like this below, what would you want to roll so you can get to 3 wholes? (The student will say, "I would want to roll a 1 and 2 for a $\frac{1}{2}$, roll a 3 and a 4 for a $\frac{3}{4}$, and roll a 3 and a 5 for $\frac{3}{5}$.")

Figure 5.26.4 Student game board

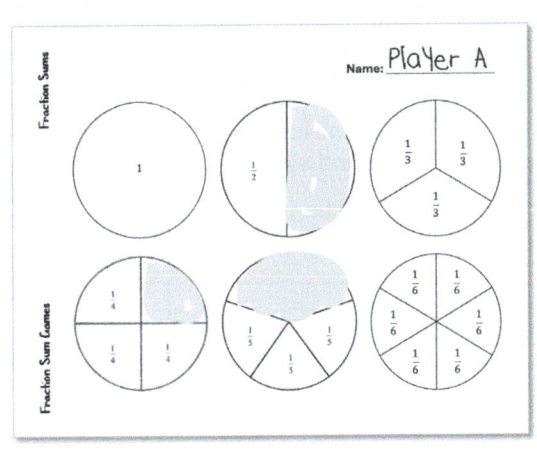

TASK 27

Fraction Multiplication: Fundraiser

General Objective:
Students will multiply fractions using a variety of strategies.

This activity might be used with:

Students who are able to (prerequisite knowledge):	Students who are Primed to (getting ready to learn):
• name and write fractions and mixed numbers represented by a model • represent fractions and mixed numbers with models and symbols • represent equivalent fractions	• interpret multiplication as scaling (resizing) • solve real-world problems involving multiplication of fractions and whole numbers and mixed numbers, e.g., by using visual fraction models or equations to represent the problem

Materials:

- Fraction circles/bars
- Open number line
- PRINTABLE: Unit Fraction Strips (tabloid sized)

  or

  PRINTABLE: Unit Fraction Strips (letter sized)
- PRINTABLE: Unit Fraction Circles
- DIGITAL RESOURCE (fraction models): https://qrs.ly/bigjc86
- DIGITAL RESOURCE (fraction strips): https://qrs.ly/2wgjc8c
- DIGITAL RESOURCE (fraction circles): https://qrs.ly/lxgjc8d

 Printables for this task are available for download at **https://companion.corwin.com/courses/ProactiveMathIntervention**.

Recommended Children's Literature:

- *Better Homes and Gardens Junior Cookbook* – 2018
- *The Math Chef: Over 60 Math Activities and Recipes for Kids* – D'Amico, 1997
- *The Complete Baking Book for Young Chefs* – America's Test Kitchen Kids, 2019

Task Overview:

Students will create a visual based on a real-world problem involving multiplication of fractions with whole numbers and mixed numbers.

Variation

Learning Target: Students will create math sketches and find solutions to the contextual problems that involve multiplication of a whole number by a fraction.

Variation Directions:

Share the context of a fundraiser where fun events happen to raise money for a school or community. Ask students if they have participated in either a walkathon or a bake sale. Use the different scenarios to make sense of the mathematics. Have students create a math sketch of a model of their choice and write a number sentence to explain their answer connecting the different representations.

After a party, we had 3 jugs that each had $\frac{2}{3}$ gallons of apple cider left. How many gallons is that in total?

To show that visually, I can use a tape diagram or a number line (see Figure 5.27.1).

Figure 5.27.1 A tape diagram and a number line showing $3 \times \frac{2}{3}$

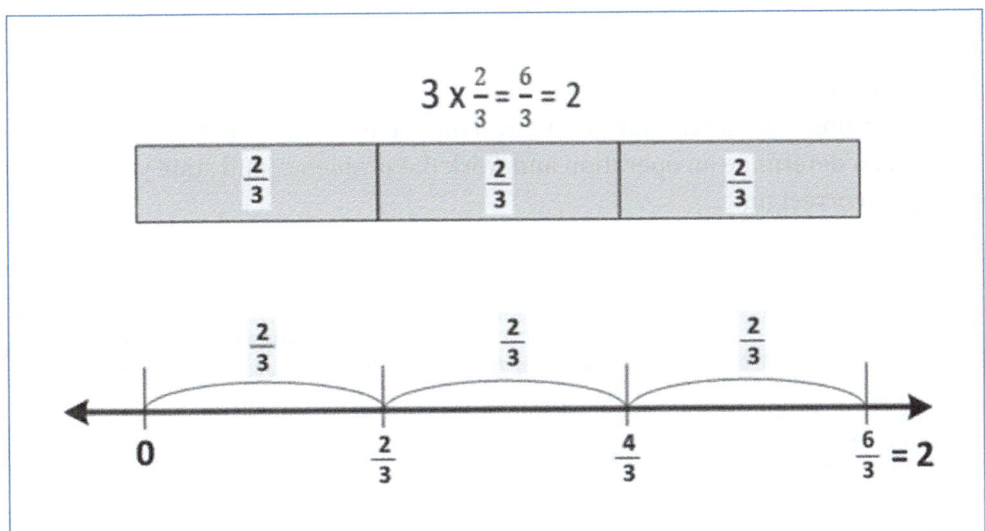

For each question, show your reasoning with a sketch of a tape diagram or a number line. Write an equation that represents the situation.

Walkathon Fundraiser

Directions: For each situation, create a model using a number line or tape diagram, then determine an operation and work the problems, and state your answer using the correct unit.

Figure 5.27.2 Walkathon stories

Situation	Drawings (Number lines, tape diagrams, or other models)	Explain your thinking
The distance of the track is $\frac{3}{4}$ mile. Use this information for each of the situations.		
If Jose walked 2 times around the track, how many miles did he walk?	←――――――――――→	
If Jessica walked 4 times around the track, how many miles did she walk?	←――――――――――→	
If Joan walked 8 times around the track, how many miles did she walk?	←――――――――――→	
If David walked 10 times around the track, how many miles did he walk?	←――――――――――→	
	Given that the track is $\frac{3}{4}$ mile, what do you notice with Jessica's and Joan's walks?	

Bake sale

Directions: For each situation, create a model using a number line or tape diagram, then determine an operation and work the problems, and state your answer using the correct unit.

Figure 5.27.3 Bake sale cookie recipe

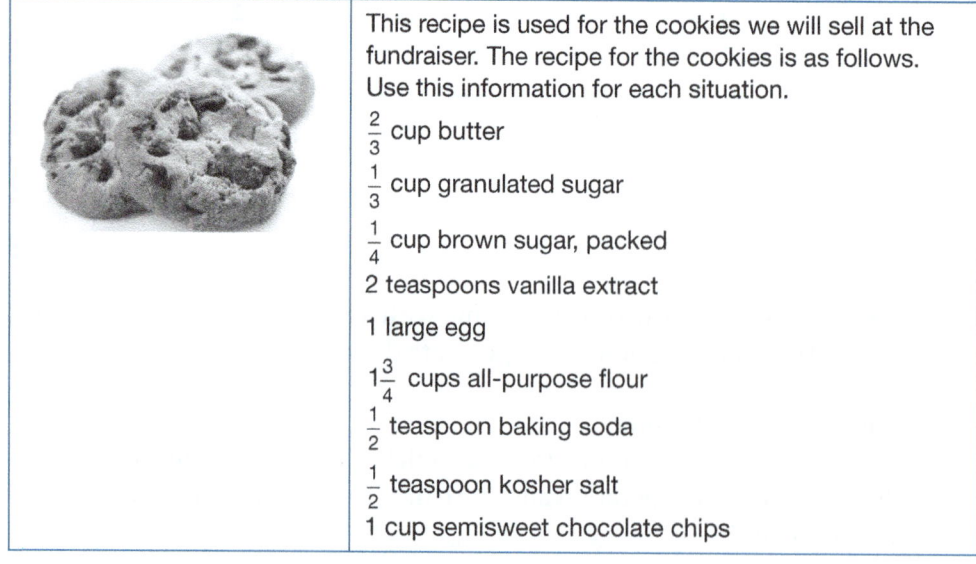

This recipe is used for the cookies we will sell at the fundraiser. The recipe for the cookies is as follows. Use this information for each situation.

$\frac{2}{3}$ cup butter

$\frac{1}{3}$ cup granulated sugar

$\frac{1}{4}$ cup brown sugar, packed

2 teaspoons vanilla extract

1 large egg

$1\frac{3}{4}$ cups all-purpose flour

$\frac{1}{2}$ teaspoon baking soda

$\frac{1}{2}$ teaspoon kosher salt

1 cup semisweet chocolate chips

Source: image of cookies from istock.com/kellyreekolibry

Figure 5.27.4 Bake sale stories

Situation	Drawings (Number lines, tape diagrams, or other models)	Explain your thinking
If Shawna made 3 batches of cookies, how much butter will she need?	←——————→	
If Shawna made 3 batches, how much brown sugar will she need?	←——————→	
If Shawna made 3 batches, how much flour will she need?	←——————→	
Challenge question: How much of each ingredient is needed to make 4 batches of this cookie recipe? Create a table to show the new recipe amounts.		

Engage in Math Discourse (Make the Mathematics Visible):

Observation: As students work, encourage them to notice important relationships and push on early and partial understandings.

- Listen for how the students are making sense of multiplying a whole number by a fraction. Ask them to explain their thinking.

Interview and **Show Me:** After the activity, bring the intervention group together for a discussion. Suggested prompts:

- What strategies did you use as you solved the problem?

- One of your friends says, "When you multiply a whole number by a fraction, you just have to multiply the numerator by the whole number." Is that true? Why? How can you explain your reasoning using models and math sketches?

Bridging Prompt (Prompt for Classroom Teacher to Use During New Lesson Content):

- Show the class the Challenge recipe with fractional measures. Have students who participated in the intervention come up and double one component of the recipe. Then, have them triple another component. Ask them to explain their thinking with the rest of the class. Blank open number lines are available for use.

TASK 28

Fraction Multiplication: Farmland

General Objective:
Students will multiply fractions using a variety of strategies.

This activity might be used with:

Students who are able to (prerequisite knowledge):	Students who are Primed to (getting ready to learn):
• name and write fractions and mixed numbers represented by a model • represent fractions and mixed numbers with models and symbols • represent equivalent fractions	• interpret multiplication of fractions as scaling (resizing) • solve real-world problems involving multiplication of fractions and whole numbers and mixed numbers, e.g., by using visual fraction models or equations to represent the problem

Materials:

- Fraction area model
- Fraction circles/bars
- Open number line
- PRINTABLE: Unit Fraction Strips (tabloid sized)

    or

    PRINTABLE: Unit Fraction Strips (letter sized)
- PRINTABLE: Unit Fraction Circles
- DIGITAL RESOURCE (fraction models): https://qrs.ly/bigjc86
- DIGITAL RESOURCE (fraction strips): https://qrs.ly/2wgjc8c
- DIGITAL RESOURCE (fraction circles): https://qrs.ly/lxgjc8d

 Printables for this task are available for download at **https://companion.corwin.com/courses/ProactiveMathIntervention**.

Recommended Children's Literature:

- *Using Alice in Wonderland to Teach Multiplication of Fractions* – Taber, 2007

Task Overview:

Students will create a visual based on a real-world problem involving multiplication of fractions with whole numbers and mixed numbers. In this activity, students will use math sketches to explore multiplying fractions through a farm-themed scenario, where they calculate portions of an acre allocated to different crops. Using area models, tape diagrams, and number lines, students will represent and solve problems, enhancing their understanding of taking a fraction of a fraction. By creating math sketches, writing equations, and explaining their reasoning, students will develop a deeper understanding of fraction multiplication and equivalent representations.

Variation

Select (or create) a variation that focuses on the prior knowledge you are working to shore up for your student group.

Learning Target: Students will use math sketches to make sense of multiplying a fraction by a fraction.

Variation Directions:

Share the context of a farmland that is measured by acres that may have different crops. Use the different scenarios to show how much crop the different friends are planting on an acre of farmland. Have students draw a math sketch, write a number sentence, and explain their answer connecting the different representations.

Farmland

Directions: For each situation, model each using a math sketch. Use an area model to start, try the other models such as the tape diagram and/or the number line, and state your answer.

Figure 5.28.1 Sample student solutions

Penny has $\frac{2}{3}$ of an acre of farmland. She wants to have $\frac{1}{2}$ of that land planted with corn. How many acres will have corn growing?	
First, I drew lines horizontally to shade in $\frac{2}{3}$ acres using the diagonal lines to show Penny's property. Then I drew a line vertically and took $\frac{1}{2}$ of that land to plant the corn and colored it yellow. When naming my answer, I looked at how much the yellow section was based on the whole acre. So, Penny planted corn on $\frac{2}{6}$ of one acre, which is the same as $\frac{1}{3}$. $\frac{2}{3} \times \frac{1}{2} = \frac{2}{6}$ or $\frac{1}{3}$ 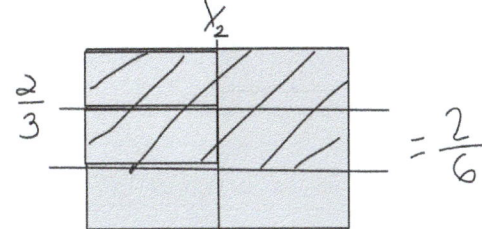	Another area model could be a tape diagram. The whole strip is one acre. Then I shaded $\frac{2}{3}$ of an acre in green to show Penny's property. Then I took $\frac{1}{2}$ of that land to plant corn. That is $\frac{1}{3}$ acre. 

Figure 5.28.2 Farmland stories

Situation	Representations (Area Model and Equation)	Explain your thinking
Kenny has $\frac{1}{4}$ of an acre of farmland. He wants to have $\frac{1}{2}$ of that land planted with blueberries. How much of an acre will have blueberries growing?	Sketch an area model and write an equation.	
Jenny has $\frac{3}{4}$ of an acre of farmland. She wants to plant $\frac{1}{3}$ of that land with tomatoes. How much of an acre will have tomatoes growing?	Sketch an area model and write an equation.	
Benny has $\frac{1}{2}$ an acre of farmland. He wants to plant $\frac{1}{3}$ of that land for lettuce. How much of an acre will have lettuce growing?	Sketch an area model and write an equation.	

Situation	Representations (Area Model and Equation)	Explain your thinking
Denny has $\frac{1}{4}$ of an acre of farmland. He wants to plant $\frac{2}{3}$ of that land with pumpkins. How much of an acre will have pumpkins growing?	Sketch an area model and write an equation.	
Use one of the other models like a tape diagram or number line to show one of these problems.		

Engage in Math Discourse (Make the Mathematics Visible):

Observation: As students work, encourage them to notice important relationships and push on early and partial understandings.

- Listen for how students are making sense of taking a fraction of a fraction. Ask them to explain their thinking.

- Students will notice that taking a fraction of a fraction gives them a smaller fraction. Ask them how that makes sense to them.

- Listen to students' discussion of units. Some students may think the product represents a fraction part "of the farmland" rather than a fractional part "of an acre." Ask students to explain their solution using the area model. Help them make sense of the 2 "wholes" that exist in the story (1 whole acre, all the farmland).

Interview and **Show Me:** After the activity, bring the intervention group together for a discussion. Suggested prompts:

- What strategies did you use as you solved the problem?

- A friend says, "When you multiply a fraction by a fraction, you just multiply straight across—numerator by numerator and denominator by denominator." Is that correct?

- What pattern are you noticing in your multiplication equations? How can we see that in the area models?

|  | **Bridging Prompt (Prompt for Classroom Teacher to Use During New Lesson Content):**

• Take one of the earlier problems that was used in the intervention. Project it to the whole class. Have students who participated in the intervention share a math sketch and describe how they are thinking about the problem. |

Fraction Races

TASK 29

General Objective:

Students will interpret and model fraction divisions to build understanding of how to divide a whole number by a unit fraction and divide a unit fraction by a whole number.

This activity might be used with:

Students who can (prerequisite knowledge):	Students who are Primed to (getting ready to learn):
• state the meaning of division (with whole numbers) • add unit fractions with common denominators • multiply a whole number by a unit fraction • partition a whole into equal parts	• partition a whole number greater than one into equal parts • interpret division with fractions as missing factor multiplication

Materials:

- Fraction tile manipulatives
- PRINTABLE: Unit Fraction Strips (tabloid sized)

  or

  PRINTABLE: Unit Fraction Strips (letter sized)
- PRINTABLE: Unit Fraction Circles
- PRINTABLE: Whole Number Tracks
- PRINTABLE: Unit Fraction Tracks
- PRINTABLE: Equal Shares Cards

 Printables for this task are available for download at **https://companion.corwin.com/courses/ProactiveMathIntervention**.

Recommended Children's Literature:

- *Wilma Unlimited* – Krull (book about running races)
- *Ready, Set, Run! The Amazing New York City Marathon* – Kimmelman & Hartland

Task Overview:

In this activity, students interpret a story of unit fractions racing along a path to make sense of division with a whole number and a unit fraction. Students will use unit fraction strips to model each situation then write and analyze the corresponding equation.

Select (or create) a variation that focuses on the prior knowledge you are working to shore up for your student group.

Variation 1

Learning Target: Students will model the division of a whole number by a unit fraction.

Variation Directions:

Begin this variation by telling students the following story:

Today, we are going to go to the unit fraction races! We will make a racetrack out of a collection of wholes that are all the same size and see how many unit fractions it takes to finish the race!

Organize students in partner pairs. Each pair of students picks a whole number and builds a racetrack of that length. Each partner picks a unit fraction "racer." Students iterate the unit fraction to see how many of the unit fraction it takes to complete the racetrack. After students have built their racetracks and modeled their unit fraction racers, they will complete the Whole Number Tracks handout. Discuss with students how the model connects to a division equation. Encourage students to notice the relationship between the number of unit fractions in 1 whole and the number of unit fractions in n wholes.

Figure 5.29.1 Fraction Races Variation 1 example

Partners A and B decide to build a racetrack that has a length of 2 (2 whole units). Both partners construct their racetracks.
Partner A:
1 | 1
Partner B:
1 | 1
Partner A chooses the racer unit fraction $\frac{1}{4}$ and begins iterating the unit fraction along the track.
$\frac{1}{4}$ $\frac{1}{4}$ $\frac{1}{4}$ $\frac{1}{4}$ $\frac{1}{4}$ $\frac{1}{4}$ $\frac{1}{4}$ $\frac{1}{4}$ 1 | 1
Partner B chooses the racer unit fraction $\frac{1}{6}$ and begins iterating the unit fraction along the track.
$\frac{1}{6}$ $\frac{1}{6}$ $\frac{1}{6}$ $\frac{1}{6}$ $\frac{1}{6}$ $\frac{1}{6}$ $\frac{1}{6}$ 1 | 1

When the partners have completed their racetracks, they fill in the Whole Number Tracks handout.

Partner A:

Racetrack Length	Unit Fraction Racer	How many unit fractions in 1 whole?	How many unit fractions in the whole racetrack?	Division Equation
2	$\frac{1}{4}$	4	8	$2 \div \frac{1}{4} = 8$

Partner B:

Racetrack Length	Unit Fraction Racer	How many unit fractions in 1 whole?	How many unit fractions in the whole racetrack?	Division Equation
2	$\frac{1}{6}$	6	12	$2 \div \frac{1}{6} = 12$

Engage in Math Discourse (Make the Mathematics Visible):

Observation and **Interview**: As students work, encourage them to notice important relationships and push on early and partial understandings.

- Which unit fraction needed more unit fraction pieces to complete the racetrack? Is that what you expected? Why?
- How many unit fractions did it take to complete 1 whole on your racetrack?
- How many unit fractions did it take to complete the whole track?
- Can you predict how many unit fractions it would take to complete the racetrack if I changed the length to ___ wholes? What strategy are you using to figure that out?
- What patterns are you noticing?
- Can you predict how many one-thirds (pick a unit fraction different from both partners) it would take to complete your racetrack? What strategy are you using to figure that out?

Interview and **Show Me**: After the activity, bring the intervention group together for a discussion. Suggested prompts:

- Have students notice relationships by highlighting models with some commonalities and some differences. For example, look for students who may have used the same unit fraction but different racetrack lengths (or students who used the same track length but different unit fraction racers). Have the students focus on one of these examples.
  - What is the same about these two races? What is different?
  - How many more unit fractions did one race need compared to the other? How did you know?

- Did any of you notice a similar pattern in the races you and your partner made?

▶ Ask students, "When we ask questions like, 'How many equal sized parts (that's your unit fraction racers) are in the whole track?' what operation does that make you think of? We are dividing the track up into equal distances! That means we can write a division equation to interpret these situations."

- When I see the equation $2 \div \frac{1}{4} = 8$ for Partner A's racetrack, what part of that equation tells me about the whole length of the track?
- What does the $\div \frac{1}{4}$ mean? Is there a way to read a division problem that would help us here? *(How many groups of $\frac{1}{4}$ are there in . . . ?)*
- What does 8 represent?

Bridging Prompt (Prompt for Classroom Teacher to Use During New Lesson Content):

- Show the class a racetrack that is 3 wholes long. Then, show the class a unit fraction racer (for example, $\frac{1}{4}$). Have students who participated in the intervention group explain the context to the class and ask, "How many $\frac{1}{4}$s will it take to complete the track?"
- Encourage the class to solve the problem using models and equations. Look for opportunities to lift up thinking from students who participated in the intervention.
- Ask, "How can thinking about the number of unit fractions in 1 whole help you think about the number of unit fractions in any other whole number?"

Variation 2

Learning Target: Students will model the division of a unit fraction by a whole number.

Variation Directions:

Begin this variation by telling students the following story:

The fraction races have decided to make racetracks out of unit fractions! These races will be a little different. This time, we will know the racetrack length and how many equal parts there are. Your challenge is to find the unit fraction runner that can do the race!

Organize students in partner pairs. Each student draws a Unit Fraction Track card and builds a racetrack to match. The card they drew gives them a goal for the number of equal parts on the track. The students determine which unit fraction "racer" matches the situation. After students have built their racetracks and modeled their unit fraction racers, they will complete the Unit Fraction Tracks handout. Discuss with students how the model corresponds to a division equation. Encourage students to notice the relationship between the denominators of the dividend and the quotient.

Figure 5.29.2 Fraction Races Variation 2 example

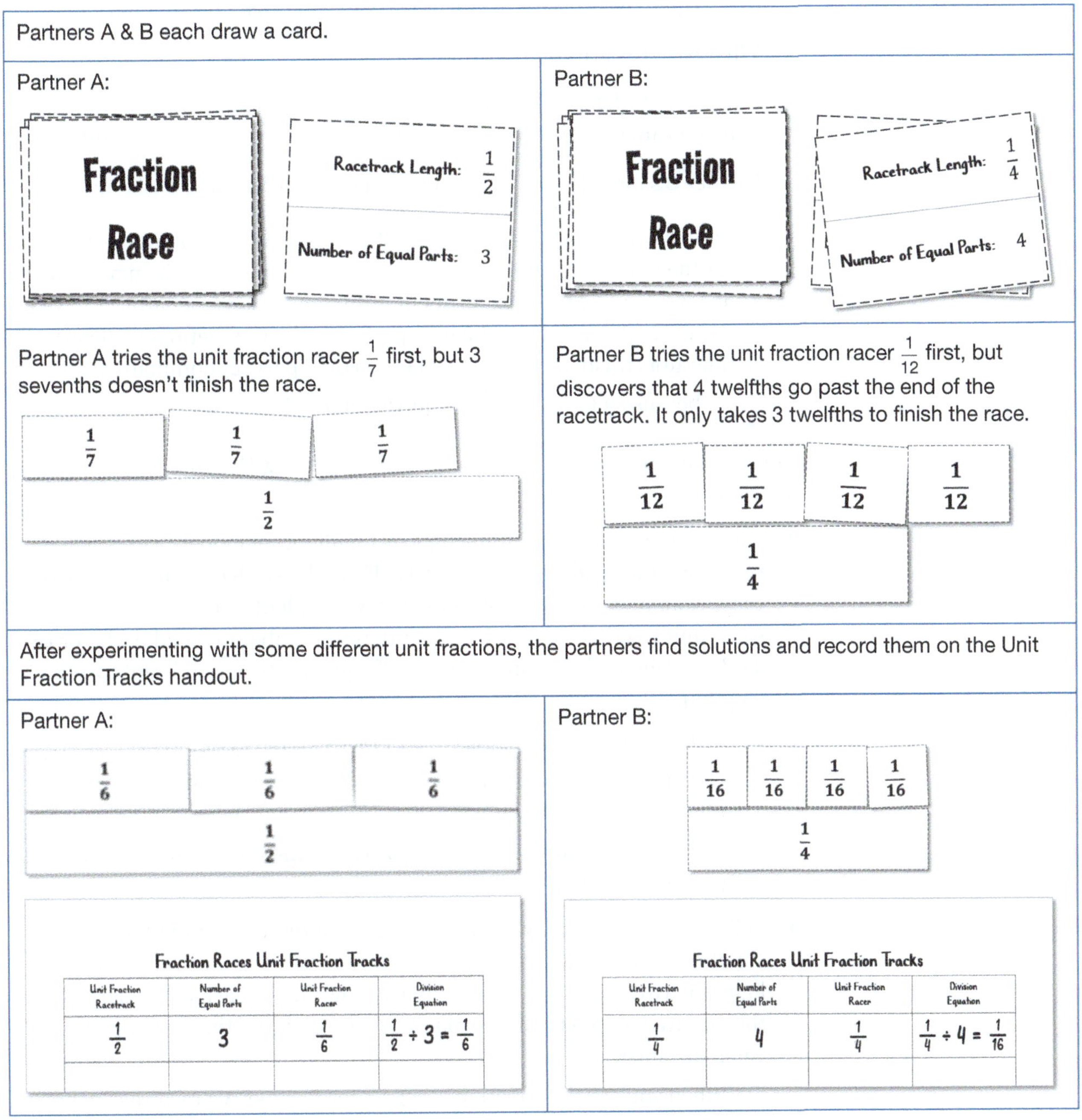

Partners A & B each draw a card.	
Partner A: Fraction Race — Racetrack Length: $\frac{1}{2}$; Number of Equal Parts: 3	**Partner B:** Fraction Race — Racetrack Length: $\frac{1}{4}$; Number of Equal Parts: 4
Partner A tries the unit fraction racer $\frac{1}{7}$ first, but 3 sevenths doesn't finish the race. [$\frac{1}{7}$, $\frac{1}{7}$, $\frac{1}{7}$ over $\frac{1}{2}$]	Partner B tries the unit fraction racer $\frac{1}{12}$ first, but discovers that 4 twelfths go past the end of the racetrack. It only takes 3 twelfths to finish the race. [$\frac{1}{12}$, $\frac{1}{12}$, $\frac{1}{12}$, $\frac{1}{12}$ over $\frac{1}{4}$]
After experimenting with some different unit fractions, the partners find solutions and record them on the Unit Fraction Tracks handout.	
Partner A: [$\frac{1}{6}$, $\frac{1}{6}$, $\frac{1}{6}$ over $\frac{1}{2}$]	**Partner B:** [$\frac{1}{16}$, $\frac{1}{16}$, $\frac{1}{16}$, $\frac{1}{16}$ over $\frac{1}{4}$]

Fraction Races Unit Fraction Tracks (Partner A)

Unit Fraction Racetrack	Number of Equal Parts	Unit Fraction Racer	Division Equation
$\frac{1}{2}$	3	$\frac{1}{6}$	$\frac{1}{2} \div 3 = \frac{1}{6}$

Fraction Races Unit Fraction Tracks (Partner B)

Unit Fraction Racetrack	Number of Equal Parts	Unit Fraction Racer	Division Equation
$\frac{1}{4}$	4	$\frac{1}{4}$	$\frac{1}{4} \div 4 = \frac{1}{16}$

Engage in Math Discourse (Make the Mathematics Visible):

Observation and **Interview:** As students work, encourage them to notice important relationships and push on early and partial understandings.

- How did you decide what unit fraction to try first?
- Were there some unit fractions you knew for sure wouldn't work? How did you know?

- Can you predict what unit fractions you would need to use if we changed it to ___ equal parts? What strategy are you using to figure that out?
- What patterns are you noticing?
- Listen for students who may notice the equivalent fraction relationships that show up in their models (e.g., in Partner A's model, $\frac{1}{2}=\frac{3}{6}$, and in Partner B's model, $\frac{1}{4}=\frac{4}{16}$). Flip over a new card for groups who notice this and ask them if they can use what they noticed to make a prediction about the size of the unit fraction racer they will need for the new card. Walk away and let them experiment until they discover a pattern . . . that the equivalent fraction can be interpreted as the number of equal parts (numerator) divided by the size of each equal part (denominator). So $\frac{1}{2}=\frac{3}{6}$ because 3 groups of $\frac{1}{6}$ is equivalent to $\frac{1}{2}$!

Interview and **Show Me:** After the activity, bring the intervention group together for a discussion. Suggested prompts:

- Have students notice relationships by highlighting models with some commonalities and some differences. Have the students focus on one of these examples. For example, look for two students who may have had the same unit fraction racetrack but a different number of equal parts (or two students who had the same number of equal parts but different racetrack lengths).
    - What is the same about these two races? What is different?
    - How many more unit fractions did one race need compared to the other? How did you know?
    - Did any of you notice a similar pattern in the races you and your partner made?
- Tell students, "We can write a division equation to interpret these situations because we can ask the question, 'When a racetrack that is ___ long is partitioned into ___ equal parts, how long is each equal part (that's the unit fraction racer that works)?' We are dividing the track up into equal distances."
    - When you see the equation $\frac{1}{2} \div 3 = \frac{1}{6}$ for Partner A's racetrack, what part of that equation tells you about the length of the track?
    - What does the ÷3 tell you?
    - What does $\frac{1}{6}$ represent?

Bridging Prompt (Prompt for Classroom Teacher to Use During New Lesson Content):

- Show the class a unit fraction racetrack (for example, $\frac{1}{2}$). Ask them to identify the unit fraction racer that would fit if the $\frac{1}{2}$ racetrack was partitioned into four equal parts. Have students who participated in the intervention explain the activity to the class.

- Ask, "When we partition a unit fraction racetrack into equal-sized parts, will the denominator of the equal parts be greater or less than the denominator of the racetrack? Why?"

TASK 30: Shape Sort

General Objective:

Identify defining and nondefining attributes of two-dimensional shapes.

This activity might be used with:

Students who are able to (prerequisite knowledge):	Students who are Primed to (getting ready to learn):
• sort shapes based on common attributes and identify the sorting categories	• describe, analyze, compare, and classify two-dimensional shapes by their number of sides and corners (angles) and connect these properties to the definitions of shapes • distinguish between defining attributes (e.g., triangles are closed and three-sided) versus nondefining attributes (e.g., color, orientation, overall size); build and draw shapes to possess defining attributes

Materials:

- PRINTABLE: Shapes for Sorting (set of assorted paper shapes, cut apart and arranged in baggies)—1 baggie per group of three students
- Large paper for sorting (several pages per group of students)
- Chart paper
- Markers

Recommended Children's Literature:

- *The Silly Story of Goldie Locks and the Three Squares* – Maccarone, 1996

Task Overview:

In this task, students sort two-dimensional shapes, analyze examples and non-examples of shapes, and identify defining and nondefining attributes of two-dimensional shapes.

Select (or create) a variation that focuses on the prior knowledge you are working to shore up for your student group.

Variation 1

Learning Target: Students will compare and classify two-dimensional shapes by their number of sides and angles (defining attributes).

Variation Directions:

If desired, read excerpts from a children's literature book about shapes, such as *The Silly Story of Goldie Locks and the Three Squares* by Grace Maccarone, to students to introduce the topic. Discuss the story details with the students. Allow time for students to discuss the attributes of the shapes Goldie Locks sorted in the story. If the book is not available, start the activity by asking students to think about different shapes they see around them every day. Say, "Shapes are everywhere! Can you think of some shapes you've seen recently, like on signs, buildings, on toys, or in our room?" Allow students to share their examples, and write a few on the board. Then, display a variety of two-dimensional shapes (either drawn on the board, on posters, or using cutouts).

Next, divide the students into groups of three and distribute the shape baggies. Allow time for students to examine the shapes. Encourage students to use descriptive language based on geometric properties. Use chart paper to record student responses. Listen for and highlight responses that include shape names (triangle, rectangle, square) and defining attributes such as number of sides and corners (angles).

Engage in Math Discourse (Make the Mathematics Visible):

Observation: As students work, encourage them to notice important relationships and push on early and partial understandings.

- What did you notice about the shapes?
- How would you describe the shapes?
- What do you notice about all of the triangles?
- What do you notice about all of the rectangles?

Interview: After the activity, bring the intervention group together for a discussion. Suggested prompts:

- Now that we've explored different shapes and their attributes, let's connect what we've noticed to a bigger idea. Think about how shapes are described—not just their names but also their sides and angles.
- How do these details help us group and compare shapes?
- How can we use what we know about shapes to sort them into categories, look for patterns, and understand what makes each shape unique?
- What do you notice about the shapes?
- How can you describe their similarities and differences?

Bridging Prompt (Prompt for Classroom Teacher to Use During New Lesson Content):
• Show the class a collection of shapes that was sorted by the intervention group. Ask the class what they notice and wonder about the shapes. Encourage the students who participated in the intervention group to try to answer the questions their classmates ask. Elicit defining attributes.

Variation 2

Learning Target: Students will compare and classify two-dimensional shapes by their number of sides and angles and identify defining and nondefining attributes of two-dimensional shapes.

Variation Directions:

This variation can be conducted over several days. Group the students into pairs or groups of three. Distribute large pieces of paper and the baggies with shapes to students to use as sorting mats. Ask the students to sort the shapes by putting all the like shapes together. Tell students to use what they know about shapes to identify characteristics of each shape.

Begin with the triangles. Ask the students to describe the triangles and record responses on a whole-group chart. Student responses might include: "Triangles have three straight sides; triangles have three corners (angles)." Some students might include color and size in their descriptions. However, ask students about nonshape attributes such as, "Does a triangle need to be red, blue, green, yellow, large, or small to be a triangle?" A shape can have a color, but it is not a mathematically defining attribute of a shape. Next, tell the student groups that they need to create a definition of what makes a shape a triangle. Write "Triangles" as a heading on a new piece of chart paper. Record the defining attributes. Allow time for students to review the full list of characteristics before concluding that there are only two defining attributes.

Figure 5.30.1 Triangle attribute chart

> Triangles
>
> Three straight sides
> Three corners (angles)

If time allows, continue with rectangles. Write "Rectangles" as a heading on a new piece of chart paper. Record the defining attributes. Allow time for students to review the full list of characteristics before concluding that there are only two defining attributes. When students notice the squares, encourage students to compare the squares to the rectangles. You can add a note to the chart that a square is a special rectangle with all four sides the same length.

Figure 5.30.2 Rectangle attribute chart

> **Rectangles**
>
> Four straight sides
> Four square corners (angles)
> Square is a special rectangle with all four sides the same length

Engage in Math Discourse (Make the Mathematics Visible):

Observation: As students work, encourage them to notice important relationships and push on early and partial understandings.

- How would you describe triangles?
- What's true about all triangles?
- How many sides does the square have? (4 sides)
- How many square corners (right angles) does the square have?
- Is a square an example of a rectangle? (Students should respond that it is because it has 4 straight sides and 4 square corners.)

Interview: After the activity, bring the intervention group together for a discussion. Suggested prompts:

- Explain the two defining attributes of a triangle in your own words.
- Why don't color or size matter when identifying a triangle or rectangle?
- What makes a rectangle a rectangle?
- How do the number of sides and angles help us identify it?
- How is a square similar to and different from other rectangles?

Bridging Prompt (Prompt for Classroom Teacher to Use During New Lesson Content):
• Show the class the Triangle attribute chart created by the intervention group. Have the students who participated in the intervention group explain to the class the chart they made. Then, present a collection of shapes. All but one of the shapes should be triangles. Challenge the class to find the shape that is not a triangle.
• Elicit reasoning from students who participated in the intervention group.

Variation 3
Learning Target: Students will use the defining attributes of two-dimensional shapes to be "shape detectives."

Variation Directions:

Introduce the role of a shape detective to the students.

- Today, you are shape detectives! Your mission is to investigate and find all the triangles and rectangles hiding in your bag of shapes. But to be a great detective, you'll need to use your knowledge of shape attributes to explain why each shape belongs in its group.
- You will use this chart paper with two large circles labeled "Triangles" and "Rectangles" to sort your shapes.
- Use the sentence frame to explain your reasoning.
  - Example: **I know this is a** <u>rectangle</u> **because** <u>it has four sides, and all the angles are right angles</u>.
  - Example: **This is not a** <u>triangle</u> **because** <u>it has more than three sides</u>.

Engage in Math Discourse (Make the Mathematics Visible):

Observation: As students work, encourage them to notice important relationships and push on early and partial understandings.

- What information did you use to decide if a shape was a triangle or a rectangle?
- Were there any shapes that were tricky to sort? Why?
- Can a square belong in the rectangle group? Why or why not?

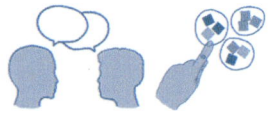

Interview and **Show Me:** After the activity, bring the intervention group together for a discussion. Suggested prompts:

- Show the group a square and a "long" rectangle (one that is not a square).
  - Ask the group to discuss why both figures should fit in the "rectangle" group.
  - Ask the group if both figures would fit in a "square" group.

> **Bridging Prompt (Prompt for Classroom Teacher to Use During New Lesson Content):**
>
> - Show the class a collection of various shapes and place them in a bag. Have students who participated in the intervention group demonstrate and explain the "shape detective" activity to the class. Then, challenge the class to find a given shape.

CHAPTER 5. Intervention Tasks 325

Alike and Different

TASK 31

General Objective:

Describe and analyze two-dimensional geometric shapes based on the number of sides and corners (angles).

This activity might be used with:

Students who are able to (prerequisite knowledge):	Students who are Primed to (getting ready to learn):
• sort two dimensional shapes based on common attributes and identify the sorting categories • describe two-dimensional shapes based on the number of sides and corners (angles)	• describe, analyze, compare, and classify two-dimensional shapes by their number of sides and corners (angles) and connect these properties to the definitions of particular shapes

Materials:

- Large copy of Image 1 to project to students
- Set of assorted paper shapes (cut apart and arranged in baggies)—1 baggie per group of three students
- Label cards (triangles, quadrilaterals, 5-sided/pentagons, 6-sided/hexagons) for each group of students OR provide students with index cards and let them create their own labels
- Large paper for sorting (several pieces of paper per group of students)
- Chart paper
- Markers
- PRINTABLE: Whole-group Charts
- PRINTABLE: Shapes for Sorting
- PRINTABLE: Shape Riddle Cards

 Printables for this task are available for download at **https://companion.corwin.com/courses/ProactiveMathIntervention**.

Recommended Children's Literature:

- *The Animals Would Not Sleep* – Levine, 2020 (sorting and classifying)

- *The Greedy Triangle* – Burns, 2008
- *A Trapezoid Is Not a Dinosaur* – Morris, 2019

Task Overview:

In this task, students sort, describe, analyze, compare, and classify two-dimensional shapes by their number of sides and corners (angles).

Select (or create) a variation that focuses on the prior knowledge you are working to shore up for your student group.

Variation 1

Learning Target: Students will sort, describe, analyze, compare, and classify two-dimensional geometric shapes by their number of sides and angles and apply their knowledge of defining and nondefining attributes of two-dimensional shapes.

Variation Directions:

Begin this variation with students together in a whole group. Show Image 1 to the students. Ask students to share what they notice about the two groups. What do they wonder? Record their ideas. Ask, "Why do you think the shapes in Group A go together? Why do you think the shapes in Group B go together?" Record students' responses on a whole-group chart. Ask, "What could we name each group?" (e.g., triangles and not triangles.) Discuss the attributes students used to label group A shapes as triangles (e.g., three sides and three angles) and label group B (e.g., four or more sides not triangles).

Figure 5.31.1 Whole-group chart

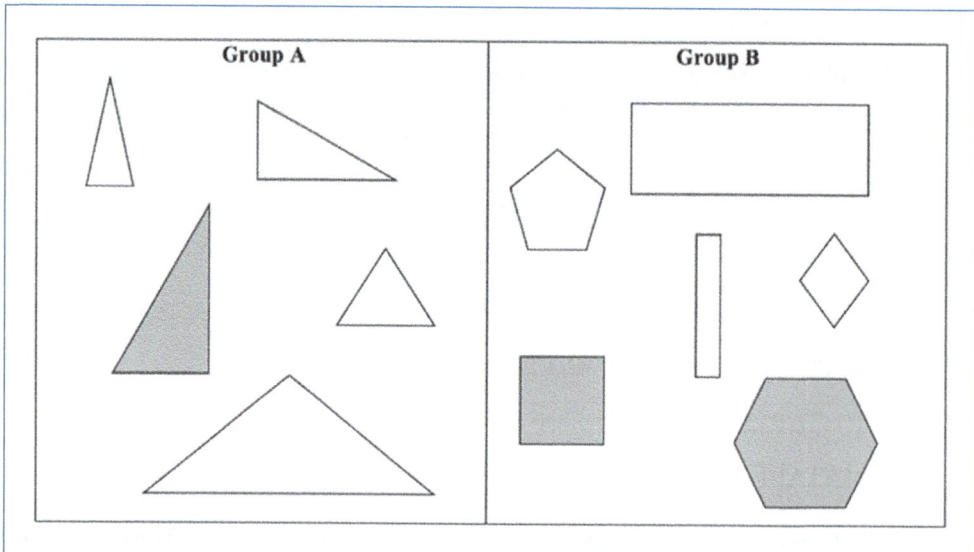

Next, organize the students into groups of three and distribute the baggies of assorted shapes and large blank pieces of paper for the sorting mats. Tell students, "Work with your group to classify the shapes. Put the shapes together that belong together and then label your groups. Be ready to share your reasoning!"

CHAPTER 5. Intervention Tasks

Engage in Math Discourse (Make the Mathematics Visible):

Observation: As students work, encourage them to notice important relationships and push on early and partial understandings.

- What do these shapes have in common? Why are they alike?
- How would you describe the shapes in this group?
- What name would you give to each group?

Interview: After the activity, bring the intervention group together for a discussion. Suggested prompts:

- Let's look at some of the ways we sorted the shapes. What do you notice about how we sorted?
- How would you describe the shapes in this group (point to group(s))?
- What name would you give to each group?
- As students share, record the categories used by students. Some students might sort shapes by type, size, number of sides or angles. Listen for language such as: side, corner, angle, triangle, rectangle, square corners, five-sided, six-sided, etc.

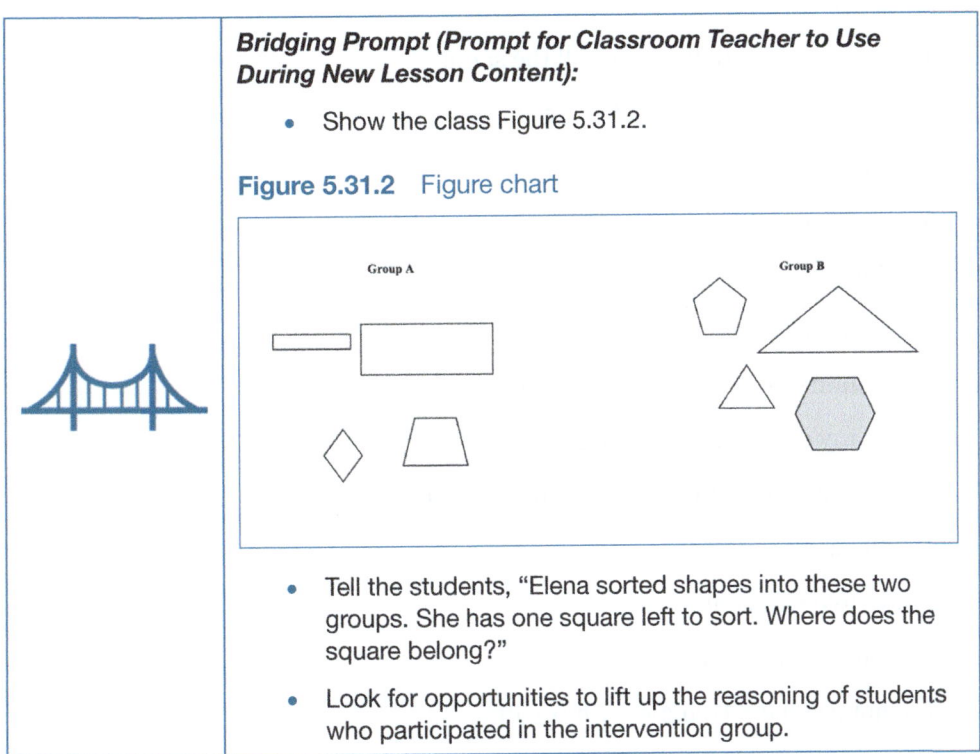

> **Bridging Prompt (Prompt for Classroom Teacher to Use During New Lesson Content):**
> - Show the class Figure 5.31.2.
>
> **Figure 5.31.2** Figure chart
>
> - Tell the students, "Elena sorted shapes into these two groups. She has one square left to sort. Where does the square belong?"
> - Look for opportunities to lift up the reasoning of students who participated in the intervention group.

Variation 2

Learning Target: Students will match shapes to clues from a riddle to identify all the shapes that match the riddle.

Variation Directions:

For this variation, students will build on their understanding of grouping shapes by their attributes. Tell the students that you are going to read a riddle (or share a Shape Riddle Card), and their task is to find all the shapes that match the riddle. As you read the riddle, ask students to find all the shapes that match the riddle by counting sides and angles, tracing their finger around the shape, and comparing the shape to other shapes. Then, have students put all those shapes into a group together. After all the shapes have been located in a group, have the students find the label card for that group.

Riddle 1:

I have three sides and three corners, no matter my size.

Sometimes, I'm tall. Sometimes, I'm wide.

What shape am I? (Triangle)

Riddle 2:

I have four sides that are all the same.

All corners are equal and share the same name.

What shape am I? (Square)

Riddle 3:

I have four sides that are not all the same.

Two sides are long; two sides are short.

But opposite sides are the same.

What shape am I? (Rectangle)

Riddle 4:

I'm a shape with six straight sides,

Like a honeycomb where bees hide.

Count my corners, one through six,

What's my name? Come take your pick! (Hexagon)

Riddle 5:

I have five sides and corners too.

Look for me in patterns, it's true!

I'm not a square, triangle, or hexagon,

What's my name? I'm a _____! (Pentagon)

Riddle 6:

I always have four sides, no less, no more.

You'll find me in shapes you adore!

I can be a square or a rectangle and other four-sided shapes too.

What's my name? Listen to this clue! (Quadrilaterals)

Have a pair of students write their own riddle on and index card and pass it to another team (let students know that the riddle does not have to rhyme).

Engage in Math Discourse (Make the Mathematics Visible):

Observation: As students work, encourage them to notice important relationships and push on early and partial understandings.

- How did you decide if the shape matches the riddle?
- What clue in the riddle helps you figure out the shape?

Interview and **Show Me:** After the activity, bring the intervention group together for a discussion. Suggested prompts:

- What are some things you were noticing as you looked for shapes to match the riddle?
- Show visual examples of the shapes discussed in the riddles (square, triangle, hexagon, pentagon, quadrilateral) and ask, "What do all these shapes have in common?" and "How are they different from one another?"
- Let's count the sides and corners (angles) of each shape together. What patterns do you see?

Bridging Prompt (Prompt for Classroom Teacher to Use During New Lesson Content):
• Have students who participated in the intervention group select a riddle they completed during the intervention, or use the riddle they created and present it to the class to solve.

Variation 3
Learning Target: Students will describe shapes using their defining attributes, such as the number of sides and corners, to help others identify them.

Variation Directions:

In this activity, students will engage in a shape-guessing game to practice identifying and describing a shape by its attributes. They will describe shapes to their peers without naming them, focusing on properties like the number of sides and corners. The goal is to strengthen their ability to use mathematical language and identify similarities and differences between shapes. To introduce the activity, explain to students that they will play a shape-guessing game. Begin by saying, "We're going to play a shape-guessing game! Each of you will choose a shape and think about how you would describe it to a partner without saying its name." Provide sentence frames to help guide their descriptions, such as: "I have ___ sides and ___ corners," "My shape looks like a ___," or "It is similar to ___ because ___. It is different because ___." Allow students to mingle around the room to find a partner, and instruct them to decide who will go first. One student will describe their shape using its attributes while the other tries to guess what the shape is. Once they finish, prompt them by saying, "After you finish, trade shapes with your partner and find someone new to explore the shapes with."

Engage in Math Discourse (Make the Mathematics Visible):

Observation: As students work, encourage them to notice important relationships and push on early and partial understandings.

- What are some geometry words you've learned that would help you describe shapes?

- How can you describe your shape using the number of sides and corners to help your partner figure out the name of your shape?

- Compare your shapes and find one similarity and one difference.

Interview: After the activity, bring the intervention group together for a discussion. Suggested prompts:

- What clues did your partner give you that made it easy to guess their shape?

- What kinds of clues did not help you guess the name of the shape?

Bridging Prompt (Prompt for Classroom Teacher to Use During New Lesson Content):
• Arrange stations with various shape cards and manipulatives such as pattern blocks and attribute blocks around the classroom.
• Challenge the class to sort the shapes based on specific attributes (e.g., number of sides, corners, or overall size).
• Have students explain their sorting choices using sentence frames:
○ I grouped these shapes because they all have __ sides.
○ This shape is different because it has __ corners.
• Call on the students from the intervention group to share their thinking.

Shape Shifting

TASK 32

General Objective:

Students will use the attributes of two-dimensional and three-dimensional shapes to consider geometric relationships and spatial thinking.

This activity might be used with:

Students who are able to (prerequisite knowledge):	Students who are Primed to (getting ready to learn):
• identify basic two-dimensional shapes • know the difference between a two-dimensional and a three-dimensional shape	• identify the essential attributes of geometric figures and begin to classify them not by appearance but by focusing on the properties of shapes • recognize defining attributes such as sides, angles, edges, faces, and vertices • put shapes in categories according to shared properties and the relationships between those properties • approximate angle sizes (eventually moving to standard angle-measuring tools such as protractors or angle rulers)

Materials:

- Loop of yarn about 1 meter long (tied)
- Angle fixer plate
- Yarn for geometry necklaces (hole punch)
- PRINTABLE: Geometry Necklace Cards
- PRINTABLE: Questions and Checklist

 Printables for this task are available for download at **https://companion.corwin.com/courses/ ProactiveMathIntervention**.

Recommended Children's Literature:

- *The Greedy Triangle* – Burns, 1994

Task Overview:

This is an activity to develop flexibility with shapes as being in different orientations and sizes but with the same properties to be called a particular name. *Select (or create) a variation that focuses on the prior knowledge you are working to shore up for your student group.*

Variation 1

Learning Target: Students will identify the attributes of geometric shapes.

Variation Directions:

This is a partner activity (can be a group of three if needed). To begin the variation, give two students each one loop of yarn to work with. They are to use their hands (some students who need support for motor skills can use a pencil grasped in their fist) to hold the yarn up as they are given directions to make various geometric figures. At this point in the instruction, it would be a good place to read excerpts from a story such as *The Greedy Triangle* (Burns, 1994) and stop along the way to act out the triangle's changes. Match the story by starting the activity with students making a triangle by holding the loop of yarn in the air with their hands as they stand in an open space in the room.

Here are questions you can pose (the comments in the brackets are for your thinking not to be shared with the students—never say anything a kid can say):

Can you show me a triangle with your loop of yarn?

How do you know it is a triangle?

Can you show me a triangle with two sides that are the same length? How can you prove to me they are the same length without using a ruler? [pinch the angles at the base of the isosceles triangle and check] Note that this is a good place to mention that it is called an isosceles triangle, but do not assess students on the name if it is not a grade-level standard. Providing opportunities for students to develop their mathematical vocabulary is important. Such geometric terms will be used throughout the grades.

Can you show me a triangle with one angle that's a square corner? (right angle)? [When the students have their "right triangle," go around with a paper-plate angle fixer (see directions below and Figure 5.29.1) to informally check that they have a right angle. Slip the plate under the angle and see if the yarn aligns with the rays of the angle. Have students adjust their triangle to make it 90 degrees if they do not have a right angle.]

Figure 5.32.1 Sketch of the checking of the suggested right angle on the triangle with an Angle Fixer

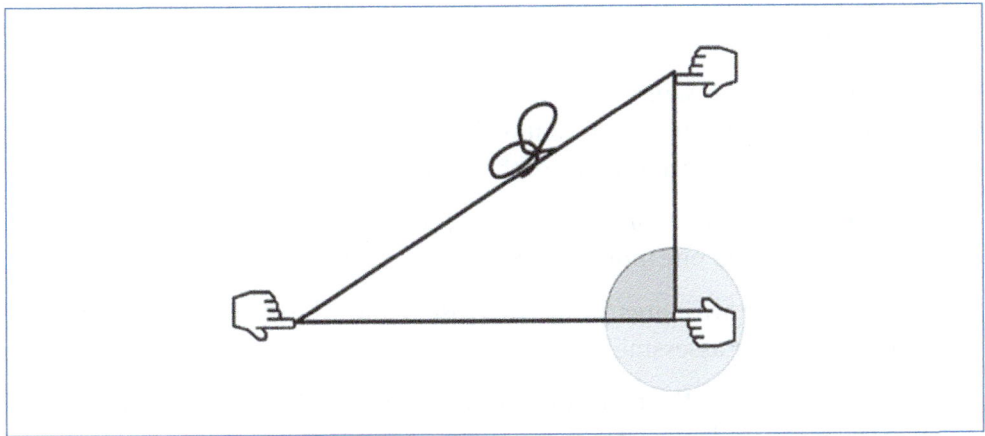

Directions for making an Angle Fixer:

You can cut on the radius two different-colored paper dessert plates that are exactly the same size and shape to make an Angle Fixer (also can be used as a fraction estimator) in the same way as you can use the hundredths wheel (see Task 24 on page 287). Merge them by slipping the cut parts together as one overlaps the other plate. (see Figure 5.32.2) You can then rotate the plates to match angles observed, create a shape with a particular angle, or estimate important benchmark angles such as 30, 45, 60, 90, 135, 180, 270, and 360 degrees.

Figure 5.32.2 Two paper plates cut to make an Angle Fixer

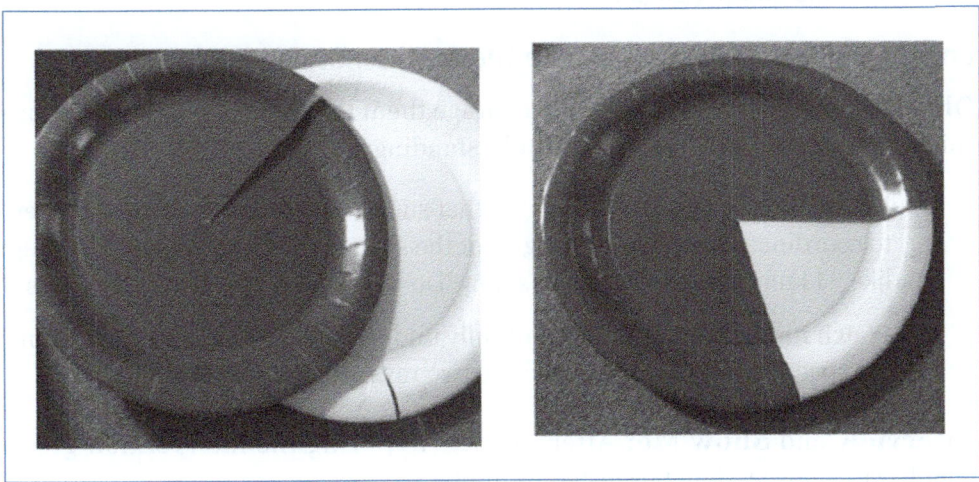

Questions continued:

- [Show an angle on the Angle Fixer plate] and then ask: Can you make a triangle with an angle of this size?
- If you keep your same triangle and move or rotate it to a new position in the air, what changes and what is the same? [The orientation may change but the properties and name of the shape will remain the same.]
- Connecting back to *The Greedy Triangle* book, the triangle wanted one more angle and one more side. Can you help make this new shape?
- What are the attributes/properties [use the word that aligns with your curriculum or assessments] of this new shape?
- How can we find the name of your new shape? [Discuss known shapes and have a student sketch it on chart paper with the name. Also draw on the chart paper shapes that are not known. These may be trapezoids, rhombuses, parallelograms, or kites. If an unknown shape, then the word *quadrilateral* can be used.] Students who have fine motor needs may be challenged to draw and can use a selection of shapes already on cards to choose from to match the new shape.
- [Ask students to switch from whatever four-sided shape they had to make a square] How do you know you have a square? How can you prove it without using a ruler? [Informal measures of lengths of sides may include pinching at the corners and folding one side over another to see if they are equal. They may also want to do an angle check with the Angle Fixer tool.]
- What would the next shape look like if the four-sided shape wants another side and another angle?

Engage in Math Discourse (Make the Mathematics Visible):

Observation: As students work, encourage them to notice important relationships and push on early and partial understandings.

- Listen for language from students that suggests they are working together to coordinate the sides and angles of the shape. They might say something like, "I know we need four sides now. Do we each need to use both hands?"
- Watch for students who may not fully understand that an angle represents the space (or rotation) between the two rays.

Interview and **Show Me:** After the activity, bring the intervention group together for a discussion. Suggested prompts:

- What were the most challenging shapes to make? Or can one group come up and show us the most challenging shape they had to make? [The pentagon and hexagon are difficult because they don't have enough hands.]
- Ask two students to come up and make a triangle. Have another child make an angle with the Angle Fixer and see if the students can reproduce it. The student with the Angle Fixer then checks the angle for accuracy.

	***Bridging Prompt (Prompt for Classroom Teacher to Use During New Lesson Content):*** • Ask two pairs of students from the intervention group to come up, and give each team a loop of yarn. Ask them to make a few shapes without naming the shape—such as, "Make a shape with four sides." If the students make the same shape, ask one team to change their shape to a different one that still fits the request for four sides. Ask the class what is the same and what is different between the two different shapes. Can anyone name the two shapes?

Variation 2

Learning Target: Students will make connections between the attributes of shapes and spatial relationships to name a secret geometric figure.

Variation Directions:

To begin this variation, the teacher needs to make Geometry Necklaces that best fit the students' needs and state standards, selecting from the provided cards or adding others. It is likely best to print them on cardstock or cut them out and glue them to cardstock. Punch two holes at the top of the card and then use a loop of yarn that is long enough to easily go over the student's head. Hang the necklace on the child's back with the shape or geometric figure not visible. Students should not try to look at their necklace, and others cannot tell them what they have on their backs. Students then walk around the room with scrap paper (or the YES/NO Checklist), preferably on clipboards, to take notes or draw sketches and can only ask YES/NO questions of each other. Review what a YES/NO question could be and also highlight a question that does not work (e.g., How many sides do I have?). You can adapt the PRINTABLE called Possible Yes/No Questions Checklist because it is an editable Word file. When they ask a classmate or teacher a question, that person carefully turns the card over and gives the answer of either yes or no, not revealing the shape or geometric figure. Students should move to another student after they have asked a person a question. They can go back to the same person after they've asked the others in the group. Students should be mentally visualizing what they might have from the information they collect, drawing sketches as they move around the room. If they are having trouble visualizing, share on a sheet the names of the various geometric figures they could have on their backs. Then, you can share pages of images of the actual cards if needed. When students are fairly certain they know their shape (and have asked at least five different questions), they should ask the teacher if they are accurate. Another approach is to show the possible shapes that could be on the cards to the students in advance of the first time you do this activity so they are aware of the options (but do not routinely do that). If the student gets theirs correct, they can give hints to others (hints that don't reveal the shape or geometric figure).

Engage in Math Discourse (Make the Mathematics Visible):

Observation: As students work, encourage them to notice important relationships and push on early and partial understandings.

- Listen for their use of the attributes of the shapes. Make a suggestion for students to ask questions about a particular number of sides. There is some challenge to this; they should ask, "Does my shape have exactly three sides?" because a four-sided shape has three sides, too, and could confuse the decision-making.

- Ask students to show you the notes they are writing and explain what they know so far. Remember, their sketches are likely not proportional. If needed for students with needs in motor skills (or other students), share a chart with many shapes/figures at this point. Place it in a plastic protector and have them circle what they think their shape/figure is. Giving them other shapes or geometric figures that you are not currently using builds interest and makes them more familiar with the terminology. They should not be assessed on any of those shapes.

- Listen for language from students that suggests that the orientation of the shape might be confusing for them.

Interview and **Show Me:** After the activity, bring the intervention group together for a discussion. Suggested prompts:

- What would be the first question you would ask if we played this game again? Why that question? [Asking if it is a two-dimensional shape eliminates a lot of options, for example.]

- What shape or geometric figure did you notice on another student's back that you thought would be hard to figure out? Why?

> *Bridging Prompt (Prompt for Classroom Teacher to Use During New Lesson Content):*
>
> - Place a hidden Geometry Necklace Card face down on the teacher's desk or attached to a whiteboard so that it is not visible. Ask students to ask only YES/NO questions to try to identify what shape it is. Call on students who participated in the intervention group to pose the initial questions and suggest important features to ask about (such as sides, faces, angles, vertices). Jot down notes from the answers. Ask, "How did you figure it out?" Place more hidden geometric figures down.

Shapes: Build It!

TASK 33

General Objective:

Compose basic shapes to create new shapes and describe the shapes used to create new shapes.

This activity might be used with:

Students who are able to (prerequisite knowledge):	Students who are Primed to (getting ready to learn):
• identify shapes and describe shapes based on similarities, differences, and identifying attributes	• compose basic shapes to create new shapes and describe the shapes used to create the new shapes • recognize shapes in combined figures from different viewpoints and orientations

Materials:

- Blank paper
- Pencils and/or crayons
- PRINTABLE: Pattern Block Cutouts

  or

  Pattern block manipulatives (one set per pair of students; need yellow, red, green, and blue)

- PRINTABLE: Triangle Outline
- PRINTABLE: Tangram Pieces

  or

  Tangram manipulatives (one set per student)

 Printables for this task are available for download at **https://companion.corwin.com/courses/ProactiveMathIntervention**.

Recommended Children's Literature:

- *Picture Pie* – Embry, 2006
- *Tangram Cat* – Rinck & Van der Linden, 2017
- *Grandfather Tang's Story* – Tompert, 1997
- *The Quest for a Tangram Dragon* – Liu-Perkins, 2024
- *Tan's Tile: A Tale of Creative Thought* – Villari, 2017

Task Overview:

In this task, students compose two-dimensional shapes to create new larger shapes and describe the shapes used to create the new shapes.

Select (or create) a variation that focuses on the prior knowledge you are working to shore up for your student group.

Variation 1

Learning Target: Students will compose basic shapes to create new shapes and describe the shapes used to create the new shapes and recognize shapes in combined figures from different viewpoints and orientations.

Variation Directions:

Begin this variation with students together in a whole group. Show the students the following image and ask, "What do you notice about all of the shapes?" For example, the students might notice:

> All the shapes are the same shape.
>
> Each shape has different shapes inside it.
>
> Shape A is made up of a triangle and a hexagon.
>
> Shape B has two trapezoids and one triangle.
>
> Shape C is made of seven triangles.
>
> Shape D has three rhombuses and one triangle.

Elicit the students' ideas that the shape is the same, but each shape is made up of or composed of different shapes. Point to the shape made up of all triangles and ask, "What shapes do you see in this shape?" Encourage them to both describe and name the shapes. If students don't remember the name of the shape, ask them to describe it by its attributes and see if another child can name it. If not, then you name it. For example, if a student says, "Shape D has three blue shapes and a triangle," you can say, "What do you notice about the sides of the blue shape?" Elicit from the students that the blue shape has four equal sides, and the opposite sides are parallel to each other. What shape(s) fits that description?

Figure 5.33.1 Composed shapes

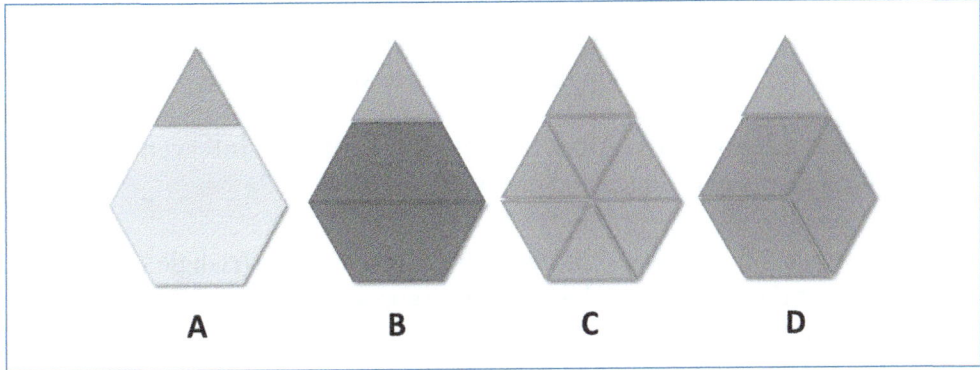

After discussing the images with the students, give the students the Triangle Outline template and pattern blocks and ask them to find as many ways as they can to make the shape using different pattern blocks. Take photos of the different combinations that the students make to project on a screen later for discussion. You can also have students record their results in "shape equations" such as Triangle A = 1Y + 4G + 2R or Triangle B = 4G + 4R (see Figure 5.33.2). Students may see patterns in these equations that guide them to the understanding that 1Y = 2R.

Figure 5.33.2 Triangle Template

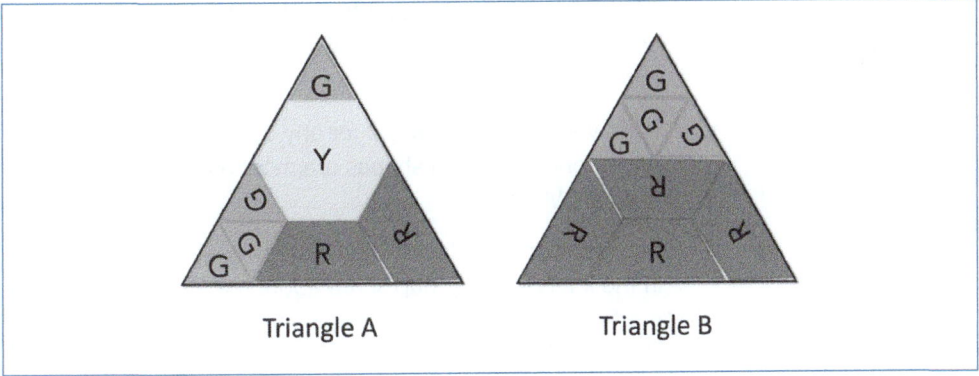

Engage in Math Discourse (Make the Mathematics Visible):

Observation: As students work, encourage them to notice important relationships and push on early and partial understandings.

- What are the different shapes you used to fill the triangle?
- How many of each shape did you use? Can you show that in a "shape equation"?
- How was the way you filled your shape the same or different from a classmate?

- Watch for students who demonstrate a strength in "shape sense" or spatial reasoning, even if they may have need for additional assistance with other areas of mathematics. Highlight these strengths and provide targeted support to further develop their confidence and skills.

Interview and **Show Me:** After the activity, bring the intervention group together for a discussion. Suggested prompts:

- Show two different ways the students represented the triangle. Ask, "What is the same about the ways they represented the triangle? What is different?"

- What patterns or relationships do you see among the shapes?

- Ask them if they can see some equal trades such as 3G is the same as 1R [3G = 1R].

Bridging Prompt (Prompt for Classroom Teacher to Use During New Lesson Content):
• Show the class several of the photos of student work from the intervention class or an image(s) from Figure 5.33.2 and project it on a whiteboard or recreate one on the document camera with pattern blocks. Ask, "What do you notice about the different combinations of shapes that fill the space? Can you suggest another combination?"
• "Can you trade a trapezoid for any of the other shapes? Which one(s)? Which shapes are most useful in making trades? Why?"
• Look for opportunities to have students who participated in the intervention group share about the shapes they made.

Variation 2

Learning Target: Students will use the same shapes to construct a new shape.

Variation Directions:

Organize the students into pairs for this activity. Provide each pair with a set of pattern blocks and ask them to create a unique shape using the blocks. Once their shape is complete, challenge the pairs to discover as many different ways as possible to recreate the same shape using various combinations of the blocks. Students can either trace their design or put other pattern block combinations right on top of it.

Example: One pair creates two attached hexagons using four trapezoids (red blocks). The pair identifies other ways to recreate the same hexagons:

- Use 12 triangles (green blocks).
- Use 6 rhombuses (blue blocks).
- Use 1 trapezoid and 9 triangles.
- Use 4 rhombuses and 4 triangles.

Engage in Math Discourse (Make the Mathematics Visible):

Observation: As students work, encourage them to notice important relationships and push on early and partial understandings.

- What shapes did you use to create your design?
- Are there any smaller shapes within the larger shape you built?
- Can you explain how the pieces fit together to form the same shape?
- Are there other ways to use fewer or more blocks to create the same shape?

Show Me: After the activity, bring the intervention group together for a discussion. Suggested prompts:

- What patterns or relationships did you notice while filling the triangle?
- Did certain shapes fit together better than others? Why do you think that is?
- Why do you think different combinations of shapes can still create the same triangle?

Variation 3

Learning Target: Students will use a set of seven shapes to construct a new shape.

Variation Directions:

Choose excerpts from a book such as *Grandfather Tang's Story* or just tell the story of a tilemaker who made a tile for the emperor that broke. The kind emperor was happy to receive the puzzle we now call a tangram. Show the class a tangram and say, "Each tangram puzzle has seven shapes. Use what you know about putting smaller shapes together to make new shapes. You can make a square, rectangle, or triangle. You don't need to use all the tangram pieces."

Engage in Math Discourse (Make the Mathematics Visible):

Observation: As students work, encourage them to notice important relationships and push on early and partial understandings.

- What shapes did you use to create your design?
- Can you find a way to make two squares with one larger than the other?

- Can you make a triangle using two shapes? Three shapes?
- Show how pieces fit together to form a trapezoid.

Show Me: After the activity, bring the intervention group together for a discussion. Suggested prompts:

- What patterns or relationships did you notice while trying to make the squares or the triangles?
- Were there any pieces that you did not use as often as others? Which pieces?
- Did certain shapes fit together better than others? Why do you think that is?
- Why do you think different combinations of shapes can still create the same triangle?

	***Bridging Prompt (Prompt for Classroom Teacher to Use During New Lesson Content):*** • Show the class a design created by students who participated in the intervention group. Ask the class to use tangram shapes to recreate the design. • Ask students to describe how they made their shapes using the correct shape names.

TASK 34

What's My Rule?

General Objective:
Students will sort and classify two-dimensional geometric figures based on their attributes.

This activity might be used with:

Students who can (prerequisite knowledge):	Students who are Primed to (getting ready to learn):
• identify some attributes of shapes (e.g., regular, number of angles, number of sides, etc.) • count sides and angles	• recognize parallel and perpendicular lines • recognize acute, right, and obtuse angles • relate two-dimensional figures hierarchically

Materials:

- Assorted sized circles to use for sorting (think life-sized Venn diagrams)
    - yarn tied off at different lengths
    - hula hoop
    - belts

    etc.

- Polygons Sorting Set
    - PRINTABLE: Shape Cards
    - PRINTABLE: Note Page (consider using with sheet protectors and dry-erase markers)
    - PRINTABLE: Group Labels

- Attributes Sorting Set
    - PRINTABLE: Shape Cards
    - PRINTABLE: Note Page (consider using with sheet protectors and dry-erase markers)
    - PRINTABLE: Group Labels

- Triangles Sorting Set
  - PRINTABLE: Shape Cards
  - PRINTABLE: Note Page (consider using with sheet protectors and dry-erase markers)
  - PRINTABLE: Group Labels
- PRINTABLE: Group Number Cards

 Printables for this task are available for download at **https://companion.corwin.com/courses/ProactiveMathIntervention**.

Recommended Children's Literature:

- *The Greedy Triangle* – Burns, 1994
- *Wing on a Flea* – Emberley, 2015
- *A Trapezoid Is Not a Dinosaur* – Morris, 2019
- *Which One Doesn't Belong? Playing With Shapes* – Danielson, 2019

Task Overview:

This activity provides students an opportunity for mathematical argumentation and discourse. They will sort geometric figures into sets based on shared attributes and work to determine which attributes their partners used to generate sets.

Select (or create) a variation that focuses on the prior knowledge you are working to shore up for your student group.

Variation 1

Learning Target: Students will sort geometric figures into two groups based on their attributes.

Variation Directions:

Begin this variation by organizing students into partner pairs or groups of four. Each group will need one set of shape cards and two large grouping circles (e.g., yarn, hula hoop, etc.). Each team will need a corresponding note page. The group should set up two grouping circles and use the group labels (or sticky notes) to number the circles Group 1 and Group 2. *Note: Allow students to set the circles up independently. It is okay if they do not overlap at the start of the game. This will provide an opportunity for reasoning and sense-making during the activity.*

Playing the game:

1. Partner A (if a group of four, two students act as Team A) begins by ***secretly*** identifying one attribute or figure name on the note page to use for a sorting rule for each of the grouping circles. Partner A marks these on their own attribute note page (Partner B cannot see this). *The goal for Partner B will be to figure out the sorting rules.*

2. Next, Partner B chooses shapes from the shape set (one at a time) for Partner A to place inside or outside the grouping circle).

- *This step in the game is an opportunity to support reasoning around shared attributes of shapes. Watch for students who get "stuck" when they have a shape that needs to fit into **both** circles. What can we do? Leverage the moment to facilitate a whole-group discussion about overlapping the circles to create an area in which geometric figures with shared attributes can be placed.*

3. As Partner A places each shape in the grouping circle, Partner B attempts to figure out the rules for each of the groups, using the note page to keep track of their thinking.

4. The game ends when Partner B accurately identifies the rules for both grouping circles.

   Repeat with Partner B choosing the sorting rules.

Figure 5.34.1 What's My Rule? Variation 1 example

A group of four is playing in teams of two. The teams set up two grouping circles and label them Group 1 and Group 2. They will play with sorting set C (shape attributes).

Team A decides to use the following sorting rules:

Group 1: *figures with at least one right angle*

Group 2: *figures with no sides congruent (the same lengths)*

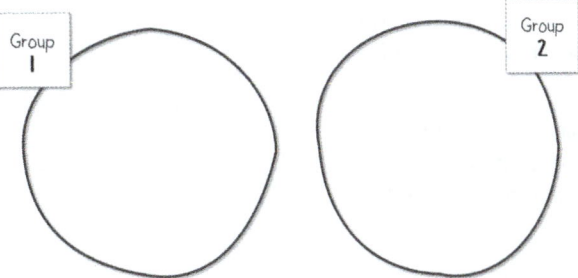

Possible Group Rule	Group 1	Group 2
all sides are congruent		
no sides are congruent		✓
all angles are congruent		
no angles are congruent		
at least one right angle	✓	
four right angles		
at least one obtuse angle		
all angles are acute		
one pair of parallel sides		
two pairs of parallel sides		

What's My Rule? — Team A — Note Page: Attributes

(Continued)

(Continued)

Team B chooses figure Q.	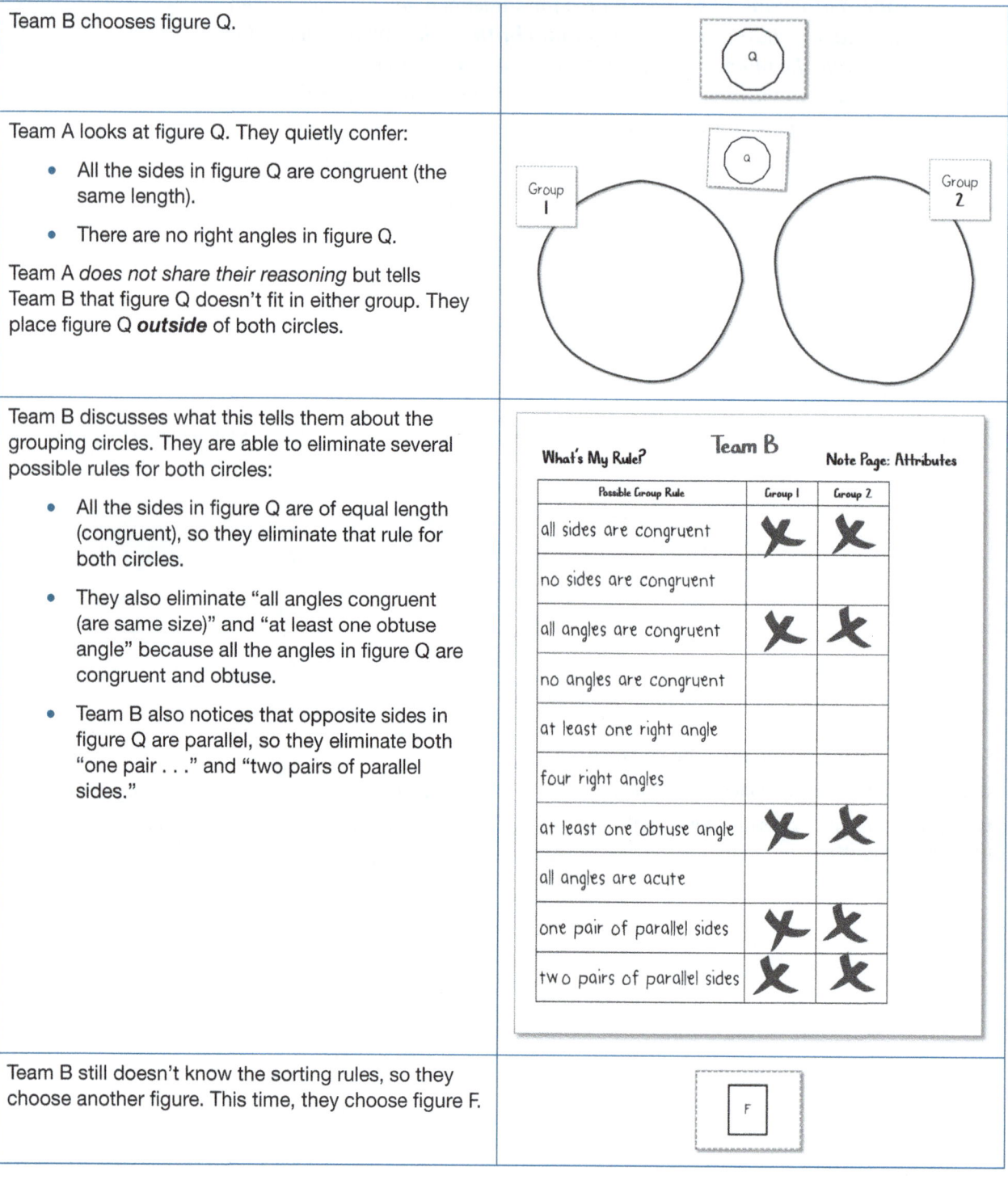
Team A looks at figure Q. They quietly confer: • All the sides in figure Q are congruent (the same length). • There are no right angles in figure Q. Team A *does not share their reasoning* but tells Team B that figure Q doesn't fit in either group. They place figure Q **outside** of both circles.	
Team B discusses what this tells them about the grouping circles. They are able to eliminate several possible rules for both circles: • All the sides in figure Q are of equal length (congruent), so they eliminate that rule for both circles. • They also eliminate "all angles congruent (are same size)" and "at least one obtuse angle" because all the angles in figure Q are congruent and obtuse. • Team B also notices that opposite sides in figure Q are parallel, so they eliminate both "one pair . . ." and "two pairs of parallel sides."	
Team B still doesn't know the sorting rules, so they choose another figure. This time, they choose figure F.	

Team A looks at figure F. They quietly confer together: • All the angles in figure F are right angles, so it fits in Group 1. • The opposite sides in figure F are of equal length, so it does not fit in Group 2. Team A *does not share their reasoning* but tells Team B that figure F fits in group 1. They place figure F *inside* the Group 1 circle in the diagram.	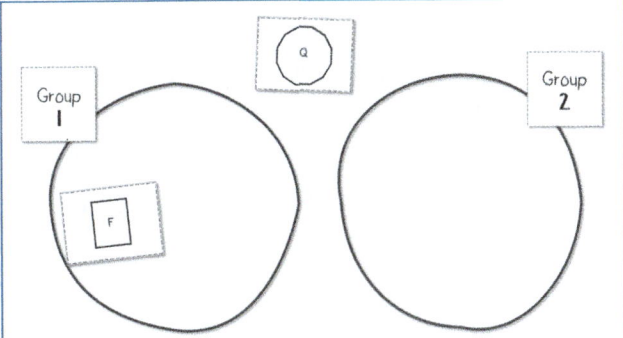
Team B discusses what this tells them about the grouping circles. They are able to eliminate two more possible rules for Group 1: • Opposite sides in figure F are of equal length, so they eliminate "no sides are congruent." • They also eliminate "all angles are acute" because figure F has all right angles. They are able to eliminate all but one possible rule for Group 2: • All the angles in figure F are 90°, so they eliminate "no angles are congruent" and "all angles are acute" as possible rules for Group 2. • Because Team A did not place figure F inside the Group 2 circle (even though it has four 90° angles), Team B also eliminates "at least one right angle" and "four right angles" for Group 2. That leaves only one possible option for Group 2: "no sides are congruent."	**What's My Rule?** **Team B** Note Page: Attributes | Possible Group Rule	Group 1	Group 2
---	---	---
all sides are congruent	✗	✗
no sides are congruent	✗	✓
all angles are congruent	✗	✗
no angles are congruent	✗	✗
at least one right angle		✗
four right angles		✗
at least one obtuse angle	✗	✗
all angles are acute	✗	✗
one pair of parallel sides	✗	✗
two pairs of parallel sides	✗	✗	
Team B looks for a shape that will help them rule out one of the last possible rules for Group 1. They choose figure DD because it has exactly one right angle.	(figure DD shown)
Team A looks at figure DD and tells Team B that it fits in group 1. They place figure DD *inside* the Group 1 circle.	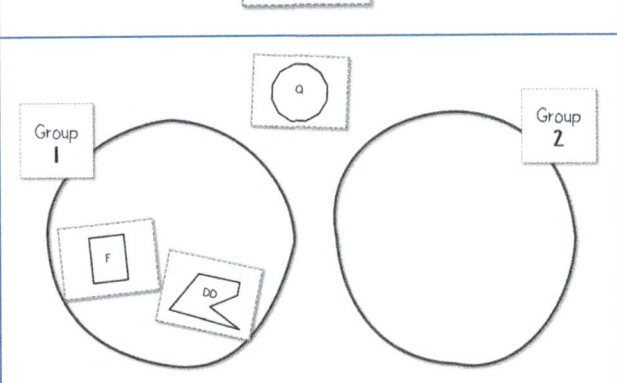

(Continued)

(Continued)

Team B is able to eliminate "four right angles" for Group 1. They tell Team A they think the rules are: • **Group 1:** *figures with at least one right angle* • **Group 2:** *figures with no sides congruent (of equal length)*	**What's My Rule?** **Team B** Note Page: Attributes | Possible Group Rule | Group 1 | Group 2 | | all sides are congruent | X | X | | no sides are congruent | X | ✓ | | all angles are congruent | X | X | | no angles are congruent | X | X | | at least one right angle | ✓ | X | | four right angles | X | X | | at least one obtuse angle | X | X | | all angles are acute | X | X | | one pair of parallel sides | X | X | | two pairs of parallel sides | X | X |

Team A confirms that Team B correctly identified the rules. They switch roles and play again.

Engage in Math Discourse (Make the Mathematics Visible):

Observation: As students work, encourage them to notice important relationships and push on early and partial understandings.

- Listen for mathematical vocabulary as students are describing the relationships they see in the sorts. Ask students to point to and show where and how they see these attributes in the geometric figures.

- Pay attention to moments when students may disagree about whether a figure belongs in or out of a circle during sorting. This is a great opportunity for mathematical argumentation. Encourage students to demonstrate their thinking using the figures.

- Notice that we use the word *congruent* interchangeably with language about the same length or the same-sized angle. The word should not be on any assessment because it isn't a grade-level standard, but we like to preview language (when it comes up in context) so students have time to learn it.

- Watch for students who may get stuck when they have a shape that can fit into both grouping circles. When this happens, stop everyone and bring the whole intervention group together for a quick discussion about how to solve this problem. Move to a discussion (see Figure 5.34.2) about overlapping the circles to create an area in which figures with shared attributes can be placed. For example, you might start the conversation like this:

Figure 5.34.2 Sample group discussion

Teacher:	Everyone pause for a moment and come take a look at this group's grouping circles. They are having an interesting discussion!

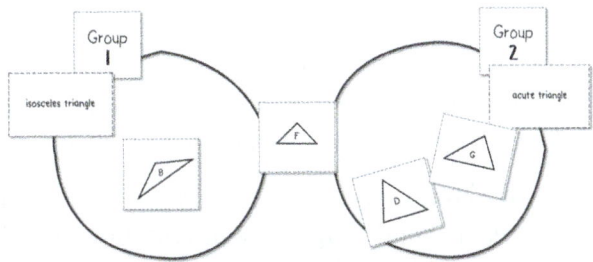

After students gather, the teacher asks the group to explain their discussion about where they should place figure F.

Student A:	We think triangle F should go in both circles.
Student B:	Yeah, we picked Isosceles for the Group 1 rule, and triangle F has two sides that are the same length.
Student A:	But it also has only acute angles, so it also fits in Group 2! We need another F card.
Teacher:	Does anyone have any ideas about what this group could do? They only have one F card. Is there a way we could arrange the grouping circles so that the F card can be in both groups?
Student C:	Can we pull the strings over?
Teacher:	Show us what you are thinking.

Student C rearranges the strings for both circles so they overlap. They put card F in the intersection of both grouping circles.

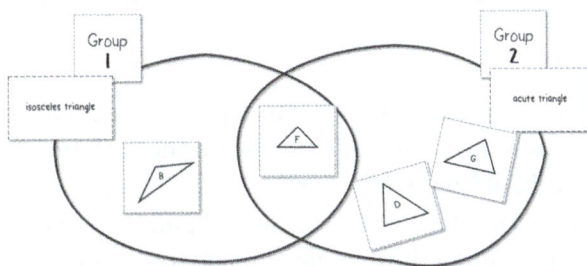

At this point, the teacher helps the group attach meaning to each part of this new arrangement of the circle. The teacher tells students to go back to their own sorting and think about whether their grouping circles should overlap.

Interview and **Show Me:** After sorts are completed, bring the intervention group together for a discussion. Suggested prompts:

- If necessary, clarify how a two-circle Venn diagram works using pattern blocks and simple attributes such as "has curves" and "is red."
- How did you figure out what your partner had picked for a sorting rule?
- Was there a specific figure your partner put into or outside of one of the grouping circles that gave it away to you? Why?
- Are there any geometric figures that you and your partner talked a lot about? Maybe they felt like they would fit in a different location in the Venn diagram? What made you decide where it should go?
- If your group overlapped their circles, what did you notice about the figures that ended up in the intersection of both circles of the Venn diagram? What did you wonder?

Bridging Prompt (Prompt for Classroom Teacher to Use During New Lesson Content):

- Play the game with the whole class. Have students who participated in the intervention group be in charge of picking the sorting rules (and keeping them secret from the rest of the class). They will call on classmates to pick shapes then place the shape in the correct location on the Venn diagram. The class will try to guess their rules.

Variation 2

Learning Target: Students will sort geometric figures in a Venn diagram based on the presence or absence of three different attributes.

Variation Directions:

This variation follows the same rules as Variation 1, except students will create *three* sorting groups this time. Begin the variation by having a whole-group discussion about how a three-circle Venn diagram works. Elicit students' ideas about how two grouping circles overlapped and what the intersection of the two circles represents. Add a third circle and ask the group to identify what each of the new intersections represents in the Venn diagram.

Engage in Math Discourse (Make the Mathematics Visible):

Observation: As students work, encourage them to notice important relationships and push on early and partial understandings.

- Watch and listen for strategic thinking as students are selecting figures to be placed in the Venn diagram. As you notice students purposefully selecting specific shapes in order to narrow down possible rules, celebrate and lift up the strengths in their reasoning.

- Listen for mathematical vocabulary as students are describing the relationships they see in the sorts. Ask students to point to and show where and how they see these attributes in the figures.

- Pay attention to moments when students may disagree about whether a figure belongs in or out of a grouping circle during sorting. This is a great opportunity for mathematical argumentation. Encourage students to demonstrate their thinking using the figures.

Interview and **Show Me:** After sorts are completed, bring the intervention group together for a discussion. Suggested prompts:

- If necessary, clarify how a three-circle Venn diagram works using pattern blocks and simple attributes such as "has curves" and "is red."

- How did you figure out what your partner had picked for a sorting rule?

- Was there a specific figure your partner put into or outside of one of the grouping circles that gave it away to you? Why?

- Are there any figures that you and your partner talked a lot about? Maybe they felt like they would fit in a different location in the Venn diagram? What made you decide?

- If your group overlapped their circles, what did you notice about the figures that ended up in the intersection of both circles of the Venn diagram?

Bridging Prompt (Prompt for Classroom Teacher to Use During New Lesson Content):

- Play the game with the whole class. Have students who participated in the intervention group be in charge of picking the sorting rules (and keeping them secret from the rest of the class). They will call on classmates to pick shapes then place the shape in the correct location on the Venn diagram. The class will try to guess their rules.

TASK 35

Counting Coins

General Objective:
Students will count money amounts using quarters, pennies, nickels, and dimes.

This activity might be used with:

Students who are able to (prerequisite knowledge):	Students who are Primed to (getting ready to learn):
• identify coins (penny, nickel, dime, and quarter) and know the value of each coin	• decompose the values of quarters, nickels, and dimes using pennies, nickels, and dimes • count to 25 cents and $1.00 using different coin combinations (pennies, nickels, dimes, and quarters)

Materials:
- Coins (real or plastic)
- Dot cube
- PRINTABLE: Coin Spinners
- PRINTABLE: Counting Coins Recording Sheet
- PRINTABLE: Make 25 Cents Game Board
- PRINTABLE: Make $1.00 Game Board
- PRINTABLE: Hundreds Chart

 Printables for this task are available for download at **https://companion.corwin.com/courses/ ProactiveMathIntervention**.

Recommended Children's Literature:
- *Peter the Pickle and His Nickels* – Anger, 2023
- *Deena's Lucky Penny* – deRubertus, 1999
- *Tim's Talking Coin* – Endeavor, 2023
- *The Penny Pot* – Murphy, 1998

Task Overview:

Second-grade students are expected to solve word problems involving money. In this task, students review coins and their values by playing the Counting Coins game to count money amounts of like coins. Students then transition to counting money amounts up to 25 cents and $1.00.

Consider introducing the task by reading a book to students. This will provide an opportunity to review coin names and values as you discuss coins featured in the story.

Select (or create) a variation that focuses on the prior knowledge you are working to shore up for your student group.

Variation 1

Learning Target: Students will count collections of quarters, nickels, and dimes.

Variation Directions:

Begin this variation with students together in a whole group. Show students a dime and ask students to identify the coin and its value. Tell the students that they will use what they know about skip counting to identify and count the value of a collection of coins. Review the value of one dime and then the value of three dimes. Ask students to show how they could skip count by tens to find the value of three dimes. Skip count as you count the dimes: "10 cents, 20 cents, 30 cents—three dimes are equal to 30 cents." Ask students how they would skip count if they were counting by nickels and quarters.

Show students the Counting Coins game materials. Model the procedure for the game by playing one round with a student in front of the whole group. Ask students which coin they should start with as they count the value of the coins (always encourage starting with the coin with the largest value). *NOTE: There are three spinners to offer opportunities to differentiate based on students' Priming needs. Spinner A has a penny and a dime, Spinner B has a penny, dime, and nickel, and Spinner C has a penny, dime, nickel, and quarter. Start students out with Spinners A and B and introduce Spinner C for students who need a little challenge and can skip count by 25.*

Students can use a paper clip spinner. *(It may be helpful to unfold the paper clip or tape a paper arrow to the end of the paper clip, as shown in Figure 5.35.1.)*

Figure 5.35.1 Paper clip spinner

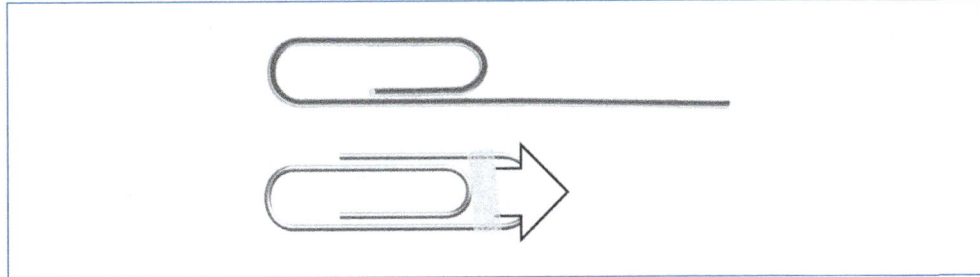

Source: iStock.com/RUSSELLTATEdotCOM

Organize students into partners and allow them to play the game. Each student should spin the spinner and count coins at least three times. Encourage student partners to help each other determine the total value of the coins collected.

Directions for Game:

1. Spin the spinner to select a coin. This is the coin you will collect and count.
2. Roll the dot cube. The number the cube lands on tells how many of that coin you will collect.
3. Collect the correct number of coins and record the total amount of money.

Figure 5.35.2 Counting Coins Variation 1 example

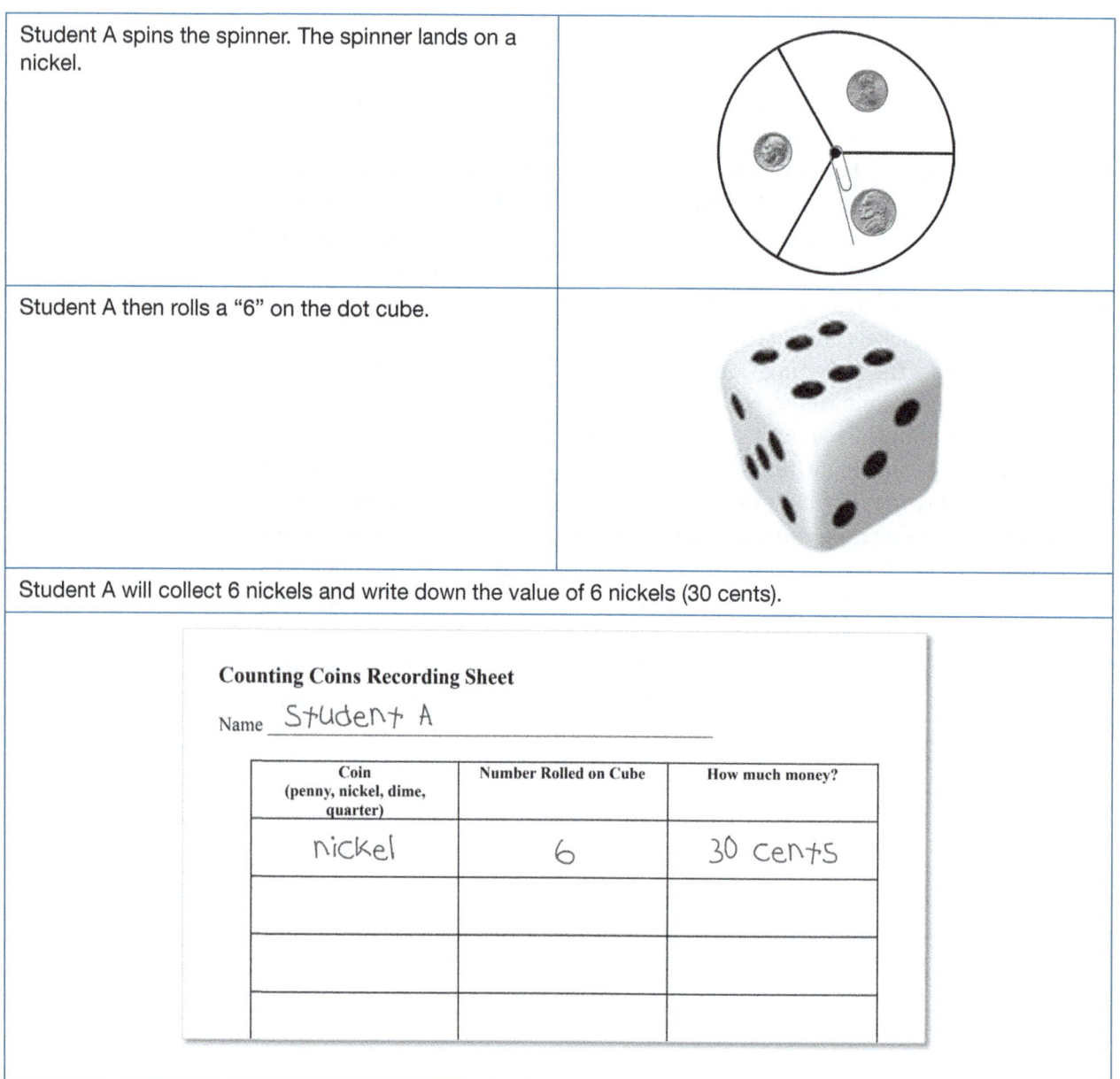

Source: US coins by istock.com/TokenPhoto; dice icon by istock.com/seanmartini

Engage in Math Discourse (Make the Mathematics Visible):

Observation: As students work, encourage them to notice important relationships and push on early and partial understandings.

- How did you determine the total amount of money for each spin?
- How did you use skip counting to find your answer?
- How does the value of the coin affect the total amount of money?
- Ask student pairs to use a sentence frame to compare the total amount for a spin. Use the prompt, "_____'s amount is greater than/less than _____'s amount of money."

Interview and **Show Me:** After the activity, bring the intervention group together for a discussion. Suggested prompts:

- Describe your actions for a spin (e.g., "My spinner landed on the nickel, and I rolled a 4 with the dot cube. Four nickels equal 20 cents.")
- How do you know you found the total amount of money for each spin?
- How do you use skip counting to find the total amount?

Bridging Prompt (Prompt for Classroom Teacher to Use During New Lesson Content):

- Have a student who participated in the intervention group show the class a coin spin from the spinner and a roll from the number cube describe how they would find the product of the two numbers.
- Then have the student who participated in the intervention group spin the spinner again and roll the number cube again. Ask the class if the total product with the new coin and roll will be greater than or less than the first amount.

Variation 2

Learning Target: Students will count to 25 cents and $1.00 using combinations of different coins (pennies, nickels, dimes, and quarters).

Variation Directions:

Bring together all the students to learn how to play the game. Model the procedure for playing the game by playing one round with a student. Or have two students play the game and provide support by asking questions. The goal of the game is to get a specific amount such as 25 cents (or a dollar).

Directions:

The object of the game is to make 25 cents. In this game, you will roll the dot cube to determine the number of pennies to place on the game board.

- Once you collect five pennies, you can trade in the pennies for one nickel.
- Once you collect five nickels, you can trade in the nickels for a quarter to reach the goal!
    1. Players take turns in this game. Roll the dot cube to determine who goes first. Players can take turns in order from highest to lowest number or lowest to highest number.
    2. Roll the dot cube to determine the number of pennies to place on your game board. For example, if the cube lands on four dots, you will place four pennies on your game board.
    3. Continue playing and trading pennies for nickels until the amount equals 25 cents.

Arrange the students in pairs and distribute the game boards, directions, and coins to the students.

Provide access to spinner A as an additional tool students can use to select the coins to count to 25 cents instead of the number cube. Using the semiconcrete representation provides an additional choice and scaffold for students. Consider allowing student choice as to which spinner they would like to use.

Engage in Math Discourse (Make the Mathematics Visible):

Observation and **Show Me:** As students work, encourage them to notice important relationships and push on early and partial understandings.

- What is your strategy for reaching 25 cents?
- What is the total value of your coins on the board?
- How many nickels do you have? What is their total value?
- How are you keeping track of the trades you've made?

Interview: After the activity, bring the intervention group together for a discussion. Suggested prompts:

- How did you use skip counting to find the total amount of money for each spin?
- What did you learn about counting coins from this activity?
- How does the value of the coin affect the total amount of money?
- When playing Counting Coins, was there a coin you wanted to get on the spinner to result in the highest possible amount of money? Why?

	***Bridging Prompt (Prompt for Classroom Teacher to Use During New Lesson Content):*** • Ask the class: "If you wanted to make 25 cents, what coins could you use? Can you think of all the possible combinations? Use your knowledge of coin values and skip counting to show me how you know your answers are correct." • Provide opportunities to have students who participated in the intervention group to share combinations and explain how they arrived at the combinations.

Variation 3

Learning Target: Students will use a hundreds chart to model different ways to find a combination to add up to 30 cents.

Variation Directions:

Bring together students in a whole group. Display a hundreds chart and explain that each square represents one cent. Have students locate key amounts such as 5, 10, 25, and 30 to help them make connections to common coin values. Place a counter or object on the "1" square of the hundreds chart. Explain that Jayla has 30 cents, and you will use the chart to represent how she could reach 30 cents using coins. Ask students how we should count forward on the chart based on coin values (e.g., move by skip counts of 10 for dimes or by 5 for nickels).

For example:

▸ A dime moves you forward 10 spaces.

▸ A nickel moves you forward 5 spaces.

Divide the students into pairs and distribute a hundreds chart and coins. Ask students to find as many ways as possible to move from 0 to 30 on the chart using movements that match the values of coins (e.g., 10 + 10 + 10 = 30 for dimes or 10 + 5 + 5 + 5 + 5 = 30 for a combination of a dime and nickels). Have them record their equations on paper. Have student pairs transfer one of their equations to a group chart. Discuss the various results and choices.

Engage in Math Discourse (Make the Mathematics Visible):

Observation and **Show Me:** As students work, encourage them to notice important relationships and push on early and partial understandings.

▸ What is your strategy for using the coins to show 30 cents?

▸ Use the sentence frame to share how you found 30 cents. (e.g., "I used three dimes because 10 + 10 + 10 equals 30." or "I used six nickels because 5 + 5 + 5 + 5 + 5 + 5 equals 30.")

 **Interview** and **Show Me:** After the activity, bring the intervention group together for a discussion. Suggested prompts:

- What did you learn about counting coins from this activity?
- What are all the ways we made 30 cents?
- Which way is your favorite way to make 30 cents? Why?
- How would your strategies change if we changed the total amount?

	**Bridging Prompt (Prompt for Classroom Teacher to Use During New Lesson Content):** • What if Jayla had 50 cents in her pocket? What combinations of coins could she have? Can you use the hundreds chart to figure out those combinations? Have students who participated in the intervention group come up and show on a hundreds chart how coin amounts can be counted. Let them explain a strategy for coin selection and writing the equation.

CHAPTER 5. Intervention Tasks

Smart About Money

TASK 36

General Objective:

Solve money problems involving dollar bills, quarters, dimes, nickels, and pennies, using $ and ¢ symbols appropriately.

This activity might be used with:

Students who are able to (prerequisite knowledge):	Students who are Primed to (getting ready to learn):
• decompose the values of quarters, nickels, and dimes using pennies, nickels, and dimes • count to $1.00 using a combination of coins (pennies, nickels, dimes, and quarters)	• solve money word problems involving a combination of dollar bills, quarters, dimes, nickels, and pennies • find exact change

Materials:

- Copy of the poem "Smart" from *Where the Sidewalk Ends* by Shel Silverstein, 1993
- $1.00 bills and assorted coins (pennies, nickels, dimes, and quarters) for each student
- Baggie (wallet) of $2.00 in coins (pennies, nickels, dimes, and quarters) for each student
- Extra coins for making exchanges
- Blank paper
- Chart paper
- PRINTABLE: SMART Slides (to be used when reading the poem for visuals)
- PRINTABLE: Is It a Fair Trade? Problem Cards
- PRINTABLE: Get Smart Recording Sheet
- PRINTABLE: Let's Go Shopping Activity Cards
- PRINTABLE: Let's Go Shopping Recording Sheet
- PRINTABLE: Counting Tower Challenge Recording Sheet

 Printables for this task are available for download at **https://companion.corwin.com/courses/ProactiveMathIntervention**.

Recommended Children's Literature:

- *My Rows and Piles of Coins* – Mollel, 1999
- *The Penny Pot* – Murphy, 1998
- *Where the Sidewalk Ends* – Shel Silverstein, 1993

Task Overview:

Students in second grade and beyond are expected to solve word problems involving money. In the task variations, students listen to a poem and use what they know about money to act out the problem and then solve a similar "trading" word problem. In another variation, students use what they know about counting money by "shopping" for items. Students are expected to demonstrate their ability to count varying amounts of money by counting coins and using money amounts to solve problems.

Select (or create) a variation that focuses on the prior knowledge you are working to shore up for your student group.

Variation 1

Learning Target: Students will solve money problems involving a combination of dollar bills, quarters, dimes, nickels, and pennies.

Variation Directions:

Arrange the students in a whole group and read the poem "Smart" by Shel Silverstein to the whole group. Encourage students to listen carefully as you read because they will be acting out the poem alongside the narration (this popular poem is easily accessible on the web). If the poem is not available, proceed directly to the word problem activity. After reading the poem once, ask students to share what they noticed and wondered, recording their ideas on a whole-group chart. Then, distribute sets of coins and dollar bills to the students. Reread the poem and show the slides that demonstrate each trade, this time allowing students to role-play with a partner, using the money to represent the amounts and trades described in the poem. Encourage them to show the money remaining after each trade as the story unfolds. Discuss the value of the coins after each trade. Ask the students what are some fair trades he could have made. Record some of these suggestions.

Next, divide the students into pairs and give them the Is It a Fair Trade? Problem Cards. Have the students work together to use coins to model the trade and check the equivalency. Students should explain their reasoning as they work together to decide if the trade is fair.

Lila's Trade

Lila has 1 dime in her hand. Her friend wants to trade her 2 nickels for the dime. Should Lila agree to the trade? Is it fair? Why or why not?

Jared's Trade

Jared has 1 quarter, and his friend offers him 20 pennies for the quarter. Is this a fair trade? Why or why not?

Sofia's Trade

Sofia has 1 dime and 1 nickel. Her friend offers to trade her 3 nickels for both coins. Should Sofia make the trade? Is it fair? Why or why not?

Kadeeja's Trade

Kadeeja has 5 nickels, and his friend offers him 1 quarter in exchange for the 5 nickels. Should Kadeeja make the trade? Is it fair? Why or why not?

Emma's Trade

Emma has 1 dime and 3 nickels. Her friend wants to trade her 1 quarter for all her coins. Should Emma make the trade? Is this a fair trade? Why or why not?

Alvaro's Trade

Alvaro has 1 nickel, and his friend offers him 6 pennies in exchange. Should Alvaro take the trade? Is it fair? Why or why not?

If some students would like a challenge, use the Get Smart sheet to have them list fair trades for the dollar bill the boy in the poem's dad gave him.

Engage in Math Discourse (Make the Mathematics Visible):

Observation: As students work, encourage them to notice important relationships and push on early and partial understandings.

- How much is the total value of the coins being traded?
- How does the value of the coins on each side of the trade compare?
- Can you use skip counting to figure out the value of the coins?
- What does it mean for a trade to be fair?

Interview and **Show Me:** After the activity, bring the intervention group together for a discussion. Suggested prompts:

- How do you know if the trade is fair or not?
- What would make an unfair trade fair?
- Explain your reasoning using prompts such as, *"I think this trade is fair/ not fair because . . ."* and *"The total value of the coins is . . ."*

Bridging Prompt (Prompt for Classroom Teacher to Use During New Lesson Content):

- Organize the class into pairs. Provide each pair of students with a bag of coins (pennies, nickels, dimes, and quarters). Write a target value on the board (e.g., 25 cents, 50 cents, or 75 cents).
- Ask students who participated in the intervention group to start with a specific combination of coins (e.g., 2 dimes and a nickel for 25 cents). Challenge them to trade coins with their partner to create a different combination that equals the same total.
  - Can you trade 1 dime for 2 nickels?
  - How about 5 pennies for a nickel?
    Are these trades fair? Why or why not?

Variation 2

Learning Target: Students will exchange dollar bills for equivalent sets of coins and determine the total value of their money using strategies such as skip counting and starting with the coin of greatest value and counting on.

Variation Directions:

Let's Go Shopping activity. Model the procedure for the game by playing one round with a student.

Directions:

Students play this game in partners or groups of no more than three students. Students will take turns selecting items to purchase on the shopping list until they spend the contents of their "wallet" or have no more than 8 cents left over.

1. Each player receives a baggie (wallet) containing a combination of coins equal to $2.00.

2. Taking turns, each player selects an item from the shopping list and "pays" for the item. Students should circle each item on their list to indicate that the item has been purchased.

3. Students spending all their money with no more than 8 cents left over meet the goal!

Organize students into partners or groups of three and allow students to play the game. After the students play the game, bring them back for a whole-group discussion.

Engage in Math Discourse (Make the Mathematics Visible):

Observation: As students work, encourage them to notice important relationships and push on early and partial understandings.

- Which coin is the best coin to start counting with?
- How can you use skip counting to help you?
- Does it make a difference which coins by value you start with to find the total?

Interview: After the activity, bring the intervention group together for a discussion. Suggested prompts:

- Did you and your partner use the same strategy to count the money? How were they similar or different?
- Which strategy do you think was the best? Why?

> ***Bridging Prompt (Prompt for Classroom Teacher to Use During New Lesson Content):***
>
> - Marisha wants to buy a snack at Field Day. She has $1.00, 1 quarter, 5 dimes, 4 nickels, and 40 pennies. Which snacks can she purchase? Can she purchase more than one of something?
>
> **Figure 5.36.1** Snack Prices
>
> | Snack | Price |
> |---|---|
> | Hot Dog | $1.25 |
> | Ice Cream Cone | $0.65 |
> | Soft Drink | $0.75 |
> | Popcorn | $0.50 |
>
> - Pose the task above to the class to complete individually. After completion, use a routine such as "Stand Up, Hand Up, Pair Up" to allow students to find a partner to share and discuss their responses. Encourage students to look for similarities and differences in their responses. Have students who participated in the intervention group share an answer with the class. Use chart paper to identify the different combinations of snacks Marisha could purchase to highlight the varying responses.

Variation 3

Learning Target: Students will explore the value of coins by creating towers while determining the total value of their coin towers.

Variation Directions:

Provide each pair or small group of students with a collection of coins (pennies, nickels, dimes, quarters) to play the Coin Tower Challenge. Distribute recording sheets for students to calculate and record the value of their towers. Introduce the Coin Tower Challenge rules:

1. Students will build a coin tower that must stand upright for at least 10 seconds to qualify.

2. For each round, groups can choose coins of any denomination to construct their tower.

3. Students must calculate the total value of the coins used in their tower and record it.

After students have played a few rounds of the Coin Tower Challenge, changing the total tower amount each time (e.g., 50¢, 75¢, and $1.00), you can add other challenge criteria such as:

- Use at least three different types of coins in the tower.
- Create a tower using the fewest or the most coins for a specific value.

Figure 5.36.2 Recording Sheet example

Coin Type	Number of Coins	Value Per Coin	Total Value
Pennies	10	$0.01 or 1 cent	$0.10 or 10 cents
Nickels	5	$0.05 or 5 cents	$0.25 or 25 cents
Total			$0.35 or 35 cents

Engage in Math Discourse (Make the Mathematics Visible):

Observation: As students work, encourage them to notice important relationships and push on early and partial understandings.

- How much is the total value of your tower?
- Does the tallest tower always have the greatest value? Why or why not?
- How did you decide which coins to use for your tower?

Interview and **Show Me:** After the activity, bring the intervention group together for a discussion. Suggested prompts:

- What did you notice about the height of your tower and the coins you used?
- What challenges did you face when building your tower?
- What's more important for this challenge: height or total value? Why?
- Do you think a taller tower will always have a greater value? Why or why not?
- What would happen if you had to use only one type of coin? How would that change your strategy?

> **Bridging Prompt (Prompt for Classroom Teacher to Use During New Lesson Content):**
>
> - Have students who participated in the intervention group show the class how a coin tower is made and how its value is calculated. Then, ask the full class to imagine they are building a coin tower for a big competition, but there's a catch: the tower needs to have the highest value and still be stable enough to stand tall. What combinations of coins could you use to make your tower both valuable and strong? Think about how the size, shape, and value of the coins affect your choices. Let's see how you can combine math and engineering to create the best tower possible!

CHAPTER 5. Intervention Tasks

TASK 37

Measure Up!

General Objective:
Measure lengths of objects indirectly and by iterating length units.

This activity might be used with:

Students who are able to (prerequisite knowledge):	Students who are Primed to (getting ready to learn):
• measure objects using a variety of measurement units (standard and nonstandard)	• describe the length of an object as a whole number of length units by laying multiple copies of the length unit end to end • understand that the length measurement of an object is the number of same-size length units that span it from beginning to end with no gaps or overlaps

Materials:

- Blank paper
- Chart paper
- PRINTABLE: Measuring Worm Units
- PRINTABLE: Cuisenaire Rods

  or

  Cuisenaire rod manipulatives (3 to 4 sets per group of three students)
- PRINTABLE: Measuring Cuisenaire Rods Recording Sheet

 Printables for this task are available for download at **https://companion.corwin.com/courses/ProactiveMathIntervention**.

Recommended Children's Literature:

- *Inch by Inch* – Lionni, 2018
- *Little One Inch and Other Japanese Children's Favorite Stories* – Sakade, 2018
- *Issun Boshi: The One-Inch Boy* – Icinori, 2014

Task Overview:

In this task, students measure the length of a variety of objects with nonstandard units of varying lengths. Students demonstrate their ability to correctly measure the length of objects by placing measurement units end to end without gaps or overlaps. They use two different units to measure identical objects to determine the relationship between the size of the unit and the number of units needed to measure the length of the object.

Variation 1

Select (or create) a variation that focuses on the prior knowledge you are working to shore up for your student group.

Learning Target: Students will accurately measure the length of objects by iterating one length unit (nonstandard unit) and describe the length of an object as a whole number of length units by laying multiple copies of the length unit end to end.

Variation Directions:

If possible, read excerpts from a children's literature book about measuring length such as *Inch by Inch* by Leo Lionni. Discuss the story details with the students. Discuss the method the worm used to measure objects in the story. The worm used himself as the measurement unit and lined himself up end to end without any gaps or overlaps to get a measurement. *Note: If a copy of the suggested book isn't available, begin with distributing worm units to each student.* Allow students to select an object and instruct students to measure the length of that object using their worms. Make note of whether students line up the measurement unit end to end without any gaps or overlaps as they measure their objects. Make it clear that although the story (if you read it) says "inchworm," this worm is not one inch in length. This is a nonstandard unit, so the answer will be in worms.

Figure 5.37.1 Measuring worm

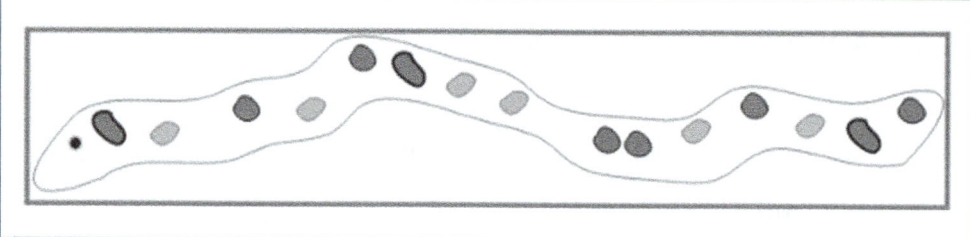

Engage in Math Discourse (Make the Mathematics Visible):

Observation: As students work, encourage them to notice important relationships and push on early and partial understandings.

- What do you notice about how your worm lines up as you measure the object?

- Are there any gaps or overlaps as you measure? Why is it important to avoid them?

- How can you make sure your measurement is accurate?

- What happens if your worm isn't long enough to reach the exact end of the object? How do you handle that? What if it is too long?

Interview: After the activity, bring the intervention group together for a discussion. Suggested prompts:

- What strategies did you use to measure your object accurately?
- Why do we need to use measurement units like the worm? What would happen if everyone used a different unit?
- What did you find challenging about measuring? How did you solve that problem?

	**Bridging Prompt (Prompt for Classroom Teacher to Use During New Lesson Content):** • Show the class an object. Have students who participated in the intervention group come up and demonstrate how they aligned the worm to the object and measure it. Have them talk about things to avoid when measuring (e.g., gaps, etc.).

Variation 2

Learning Target: Students will accurately measure the length of objects by iterating one length unit and describe the length of an object as a whole number of length units by laying multiple copies of the length unit end to end.

Variation Directions:

Organize students into groups of three and distribute Cuisenaire rods to each group. Allow time for students to examine the rods and notice that each rod is a different length. Ask students to think about how to measure accurately, emphasizing the need to line up units end to end without any gaps or overlaps.

Explain the rules for measuring length:

1. Units must be aligned end to end with no gaps or overlaps.
2. Use a number and a unit to describe the length.

Tell the students that they are going to measure the other rods using the red rod as the unit. Begin by having students measure the purple rod. Ask them to work with their group to first estimate and then measure to determine the number of red rods it takes to equal the length of the purple rod. Observe as students estimate and then measure. Allow students to estimate and then measure additional rods (dark green, brown, and orange) using the red rod as the unit. Encourage students to record their results on blank paper. The measurements include:

- Purple rod = 2 red rods
- Dark-green rod = 3 red rods
- Brown rod = 4 red rods
- Orange rod = 5 red rods

Engage in Math Discourse (Make the Mathematics Visible):

Observation: As students work, encourage them to notice important relationships and push on early and partial understandings.

- What do you notice about the relationship between the lengths of the rods?
- How does using the red rod help us measure accurately?
- Why is it important to line up the rods end to end without gaps or overlaps?
- What strategies worked well when measuring with the red rod?
- What patterns did you notice in the lengths of the rods?

Interview and **Show Me:** After the activity, bring the intervention group together for a discussion. Suggested prompts:

- How would your measurements change if we used a different rod as the unit?
- What would you do if the object you're measuring is longer than the rods? How could you solve that problem?
- Predict how many red rods it would take to measure a rod that's twice as long as the orange rod.

Bridging Prompt (Prompt for Classroom Teacher to Use During New Lesson Content):

- Organize the class into small groups. Provide each group with a set of rods. Ask them to estimate and then measure the length of each rod (e.g., purple, green, brown, orange) using the red rod as the unit.
- Encourage students to record their measurements in a table, such as:

Figure 5.37.2 Rod lengths in units of red rods

Rod Color	Length in Red Rods
Purple	2
Dark Green	3
Brown	4
Orange	5

- After measuring a few rods, ask students, "What patterns do you notice in the measurements? How can you predict how many red rods long the blue rod or the black rod will be?" Call on the students who participated in the intervention group to share their thinking.
- How does the length of the red rod compare to the other rods?
- What happens as the rods get longer? How do the numbers in the chart change?

Comparing With Measurement

TASK 38

General Objective:

Measure lengths of objects indirectly and by iterating length units.

This activity might be used with:

Students who are able to (prerequisite knowledge):	Students who are Primed to (getting ready to learn):
• measure objects using a variety of measurement units (standard and nonstandard)	• recognize that as the size of a measurement unit increases, fewer units are needed to measure an object, and as the unit size decreases, more units are required

Materials:

- Large and small paper clips (one baggie of each for each pair of students)
- Nonstandard units of varying lengths, such as inch tiles, connecting cubes, centimeter cubes, teddy bear counters (You'll need a large enough quantity of each item so that students can choose two different measurement units for the measurement task.)
- Measurement bags: identical baggies of objects such as markers, crayons, unsharpened pencils, glue sticks, toy cars, etc.—one per pair of students
- Chart paper
- Markers
- PRINTABLE: Comparing With Measurement Recording Sheet (one per pair of students)

 Printable for this task available for download at **https://companion.corwin.com/courses/ ProactiveMathIntervention**.

Recommended Children's Literature:

- *Twelve Snails to One Lizard: A Tale of Mischief and Measurement* – Hightower, 1997

Task Overview:

In this task, students measure the length of objects with nonstandard units of varying lengths. Students correctly measure the length of objects by placing equal measurement units end to end without gaps. They use two different units to measure identical objects to determine the relationship between the size of the unit and the number of units needed to measure the length of the object.

Variation 3

Learning Target: Students will measure the lengths of objects by iterating (placing end to end) length units and recognize the relationship between the size of the measurement unit and the number of units needed to measure the length of an object.

Variation Directions:

Introduce the activity by reading excerpts from the children's literature book *Twelve Snails to One Lizard* by Susan Hightower. Discuss the story details with the students. Discuss the idea that Milo measured the length of the dam using snails, lizards, and snakes, and the number of animals needed to measure the dam was different. Students should notice that Milo needs more (36) snails to measure the length of the dam than the two other units (lizards and snake) because the snails were a shorter length. When the length of the unit was longer (lizard), they needed fewer units to equal the length of the dam.

(Begin here if the book is unavailable.) Show the students an unsharpened pencil and large and small paper clips. Ask the group to estimate the number of small and large paper clips that will be needed to measure the pencil. Record the estimates.

Next, ask the students to share the rules for measuring the length of the object. Elicit from the students that the units need to be lined up end to end with no gaps or overlaps between the units. First, ask the students to estimate what they think the measurement will be. Select a student volunteer to estimate and then measure the length of the pencil using large paper clips, making sure there are no gaps and overlaps. Record the measurement, emphasizing the correct way to record the number and unit (e.g., 7 large paper clips). Ask another student volunteer to estimate and then measure the pencil using small paper clips, making sure there are no gaps and overlaps. Record the measurement, emphasizing the correct way to record the number and unit (e.g., 12 small paper clips).

Display both paper clip train measurements under a document camera with the pencil beside it so the students can see both measurements. Facilitate a discussion about the difference between the measurements when using the large and small paper clips. Create a whole-group chart together about measuring and comparing units of measure.

Figure 5.38.1 Chart prompts

Suggested prompt:	Related ideas to elicit:
Why does it take more small paper clips than large ones to measure the same pencil?	• The smaller the unit, the more units are needed to cover the same length. • The size of the unit directly affects the number of units required to measure an object.

Suggested prompt:	Related ideas to elicit:
What does this tell us about the relationship between the size of the unit and the number of units needed?	• There is an inverse relationship between unit size and the number of units used. • As the size of the unit increases, the number of units decreases and vice versa.
Does the pencil's length change depending on which unit we use? Why or why not?	• The pencil's length remains constant regardless of the unit used for measurement. • Units are a tool to describe the same length in different terms, but the object itself does not change.
Why is it important to measure with no gaps or overlaps?	• Gaps or overlaps lead to inaccurate measurements. • Accurate measurements require units to be aligned consistently, without extra space or overlapping portions.

Next, organize students into partner pairs and provide each pair with the measurement bag, recording sheet, and two different nonstandard measurement units. Make sure each pair of students receives measurement units of two different lengths. Bring the whole group together to discuss the measurements.

Engage in Math Discourse (Make the Mathematics Visible):

Observation: As students work, encourage them to notice important relationships and push on early and partial understandings.

- How are you lining up your units? Are there any gaps or overlaps?
- What's different about measuring with the smaller unit compared to measuring with the larger one?
- What are you noticing as you record your measurements? Do the results make sense?
- How does the size of your measurement unit affect the number of units you need?
- Are your estimates close to the actual measurements? Why or why not?

Interview and **Show Me:** After the activity, bring the intervention group together for a discussion. Suggested prompts:

- How did the size of the measurement unit affect the number of units you needed?
- Which unit required more pieces to measure the same object? Why do you think that is?

- If you used an even smaller unit, what do you predict would happen to the number of units needed?

- If you had a larger object to measure, how would the size of the unit affect the number of units required?

- Does the length of the object you measure change depending on the unit used? Why or why not?

- What does this activity teach us about the relationship between the size of a unit and the number of units required to measure something?

	**Bridging Prompt (Prompt for Classroom Teacher to Use During New Lesson Content):** • Ask students to estimate and then measure the length of one side of a book using two different measurement units (large paper clips and small paper clips). • What can you say about the relationship between the size of the unit and the number of units it takes to measure the length of an object? How do you know? Call on the students who participated in the intervention group to share their thinking.

On the Edge

TASK 39

General Objective:

Students will explore the meaning of perimeter as an attribute of two-dimensional figures.

This activity might be used with:

Students who are able to (prerequisite knowledge):	Students who are Primed to (getting ready to learn):
• measure to the nearest inch using a ruler • add two-digit numbers or three single-digit addends	• recognize perimeter as an attribute of two-dimensional figures • find perimeters, given side lengths or with an unknown side length and generate rectangles with a given or the same perimeters

Materials:

- Chart paper
- Ruler/tape measure in inch units
- PRINTABLE: 1-centimeter grid paper
- PRINTABLE: 1-inch grid paper

 Printables for this task are available for download at **https://companion.corwin.com/courses/ ProactiveMathIntervention**.

Recommended Children's Literature:

- *Super Sand Castle Saturday* – Muphy, 1999 (nonstandard to standard length units)
- *Spaghetti and Meatballs for All: A Mathematical Story* – Burns, 2000
- *Zachary Zormer* – Reisberg, 2006
- *Tia Tape Measure* – Klein, 2011
- *How Big Is a Foot?* – Myller, 1991
- *Racing Around* – Murphy, 2001 (perimeter)
- *Perimeter, Area, and Volume: A Monster Book of Dimensions* – Adler, 2013 (beyond just perimeter but focuses on attributes)
- *Chickens on the Move* – Belviso & Pollack, 2006 (perimeter)
- *The Pigpen Problem: How to Calculate Area and Perimeter* – Wise & Lynch, 2024 (perimeter)

Task Overview:

Students will explore the foundational components of learning about perimeter of a rectangular figure by exploring length measures.

Select (or create) a variation that focuses on the prior knowledge you are working to shore up for your student group.

Variation 1

Learning Target: The students will use concrete and semiconcrete representations to develop and expand their understanding of length measurement using lines and objects in a variety of orientations and combinations.

Variation Directions:

This variation can be delivered over several days. Begin with a story such as Loni and Jasmine are building a fence that came in different pieces. How can they measure the sections of the fence so it fits their yard? Begin with students measuring one line segment (fence section) on the chart paper. When Priming, we don't give them a whole rectangle, as commonly shown in perimeter, but just give them parts at this point. Have several segments drawn on chart paper, and students can come up as teams of two to measure the lines (see Figure 5.39.1). Make sure the lines are a precise number of inches, so the students are not measuring fractional parts of a unit at this time. Now, we review some of the main points of length measurement. Ask students, "How should you align the ruler to count the units?" Check to see that they are all starting with their ruler or tape measure in the correct position. Remind students that there are no gaps or overlaps when we measure if we are using individual inches. This may be the first time that students are measuring with continuous units this year, so connect this work to their use of the number line to highlight similarities such as how to read the unit at the end of its length by counting units (you may want them to position their finger there for clarity). Students need to move away from counting tick marks or numbers on the ruler or number line, because that is a common hiccup.

Figure 5.39.1 Chart paper with lines

Now, move to two lines (still no fractional measures at this time) that are connected at a right angle or other acute angle. Here, we are asking the students to measure the total length of the two lines that are connected. Ask them how they can do that. Have students record the measures of the lines on the chart paper in inches next to each line. They should find the total length by adding together these two lengths and recording the totals on the chart. Leave the charts displayed, or keep them to use again for another variation.

Figure 5.39.2 Chart paper with two lines connected by an angle

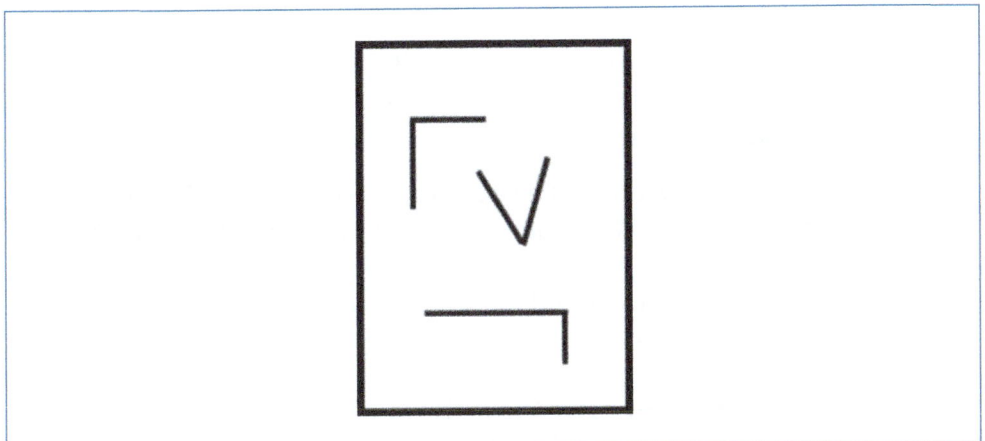

During the next intervention session, share the first of the following figures (see Figure 5.39.3) on chart paper. Ask students how these lines are different from the previous chart. They should note how they see the same two lines as they already measured. Now, ask them to use a ruler to add another line connected to the original two lines. Again, ask them how we could find out the total length of the three lines. [Find the two lines' measure and add the measure of the third line.] What lines do you need to measure to do that? Have students measure and record the length of the third line on the chart paper. Then, post the final chart with three lines attached and ask them to calculate the total length (see Figure 5.39.3).

Figure 5.39.3 Three lines: one chart with students' additions; the other with three lines in place

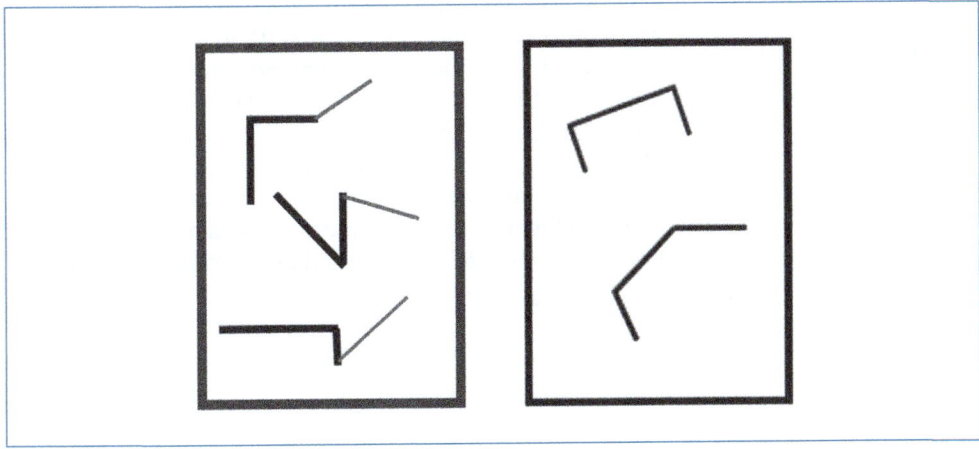

As we add on one more line (or fence section) for measuring each time, we can highlight that addition using a number line connects to continuous units. So it's a good opportunity to talk about a longer number line to add, for example, five units and then add on the next seven units and then add on the last five units. Students should have access to tape measures to use as needed.

Engage in Math Discourse (Make the Mathematics Visible):

Observation: As students work, encourage them to notice important relationships and push on early and partial understandings.

- Look for ruler/tape measure to be carefully placed at the beginning of the line. Note if the scale is being read accurately as a count of the units—not merely looking at the number.

- When the students start working on the three-line figure where they added a third line, avoid a common hiccup by asking, "Should the first two lines be remeasured?" Have students discuss what to do instead. This goes back to their knowledge of addition and the approach to "adding on" from the initial measurement.

Interview and **Show Me:** After the activity, bring the intervention group together for a discussion. Suggested prompts:

- Can you show me what strategies you used to measure the lines?

- How did you keep track of the total when you had two or three lines to measure?

- Choose one or two examples from student work and ask, "Is it possible that you would get different total measures from another team? How can we check our answers?"

Bridging Prompt (Prompt for Classroom Teacher to Use During New Lesson Content):

- Introduce the actual lesson on perimeter by using a launch in which the same chart paper from the intervention session with the two line segments connected by the angle is printed (see Figure 5.39.2). Then ask, "How could I measure the combined length of these two segments together?" Students who participated in the intervention group would then have the opportunity to recognize that chart paper right away. Hopefully, they say, "Oh, I've seen that before!" Let the students explain how they might measure them while asking them to explain as they work. Encourage them to use the words they've just practiced such as *length*, *line*, *measure*, *unit*, *inch*, and *combine*.

CHAPTER 5. Intervention Tasks

Field Day

TASK 40

General Objective:

This task adapted from the AIMS EDUCATION FOUNDATION "Metric Olympics" task.

Students will use concrete representations to make sense of concepts related to measuring and converting measurements of length (metric and customary).

This activity might be used with:

Students who can (prerequisite knowledge):	Students who are Primed to (getting ready to learn):
• organize measuring tools (inch tiles, unit cubes, strips of cardstock) into rows of equal-sized units • collaborate with peers	• measure lengths with rulers and meter/yardsticks • compare and relate lengths with different units (within the same system of measurement) • convert between units of length (within the same system of measurement)

Materials:

General:

- PRINTABLE: Station Directions
- PRINTABLE: Unit Spinners

For metric measurement:

- base ten unit cubes to represent centimeters
- base ten rods to represent decimeters
- meter sticks
- PRINTABLE: Field Day Recording Sheet—Metric

For customary measurement:

- inch tiles
- 12-inch strips of cardstock or 1-foot rulers

- yardsticks *must be **yardsticks**, not meter sticks*
- PRINTABLE: Field Day Recording Sheet—Customary

 Printable for this task is available for download at **https://companion.corwin.com/courses/ProactiveMathIntervention**.

Recommended Children's Literature:
- *Wilma Unlimited* – Krull, 1996
- *Millions to Measure* – Schwartz, 2006
- *Inchworm and a Half* – Pinczes, 2001

Task Overview:

Student groups will rotate through various Field Day Event stations (this will likely take multiple intervention sessions). As students complete each station, they will measure and record their results (distances).

Stations: (Suggestions here are based on several activities from the AIMS EDUCATION FOUNDATION (1987) "Mini Metric Olympics," but you may want to create your own based on your students' interests. Use the Station Directions cards at each location.)

- Cotton Ball Shot Put: Stand at the tape line on the floor and throw one cotton ball as far as you can.
- Paper Plate Discus Throw: Throw a paper plate like a discus (or Frisbee) from the tape line on the floor.
- Side Step: Put both feet together next to a line on the floor and take one giant step to the side.
- Giant Step: Put both feet at a starting line on the floor and take one giant step forward.
- Straw Javelin: Stand at the tape line on the floor and throw a straw as far forward as you can.

Select (or create) a variation that focuses on the prior knowledge you are working to shore up for your student group.

Variation 1

Learning Target: Students will understand that changing from a larger to a smaller unit will result in a greater number of units needed (e.g., "I need more inches than feet to measure the same distance") and changing from a smaller to a larger unit will result in a lesser number of units needed (e.g., "I need fewer meters than centimeters to measure the same distance").

Variation Directions:

Place the directions for activities (see handout Field Day Activities) along with a unit spinner at each station. Students will spin the unit spinner to determine which unit they will use to measure. For example, when Student A takes a turn at the Straw Javelin, she spins the spinner and gets "cm" for centimeters. The group will use centimeter cubes to measure the distance Student A throws the straw. Next, Student B takes a turn, and he spins "meters." The group will use meter sticks to measure the distance Student B throws the straw.

Engage in Math Discourse (Make the Mathematics Visible):

Observation: As students are working through the stations, attend to their reactions to spinning various units to make the measure.

- You may begin to hear students groan or cheer if they spin centimeters (or inches) when they need to measure a long distance. Ask them to explain why.

- Be on the lookout for students who begin to use shortcuts for measuring with some of the smaller units, like measuring a long distance with a meter stick but recording the length in centimeters.

- Identify a couple examples of students who have measured the same event with different units. Discuss the differences and why that happened.

Interview and **Show Me:** After the activity, bring the intervention group together for a discussion. Suggested prompts:

- Did you hope for a certain unit when you were spinning the spinner? Why?

- What difference did it make when you measured distances with different units?

- I noticed that Student A measured 145 on the Straw Javelin throw, but Student B got a measure of 2. Wow. Is Student A just really, really strong? What happened there?

- Some of you were using the meter stick (yardstick) to measure even if you got centimeters or decimeters (inches or feet) on the spinner. How did that work?

Bridging Prompt (Prompt for Classroom Teacher to Use During New Lesson Content):

- Show the class three different-sized measurement tools (for metric measurement: meter sticks, base ten decimeter rods, and base ten centimeter cubes; for customary measurement: yardsticks, 12-inch strips of cardstock/ruler, and inch tiles). Ask: "Which unit would you choose to measure the length of this room? Why? Would it take more or fewer of that unit than the other units?" Call on the students who participated in the intervention group to share their thinking.

Variation 2

Learning Target: Students will understand the relationship between two given units of measure (for example, centimeters and meters or inches and yards).

Variation Directions:

Place both measurement units at every station (e.g., centimeter cubes and meter sticks or inch tiles and yardsticks). Each student group will designate one "competitor" per station. After the competitor completes the activity, the group will work together to measure the distance simultaneously with *both* units used side by side and record their results.

Figure 5.40.1 Inch tiles lined up next to yardstick

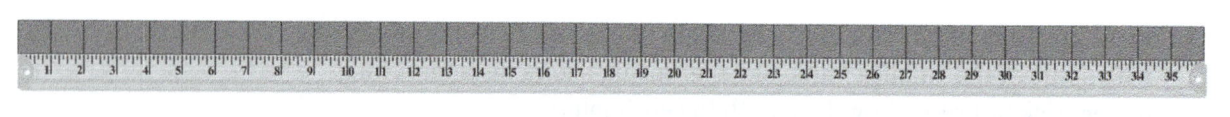

Engage in Math Discourse (Make the Mathematics Visible):

Observation: As students are working through the stations, attend to their reactions to spinning various units.

- You may begin to hear students start to say they don't need to match the individual units to every meter/yardstick. Ask them to explain.

- Listen/look for the different ways students work to make sense of partial units (such as half a yard).

Interview and **Show Me:** After the activity, bring the intervention group together for a discussion. Suggested prompts:

- Did it take you more inches or more yards to measure the distances? Why do you think that is?

- What other relationships did you notice when you were measuring?

- Some of you decided you didn't need to always lay out inch tiles for the whole distance you were measuring. Why was that?

- What did you do when the distance was only part of a yard? *If possible, use an actual example of student work here*. What are some different ways that might make sense to name this distance?

CHAPTER 5. Intervention Tasks

Bridging Prompt (Prompt for Classroom Teacher to Use During New Lesson Content):

- Show the class a distance measured with meter sticks (or yardsticks) where the meter sticks (yardsticks) are physically laid out in a row to demonstrate the measurement. Ask, "How could we figure out how many decimeters (feet) it would take to measure this same distance?" Call on the students who participated in the intervention group to share their thinking.

TASK 41

Broken Clock

General Objective:

Students will understand how to tell time on an analog clock.

This activity might be used with:

Students who can (prerequisite knowledge):	Students who are Primed to (getting ready to learn):
• describe positional relationships (before, after, etc.) • skip count by 5s	• tell time to the hour, half hour, or minute • describe time as before or after the hour • recognize how the hour hand on a clock moves as the minute hand moves

Materials:

- Geared demonstration clock
- Paper clip
- Counters
- PRINTABLE: Clock Spinner
- PRINTABLE: Student Timeline Game Boards
- PRINTABLE: What's the Hour? Cards
- PRINTABLE: 1–12 number cards
- PRINTABLE: Student Paper Clocks
- DIGITAL RESOURCE (clock): https://qrs.ly/tugjc8f (includes an option to show the digital time simultaneously)
- DIGITAL RESOURCE (clock): https://qrs.ly/fdgjc8h

 Printables for this task are available for download at **https://companion.corwin.com/courses/ProactiveMathIntervention**.

Recommended Children's Literature:

- *It's About Time* – Murphy, 2005
- *A Story About Time* – Yeung, 2023

- *Bunny Day: Telling Time From Breakfast to Bedtime* – Walton, 2002
- *Grouchy Ladybug* – Carle, 1996

Task Overview:

This is a partner game. Students will use a paper clip spinner on a blank clock face. The paper clip will represent one hand of the clock. *(It may be helpful to unfold the paper clip or tape a paper arrow to the end of the paper clip, as shown in Figure 5.41.1.)* Students take turns spinning the spinner to reveal the hour or minutes to be shown on the clock.

Figure 5.41.1 Paper clip clock hands

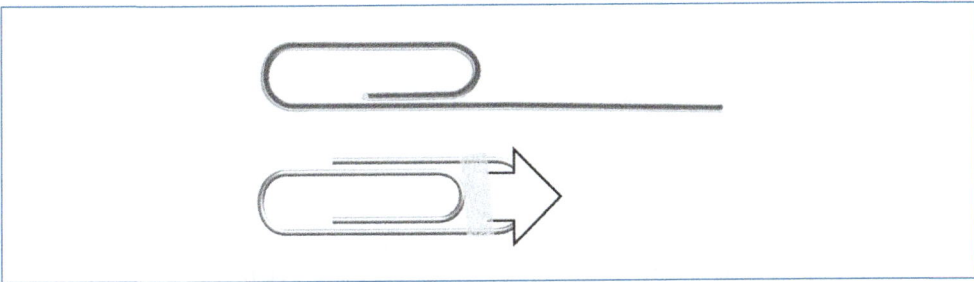

Source: iStock.com/RUSSELLTATEdotCOM

To introduce this task, tell students a story about a clock that is broken and only has *one hand*. If possible, show them a clock with only an hour hand or only a minute hand, depending on the variation of the game *(the free Math Learning Center digital clock linked above can be set to have just one hand)*. Ask them what they could know about the time if they had a broken clock with only an *hour/minute* hand.

Select (or create) a variation that focuses on the prior knowledge you are working to shore up for your student group.

Variation 1

Learning Target: Students will understand that the hour hand moves throughout the hour and be able to identify the current hour by looking at the hour hand of an analog clock.

Variation Directions:

Begin by having students show a time to the hour, such as 2 o'clock, on a geared demonstration clock. Ask students to watch the *hour hand* on the clock while they move the minute hand to change the time to the next hour. What pattern do they notice? Have students demonstrate this with several different examples until you are able to elicit from students that the hour hand is moving throughout the hour. Draw students' attention to *where* the hour hand moves (e.g., whenever the time is 2:__, the hour hand will be somewhere between the 2 and the 3).

After this discussion, introduce the game (see Task Overview). In this variation, the paper clip on the spinner represents the *hour hand* of the clock. After each spin, students should use the sentence frame, "The hour is _ because . . ." and place a counter on their student timeline (if the time is open). If the spinner lands directly on an hour, the student places the counter under that hour on the timeline.

Set a goal for winning the game based on the time available. If time allows, students could race to finish their whole timeline. If time is short, students could race to be the first with five counters in a row or three on each side (top and bottom) of the timeline.

Figure 5.41.2 Broken Clock Variation 1 example

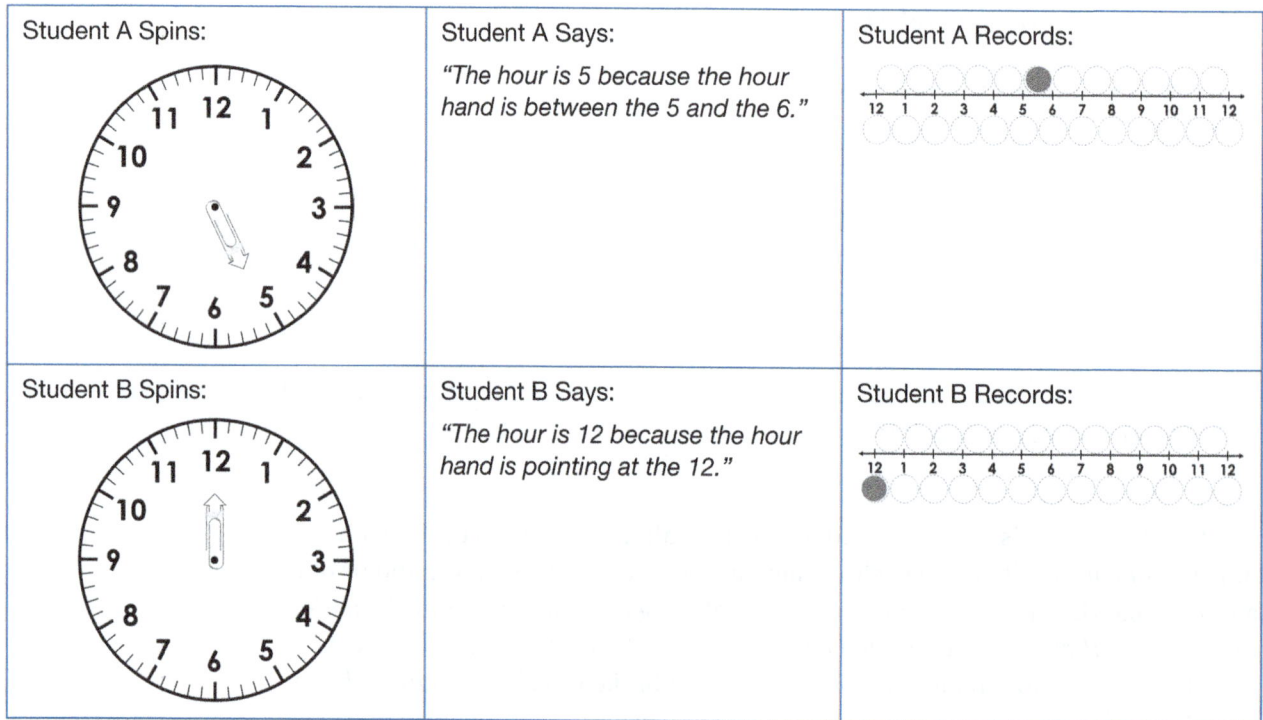

Engage in Math Discourse (Make the Mathematics Visible):

Observation: As students are playing the game:

- Watch/listen for students to have debates about whether the hour hand is "on the hour." This is an opportunity to practice mathematical argumentation. Ask each student to defend their position, telling where *they* would place a counter on the timeline and why.

- Listen for students who are saying things like, "It's *almost* 4 o'clock" or "It's *really close* to 11 o'clock." Ask them how they know.

Interview and **Show Me:** After the activity, bring the intervention group together for a discussion. Suggested prompts:

- When is it easiest to tell what the hour is? Why?

- I heard a student say that it was *almost* 4 o'clock when his spinner looked like this (show spinner). What do you think he meant? What would the hour hand look like if it was closer to 3 o'clock?

- Show students a variety of "What's the Hour?" cards (see examples) and ask them to tell the time shown on the clock.

Figure 5.41.3 What's the Hour? cards

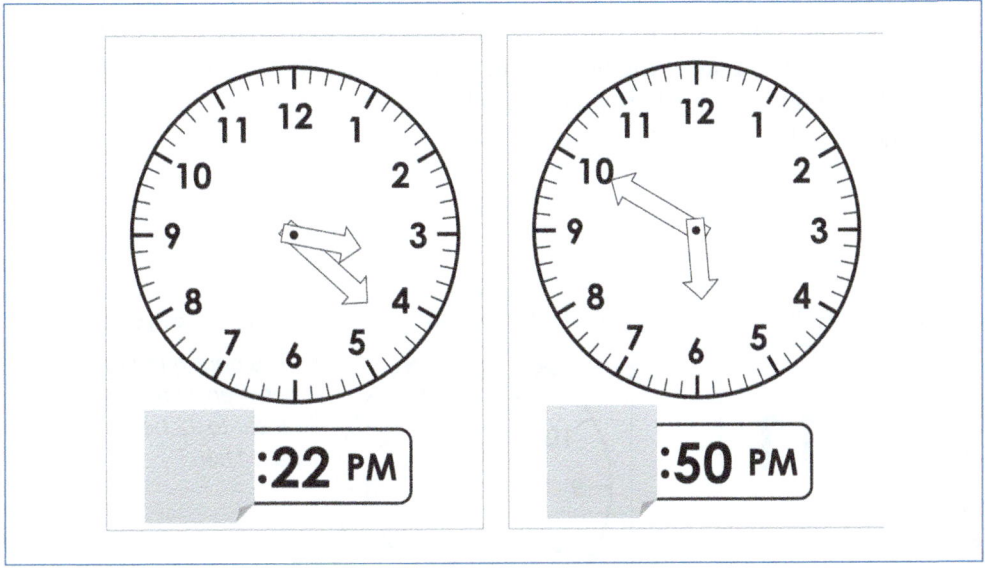

	Bridging Prompt (Prompt for Classroom Teacher to Use During New Lesson Content):
	• Show the class a geared clock set to an hour (e.g., 3:00). Ask, "How does the hour hand move on a clock as the time changes from 3 o'clock to 4 o'clock?" Call on the students who participated in the intervention group to share their thinking.

Variation 2

Learning Target: Students will understand that the hour hand is closer to the hour when the minutes are less than 30, closer to the next hour when the minutes are greater than 30, and exactly halfway between the hours when the minute hand is at the 30-minute mark.

Variation Directions:

Begin by having the students find a time to the half hour, such as 2:30, on a geared or virtual demonstration clock. Ask students to notice where the *hour hand* on the clock is. Show students several different times to the half hour and elicit from students that the hour hand is always *halfway* between the hours when the time is _:30. Next, ask students where they think the hour hand would be if the time is between 2:00 and 2:30. Ask them to watch the *hour hand* on the clock while you move the minute hand to check their conjectures. Repeat with several different examples until you are able to elicit from students that the hour hand is closer to the hour when the time is between _:00 and _:30 (e.g., the hour hand will be closer to 2 than 3 when the time is between 2:00 and 2:30). Repeat for times between _:30 and the next hour.

After this discussion, introduce the game. In this variation, the paper clip on the spinner represents the *minute hand* of the clock. After a student spins the spinner to reveal the *minutes* after the hour, they will draw a number card to reveal the *hour*. The student will set their own clock to match and then tell their partner where they placed the hour hand and why. Students can use the sentence frame, "The hour hand is between ___ and ___. It is closer to ___ because . . ." (or "The hour hand is halfway between ___ and ___ because . . .").

Figure 5.41.4 Broken Clock Variation 2 example

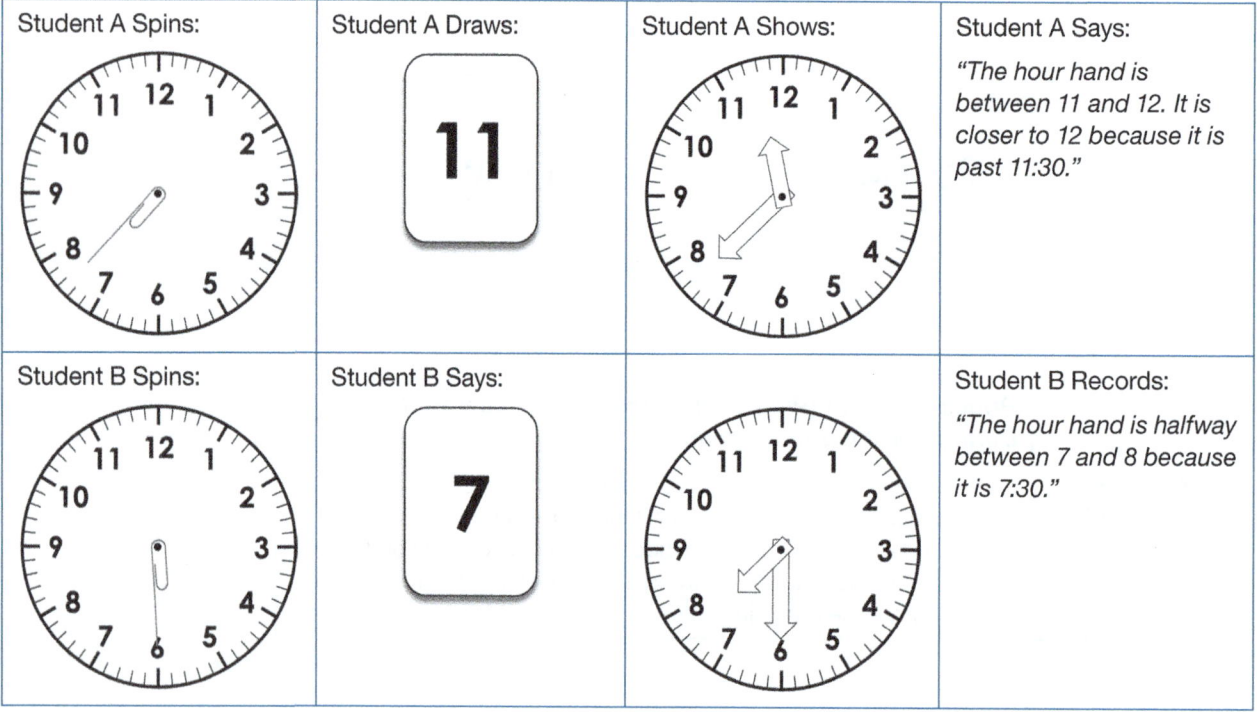

Engage in Math Discourse (Make the Mathematics Visible):

Observation: As students are playing the game:

- Be comfortable with some imprecision from students, as there is always error in measurement. It's okay if their positioning of the hour hand is not as precise as a geared clock would be. The aim here is for students to focus on the general relative distance of the hour hand between a given set of numbers.

- Listen for students who may be confusing the minute and hour hands. Remind them that the spinner represents the minute hand in this version of the game.

Interview and **Show Me:** After the activity, bring the intervention group together for a discussion. Suggested prompts:

- Are there times that seem challenging to represent? Why?

- Gina said, "If I spin 30 minutes after the hour and draw 6 for my hour, my hour and minute hand would both be pointing right at the 6." Is Gina correct?

	**Bridging Prompt (Prompt for Classroom Teacher to Use During New Lesson Content):** • Ask the class, "If the hour hand on a clock is *exactly halfway* between 1 and 2, where would you expect the minute hand to be? Why?" Call on the students who participated in the intervention group to share their thinking.

TASK 42

Tip and Plot

General Objective:

Students will measure and organize collections of real-world objects and generate line plots to represent, analyze, and interpret the collections.

This activity might be used with:

Students who can (prerequisite knowledge):	Students who are Primed to (getting ready to learn):
• measure lengths to the nearest inch • measure lengths to the nearest half inch • organize collections by relative length • count sets of objects • identify and plot points on a number line	• graph collections on a line plot • analyze and interpret data from a line plot

Materials:

- Collections of connecting cubes (make stacks of assorted lengths, with multiple copies of some lengths)
- Collections of items (bagged, assorted in length, at least 20 to 25 items per bag)
- Small sticky notes
- Adding machine tape

Recommended Children's Literature:

- *Kenley's Line Plot Graph* – Stone, 2015
- *Friends Beyond Measure* – Fisher, 2023

Task Overview:

This activity can be done individually, with partners, or in a small group. Create collections of items that can be measured for length (e.g., straws, pencils, paintbrushes, craft sticks, pipe cleaners, etc.). Each collection should contain at least 25 to 30 items and should include a variety of lengths, with multiple copies of some or all lengths. Cut items such as straws, for example, should be cut exactly to the full unit unless the students are measuring to halves, fourths, or eighths (or unless an additional goal is working on measuring to the *nearest* unit).

Figure 5.42.1 Collections of items that can be measured for length

In each variation of this task, students will be asked to tip out a portion of the collection, organize the tipped-out items by length, and then measure the items (precision of measurement will depend on grade level and task variation). Once the items have been measured, students will determine an appropriate scale and range to use and create a line plot of the data set. In some variations, students may arrange the collection items themselves in the line plot.

Select (or create) a variation that focuses on the prior knowledge you are working to shore up for your student group.

Variation 1

Learning Target: Students will organize stacks of connecting cubes by length and create a concrete, physical line plot from the data set.

Variation Directions:

Begin this variation by providing students with a box or bag filled with a collection of cube stacks (assorted lengths, with repetitions). Have students tip out a portion of the collection (one or two dozen cube stacks) onto the table. Ask students to organize the stacks by length (from shortest to longest) and label each stack with a sticky note telling its length. Students will create a physical line plot using the cube stacks, connect that to a semiconcrete line plot using the sticky-note labels, and finally generate sketched line plots to represent the data set.

Figure 5.42.2 Tip and Plot Variation 1 example

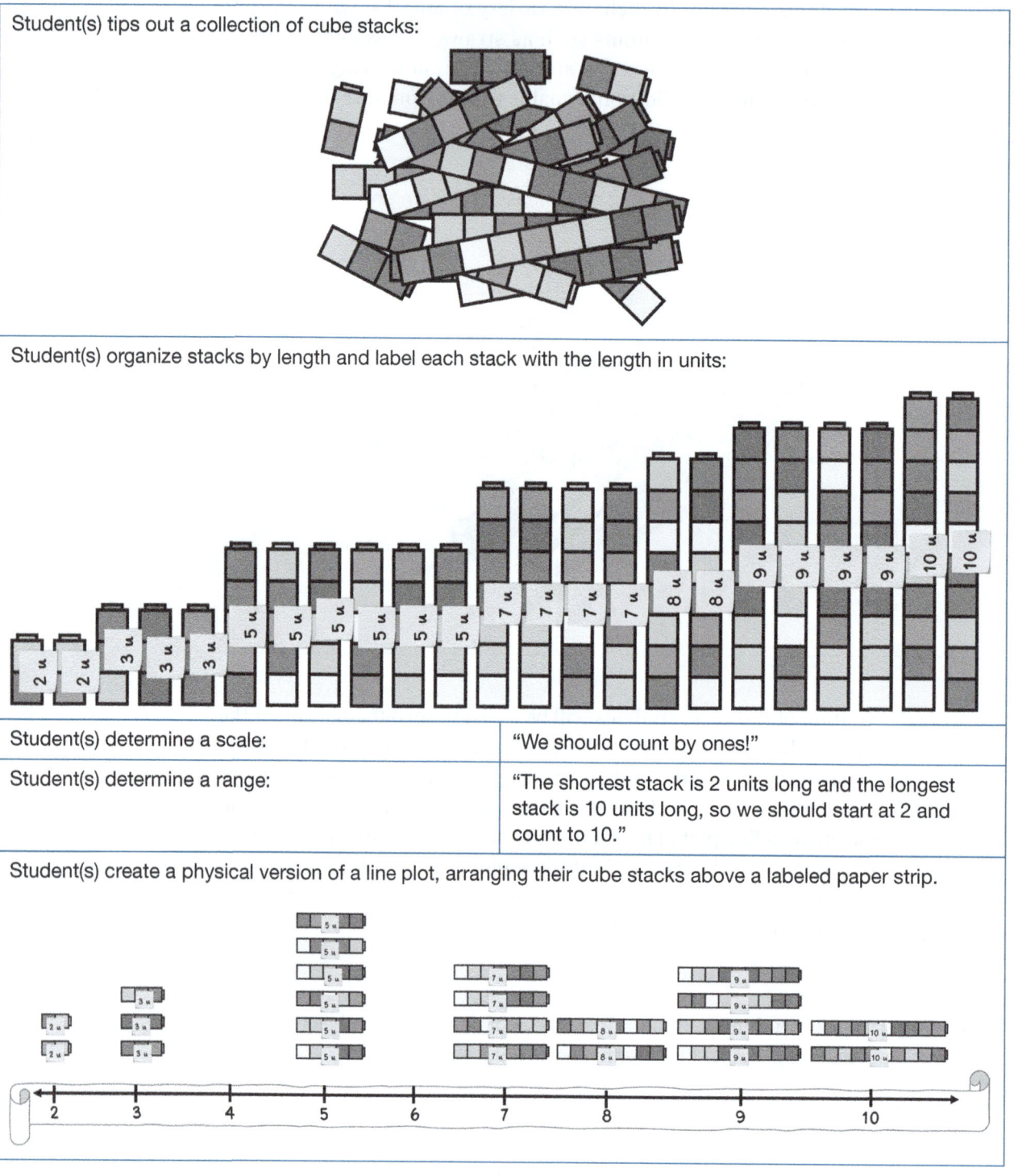

Students create a semiconcrete version of the same line plot by transferring the sticky notes to a new labeled paper strip.

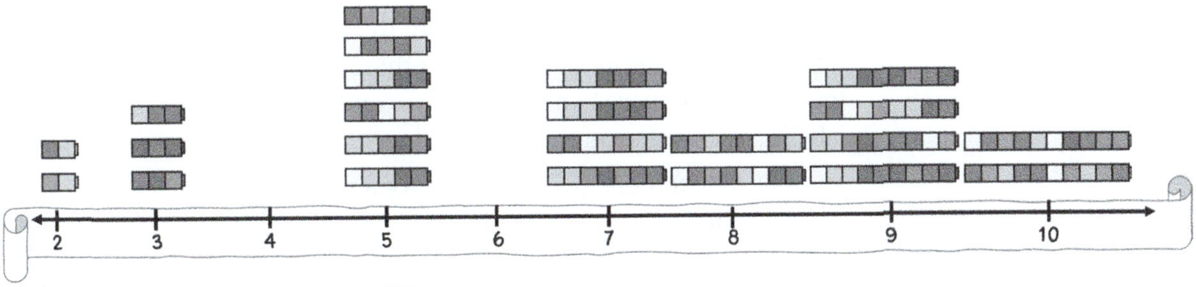

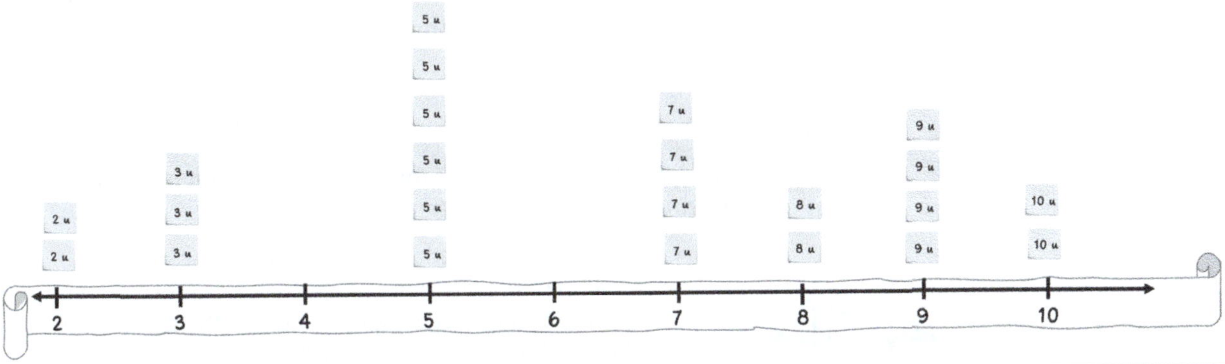

Student(s) sketch and label a dot or line plot on paper:

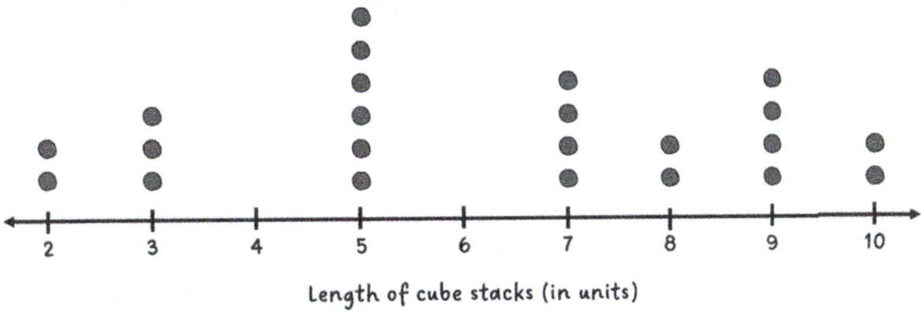

Length of cube stacks (in units)

Engage in Math Discourse (Make the Mathematics Visible):

Observation: As students work, encourage them to notice important relationships and push on early and partial understandings.

- Watch for students who want to count the number *of stacks* (think of groups) instead of the number of data points *in a stack*.

- Ask questions to push on students' understanding of their line plots:
  - Could this be a graph of the height of the students in this group? Why or why not?
  - How would your line plot change if I took these cube stacks away? (remove or indicate specific cube stacks)

Interview and **Show Me:** After line plots are constructed, bring the intervention group together for a discussion. Suggested prompts:

- Watch for students who want to omit the number-line labels for the missing quantities in the data set. If that happens, bring the group together to compare two number lines and discuss the importance of labeling each value on the number line.
  - Ask: When there are no dots above a number in the line plot, what does that mean?
- What if each cube represents a pet and each stack represents the number of pets each student has at home? What do you notice? What questions would you have?

	*Bridging Prompt (Prompt for Classroom Teacher to Use During New Lesson Content):* • Take photos of students' cube stack line plots and finished dot or line plots. Print these out and cut them apart to create matching cards. Begin the grade-level class by having students find matches. • Facilitate a discussion, encouraging students who participated in the intervention to share how they know they match.

Variation 2

Learning Target: Students will organize a collection of items by length (measured to the nearest ___) and create a concrete, physical line plot from the data set.

Variation Directions:

Begin this variation by providing students with a box or bag filled with a collection of items (assorted lengths, with repetitions—if items like paintbrushes or pencils are unavailable, straws or pipe cleaners could be cut to various lengths). Have students tip out a portion of the collection (one or two dozen of the items) onto the table. Ask them to organize the items by length (from shortest to longest) and then measure and label each item with a sticky note telling its length. Discuss with students how to measure (this can vary depending on your instructional goal). For example, place one of the paintbrushes next to a ruler but not lined up with the end of the ruler. Ask students what they notice about how the ruler is positioned. Students will create a physical line plot using the items, connect that to a semiconcrete line plot using the sticky-note labels, and finally generate sketched line plots to represent the data set.

Figure 5.42.3 Tip and Plot Variation 2 example (collection of paintbrushes, measuring to the nearest half inch)

Student(s) tips out a collection of paintbrushes:

Student(s) organize brushes by length, measure to the nearest half inch, and label each brush:

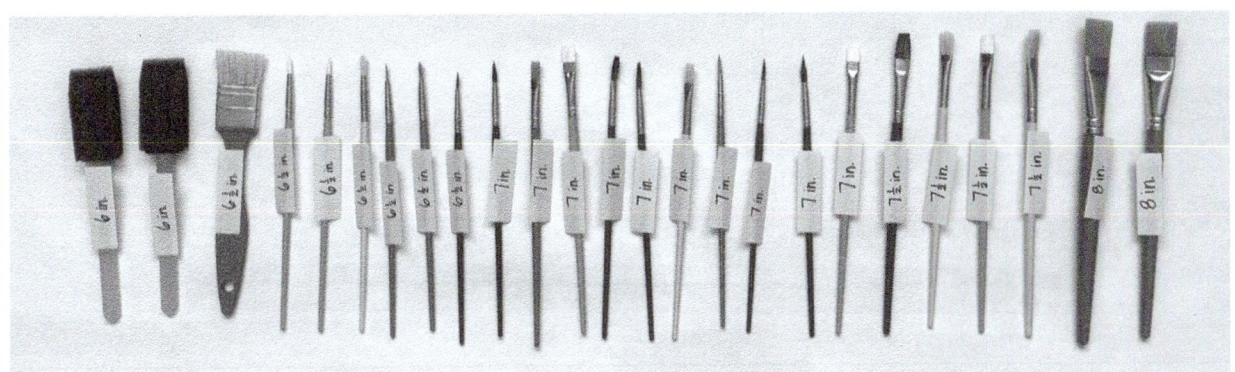

Student(s) determine a scale:	"We should count by half-inches!"
Student(s) determine a range:	"The shortest brush is about 6 inches long and the longest brush is about 8 inches long, so we should start at 6 and count to 8."

Student(s) create a physical version of a line plot, arranging their objects above a labeled paper strip.

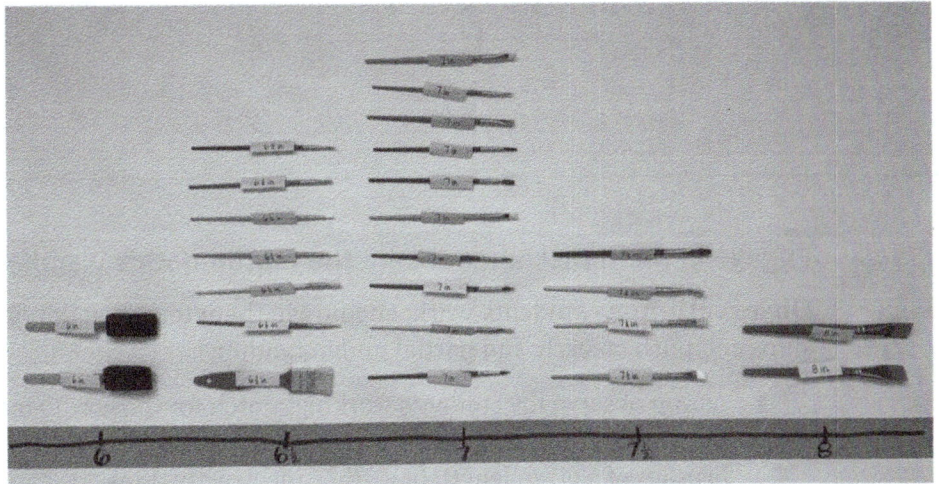

(Continued)

(Continued)

Students create a semiconcrete version of the same line plot by transferring the sticky notes to a new labeled paper strip.

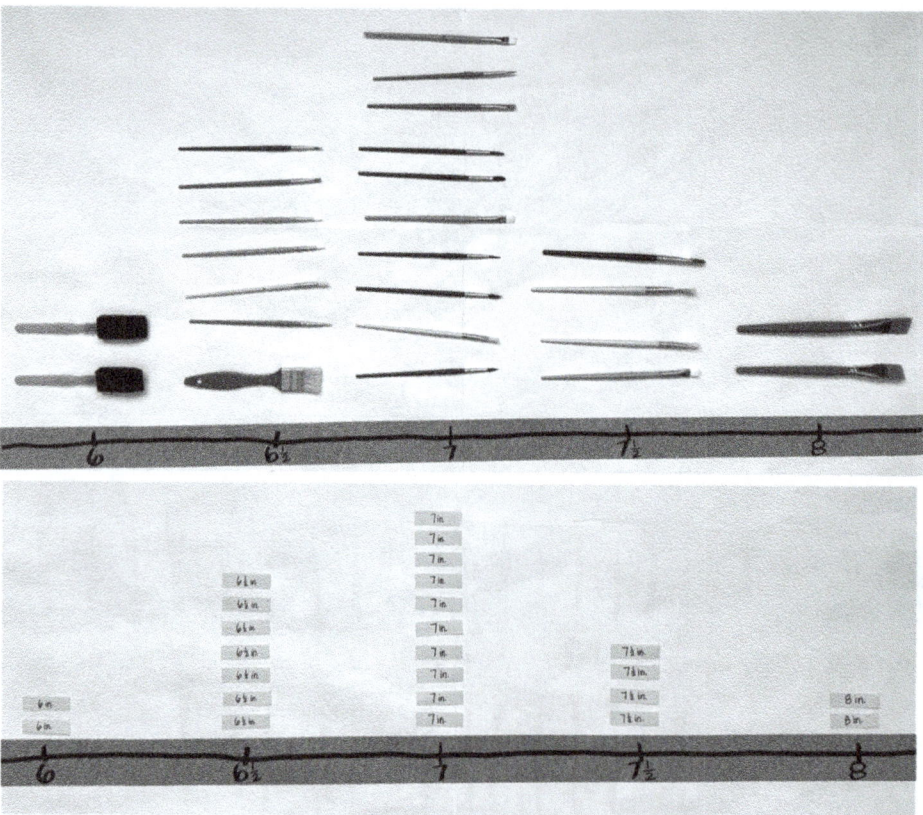

Student(s) sketch and label a dot (or line) plot on paper:

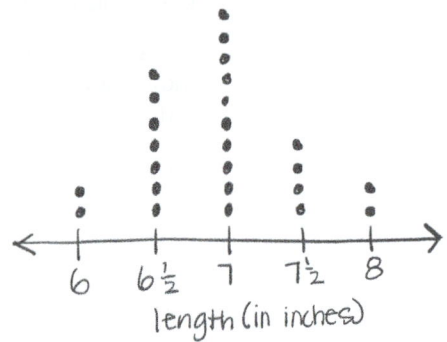

Engage in Math Discourse (Make the Mathematics Visible):

Observation: As students work, encourage them to notice important relationships and push on early and partial understandings.

- As you observe the students working, watch for those who may confuse the number indicating the *length* of the brushes with the *number of brushes* of a given length (e.g., they incorrectly think "$6\frac{1}{2}$" represents $6\frac{1}{2}$ brushes instead of the number of brushes that are $6\frac{1}{2}$ inches long).

- Ask questions to push on students' understanding of their line plots:
  - What does this graph tell us about the paintbrushes in this package?
  - Does this line plot tell me how wide the tips of the brushes are? What does it tell me?
  - How would your line plot change if I took these brushes away? (remove or indicate specific brushes)

Interview and **Show Me:** After line plots are constructed, bring the intervention group together for a discussion. Suggested prompts:

- Which brushes seem to be most popular? Why do you think so?
- If we are going to buy some more brushes, what length brush should we get? Why?

	***Bridging Prompt (Prompt for Classroom Teacher to Use During New Lesson Content):*** • Take photos of students' brushes plots and finished dot line plots. Print these out and cut them apart to create matching cards. Begin the grade-level class by having students find matches and discuss how they know they match. • Look for opportunities to partner students who participated in the intervention group with other students in the class so they can be positioned as the "expert" in the activity.

TASK 43

Picture Perfect

General Objective:

Students will create physical models of scaled pictographs and use those to make sense of the scale and multiplicative relationships within a pictograph. They will then generate semiconcrete models to correspond to their physical models.

This activity might be used with:

Students who can (prerequisite knowledge):	Students who are Primed to (getting ready to learn):
• skip count • count on • tally items • create a simple bar graph • interpret a bar graph	• use a scale • multiply with an equal groups representation • reason about division with equal groups

Materials:

▶ **Lots** of old egg cartons cut into two-egg sections (be sure to leave the lid attached to the sections)

○ *Use a dark marker to label each two-section container with a simple symbol that students can replicate later when drawing their semiconcrete versions of the pictographs. Draw the symbol in the center of the lid so you can make a few "half" containers with the symbol cut in half. See Figure 5.43.1 for example images. (Consider creating multiple sets with different symbols for each set if sufficient supplies are available.)*

Figure 5.43.1   Two-section egg cartons and "half" containers

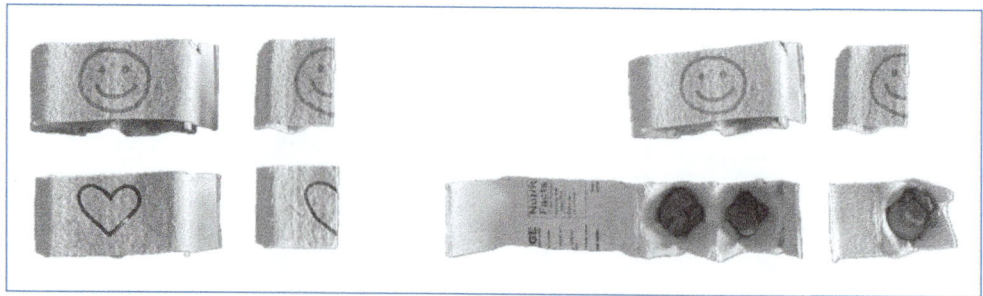

- Painters tape and/or or butcher paper to create axes and labels for pictograph
- Collections of small items that can be sorted and used to fill containers for graphing (e.g., buttons, counters, beads, coins, stickers, etc.)
- Pictograph images (create these to correspond to the collections of items you have available; see Variations 2 and 4 for more information)
- PRINTABLE – 1-inch grid paper

Recommended Children's Literature:
- *Pictographs* – Edgar, 2018 (more info about pictographs)
- *A Week of Weather* – Cortland, 2018
- *Show and Tell! Great Graphs and Smart Charts: An Introduction to Infographics* – Murphy, 2022
- *Lemonade for Sale* – Murphy, 1998

Task Overview:

This activity is designed to help students make sense of the multiplicative relationships within a pictograph. Students will fill two-section containers (egg cartons) with items according to the given scale. They will use the filled and closed containers to create a physical pictograph representing categorical data. Following a discussion, students will create a semiconcrete pictograph representing as a math sketch of the same data.

In each variation of this task, students will create physical representations of pictographs using two-section containers (egg cartons) to group items.

Select (or create) a variation that focuses on the prior knowledge you are working to shore up for your student group.

Variation 1

Learning Target: Students will sort a collection of items by a single attribute, grouping into sets according to a given scale, and then create a physical pictograph of the items. In this variation, all sets will be complete (so no partial images would need to be created to represent half a unit).

Variation Directions:

Organize students in small groups of three or four. Begin this variation by providing students with a collection of items that can be sorted by a single attribute, such as color. Precount (but then mix up) the collections in advance based on the scale you wish to use to ensure that each category will have a multiple of the chosen factor. For example, if you decide to sort buttons by color using a scale of 10 (☺ = 10), provide students collections of buttons in which the number of buttons of each color is a multiple of 10 (e.g., 10 red, 30 blue, 20 green, 10 white). Give each group a different scale to facilitate group discussion.

Give students the mixed-up collection of items and show them the scale. Have students sort the items according to the selected attribute and then fill containers according to the scale, using the filled containers to create their graph.

Figure 5.43.2 Picture Perfect Variation 1 example

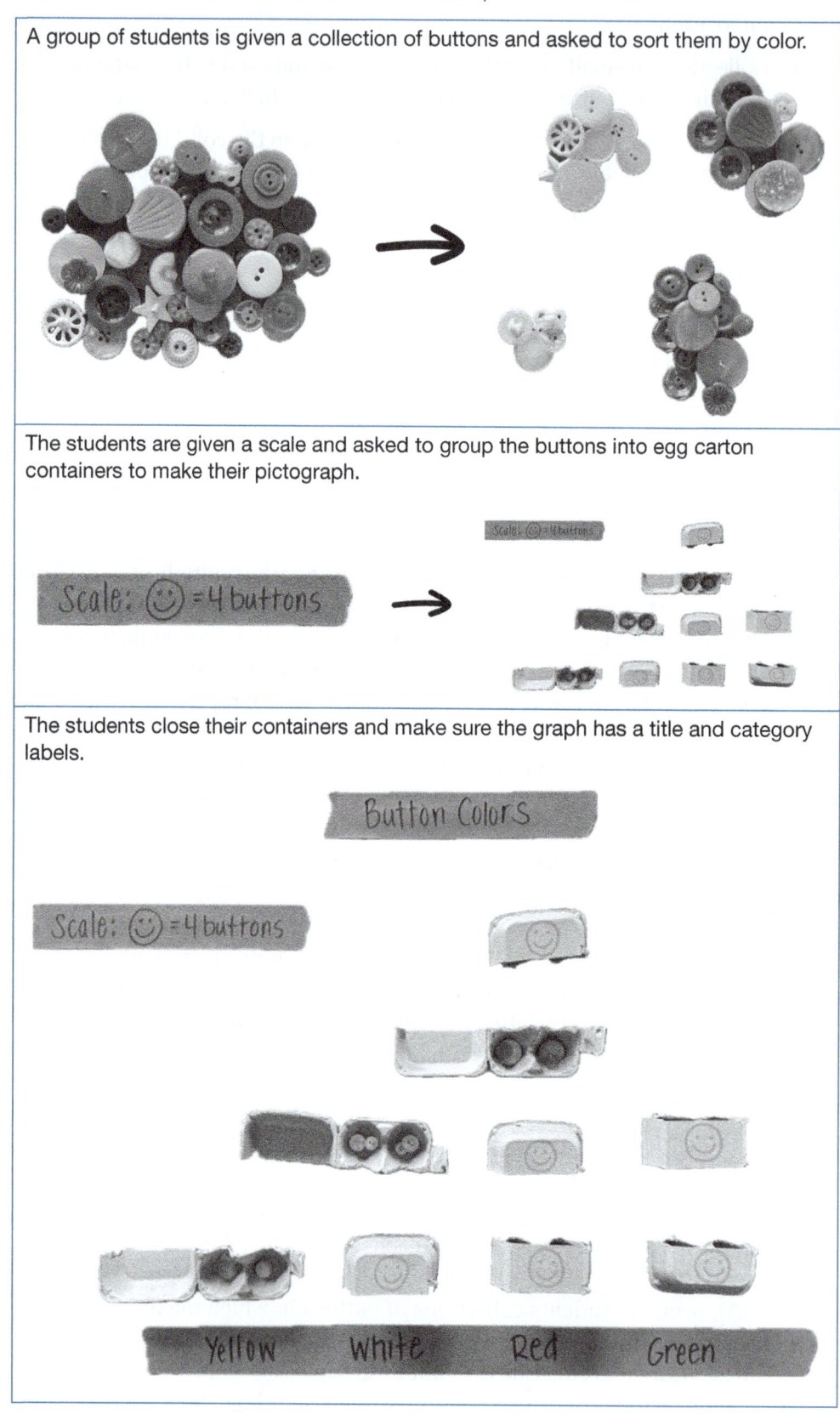

Engage in Math Discourse (Make the Mathematics Visible):

Observation: As students work, encourage them to notice important relationships and push on early and partial understandings.

- For which color of buttons do you think you'll need the most containers? Why?
- Which color will need the fewest containers? Why?
- Can you predict how many containers you will use for this color? How do you know?
- I see you have four containers of red buttons. How many red buttons do you have? How did you figure that out?
  - Look for strategies students use here. Some may need to open the containers and count all the buttons, some may be comfortable skip counting, and others may notice the multiplicative relationship and be able to multiply. Make note of these different approaches to highlight during the whole-group discussion.
- Watch for students who may lose track of which color buttons are in which containers. Provide a "just-in-time" scaffold to students by reminding them that they can open the containers to organize them and then close them back up. Be sure to ask students to make a connection to the scale of the graph after opening containers to check the quantities inside.

Interview and **Show Me:** After students have completed their pictographs, bring the intervention group together for a discussion. Suggested prompts:

- (Indicate a group's pictograph) How can we figure out how many red buttons this group had?
- (Indicate two pictographs with different scales) I notice that both of these groups have two "smiley faces" of yellow buttons. Do these two groups have the same number of yellow buttons? How could we figure that out? What happened there?
- Can you make a math sketch of your group's pictograph? Use a smiley face (heart, etc.) to show each container. (Model how students would draw one container on their picture. Provide grid paper to help students keep drawings equally distributed on their graphs.)

> ***Bridging Prompt (Prompt for Classroom Teacher to Use During New Lesson Content):***
>
> - When you see a symbol like this smiley face/heart/etc. on a pictograph, what do you wonder about? Call on the students who participated in the intervention group to share their thinking.

Variation 2

Learning Target: Students will translate an image of a pictograph into a physical version of the pictograph. In this variation, all images will be complete (so no partial containers would need to be included in the physical pictograph).

Variation Directions:

Organize students into pairs or groups of three. Begin this variation by providing students with pictograph images. Give each group a different graph to facilitate group discussion. Provide access to the collections of items (unsorted and uncounted) students will use, but do not distribute the collections to groups.

Each group is challenged to interpret their pictograph and create a physical version of the same pictograph. To do this, they will need to:

1. determine what collection of items they should use to create the graph
2. decide how to sort and count out items from that collection based on the graph
3. determine how many of each item to put in their containers

Figure 5.43.3 Picture Perfect Variation 2 example

A group of students is given the following pictograph.

Scale ☺ = 3 buttons

Button Colors			
red	☺	☺	☺
blue	☺		
yellow	☺	☺	
white	☺		

The students decide they will need to gather some red, some blue, some yellow, and some white buttons from the button collection.

They get out enough containers to match the graph and then use the scale to decide they should put three buttons in each container.

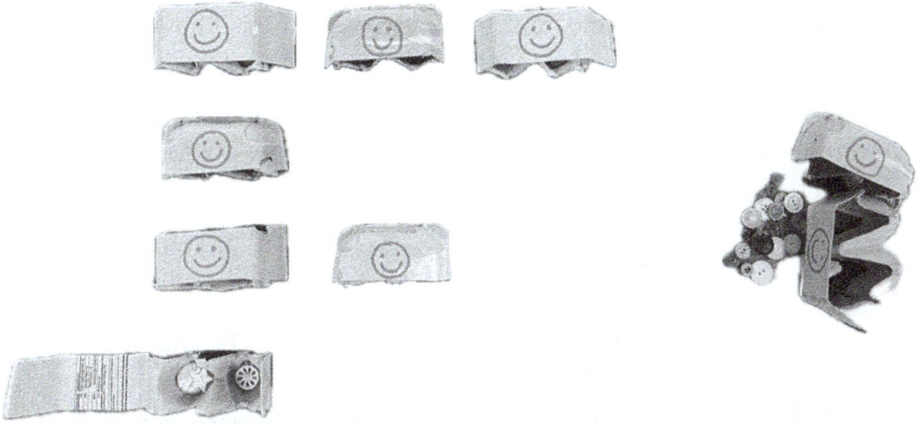

Finally, they label their pictograph.

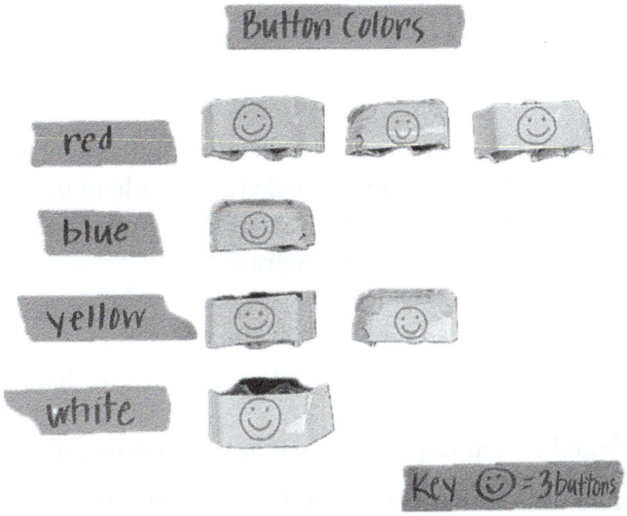

Engage in Math Discourse (Make the Mathematics Visible):

Observation: As students work, encourage them to notice important relationships and push on early and partial understandings.

- How did you decide how many buttons of each color to get?
- Which color did you get the most of? the fewest of? Why?
- Do you have a prediction about how many buttons you will need for this color? How are you predicting?
- I see you have four containers for red buttons. How many red buttons will you need? How did you figure that out?
    - Look for common hiccups here. Students may confuse containers with buttons and think each container represents a single button. Ask them to talk with their partners (or perhaps another group) about what the scale represents.

 **Interview** and **Show Me:** After students have completed their pictographs, bring the intervention group together for a discussion. Suggested prompts:

- (Indicate a group's pictograph) How can we figure out how many red buttons this group needed to get?
- (Indicate two pictographs with different scales) I notice that both of these groups have pictographs that show two "smiley faces" of yellow buttons. Did these two groups need to get the same number of yellow buttons? How could we figure that out? What happened there?

 **Bridging Prompt (Prompt for Classroom Teacher to Use During New Lesson Content):**
- How does the scale help you interpret a pictograph? Call on the students who participated in the intervention group to share their thinking.

Variation 3

Learning Target: Students will sort a collection of items by a single attribute, grouping into sets according to a given scale (a multiple of 2), and then create a physical pictograph of the items. In this variation, include some half sets (so half images would need to be created in a pictograph).

Variation Directions:

This variation works the same way as Variation 1, with two changes:

- Scales should be a multiple of 2 (so that values can easily be halved).
- Precounted collections given to students to graph should include at least one category which will need a "half container." For example, if you decide to sort a collection using a scale of 10 (e.g., ☺ = 10), provide students with collections in which the number of items for each selected attribute is a multiple of **5 or 10** (e.g., 10, 25, 20, 5).

Give students the mixed-up collection of items and show them the scale. Have students sort the items according to the selected attribute and then fill containers according to the scale, using the filled containers to create their graph. Wait to introduce the half-sized containers until students start asking questions about what to do with "extra" or "missing" items when they can't fill a whole container to match the scale. When those questions come up, bring the whole group together and facilitate a quick discussion to elicit the idea of cutting a container in half. *Then* provide students with half-sized containers to add to their graphs (or, if desired, and students have the cutting skills, allow them to cut a container in half themselves!).

CHAPTER 5. Intervention Tasks 403

Figure 5.43.4 Picture Perfect Variation 3 example

A group of students is given a collection of coins and asked to sort them by denomination.

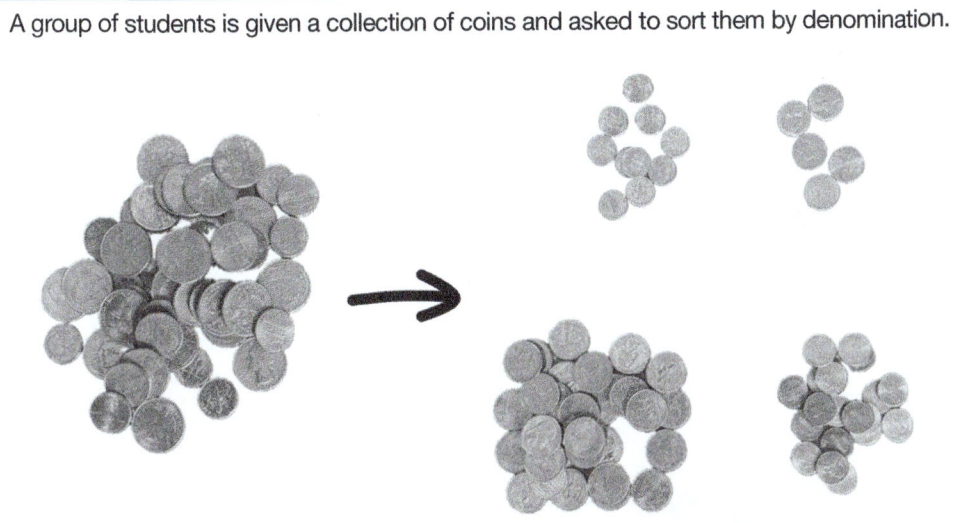

The students are given a scale and asked to group the coins into egg carton containers to make their pictograph, but they discover that they do not have enough nickels to fill up a container.

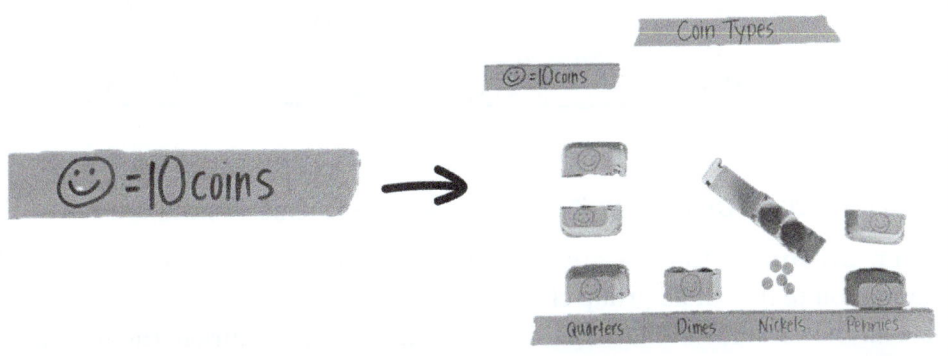

After some discussion with the teacher, the students realize they cannot leave the nickels out of the pictograph, and they cannot put the nickels in a 10-coin container. The teacher calls the whole group together for a discussion and elicits the idea that the container could be cut in half. When students return to their work, this group cuts the nickel container in half.

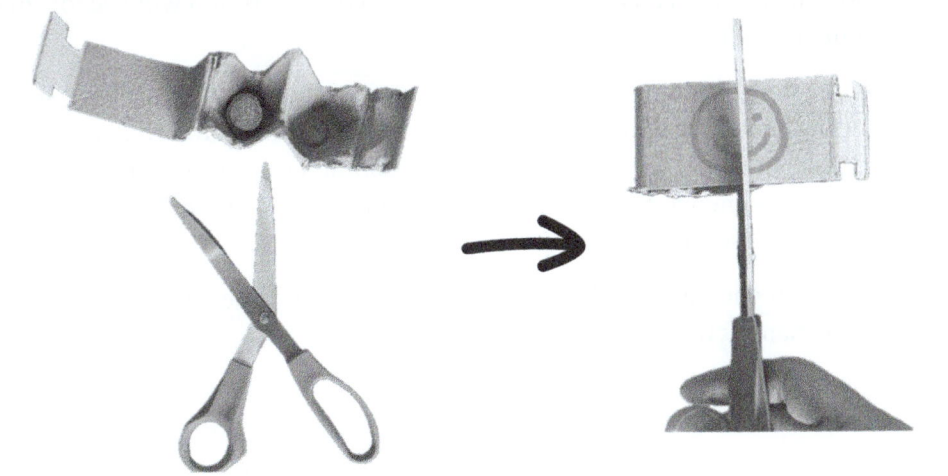

(Continued)

(Continued)

The students are now able to complete their graph, making sure the graph has a title and category labels.

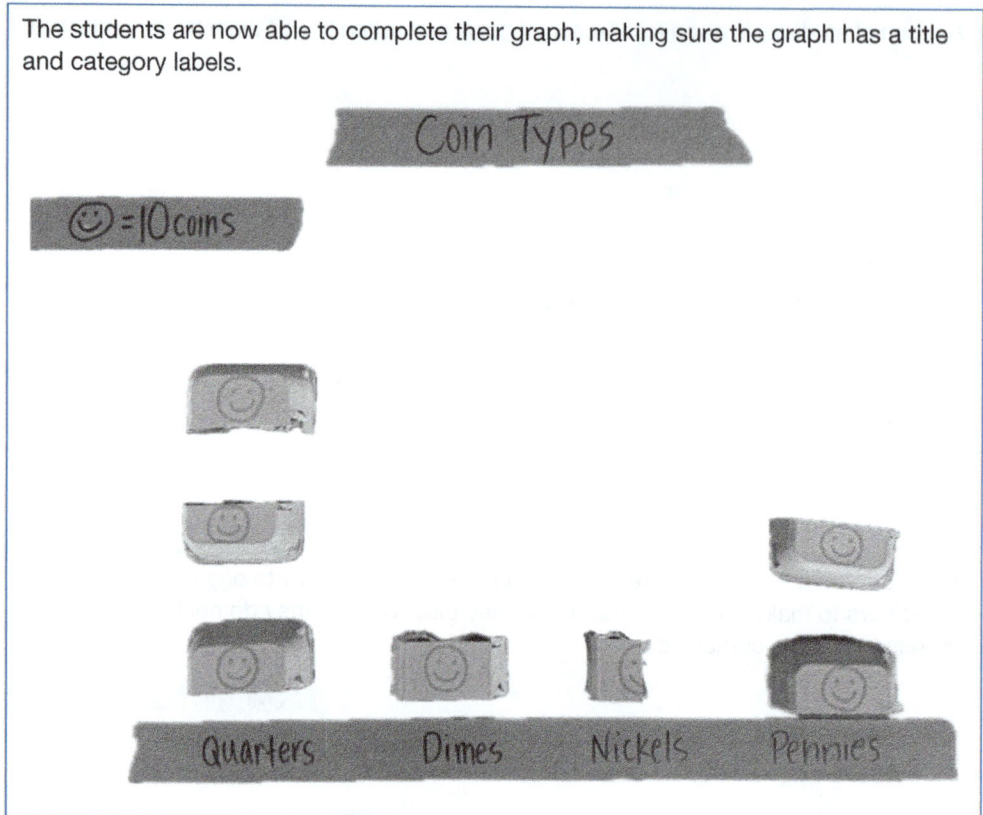

Engage in Math Discourse (Make the Mathematics Visible):

Observation: As students are finding relationships, encourage them to add information to the whole-group chart as they discover it. Students need not complete all the columns at the same time. As they work, encourage them to notice important relationships and push on early and partial understandings.

- For which coin do you think you'll need the most containers? Why?

- Which coin will need the least number of containers? Why?

- Do you have a prediction about how many containers you will use for this coin? How are you predicting?

- Watch for the moment when students begin to notice that they have some sets of items that don't quite match the scale. Ask them to brainstorm ideas about what they should do. If that container is 10 coins, how could we show someone that there are only 5 coins here? How will they know once we close the container?

- I see that you have a half of a container here. What does that mean? How many coins is that?

Interview and **Show Me:** After students have completed their pictographs, bring the intervention group together for a discussion. Suggested prompts:

- (Indicate a group's pictograph) How can we figure out how many nickels this group had?

- (Indicate a pictograph that has a "half container more" relationship between two categories) I notice that in this pictograph, there is one "smiley face" of dimes, but the nickels only have half of a smiley face. What does that tell us about the number of dimes and nickels? How could we figure that out?

- Can you make a math sketch of your group's pictograph? Draw a smiley face (or heart, etc.) to show each container. (Model how students would draw a half container on their pictograph. Provide grid paper to help students keep drawings equally distributed on their graphs.)

Bridging Prompt (Prompt for Classroom Teacher to Use During New Lesson Content):
• When you see a half symbol like this smiley face/heart/ etc., on a pictograph, what do you wonder about? Call on the students who participated in the intervention group to share their thinking.

Variation 4

Learning Target: Students will translate an image of a pictograph into a physical version of the pictograph. In this variation, pictographs will include some half images (so half sets would need to be created in the physical pictograph).

Variation Directions:

This variation works the same way as Variation 2, with two changes:

- Scales should be a multiple of 2 (so that values can easily be halved).

- Pictographs given to students to recreate should include at least one category with a half image (representing half of the scale amount).

Figure 5.43.5 Picture Perfect Variation 4 example

A group of students is given the following pictograph.

Scale ☺ = 12 coins

Coin Types			
quarter	☺	☺	☺
nickel	☺	☹	
dime	☺	☺	☹
penny	☺		

The students decide they will need to gather some quarters, nickels, dimes, and pennies from the coin collection.

They get out enough containers to match the graph, including two half containers, and then use the scale to decide they should put 12 coins in each container. They reason that 6 is half of 12, so they should put 6 coins in each half container.

Finally, they label their pictograph.

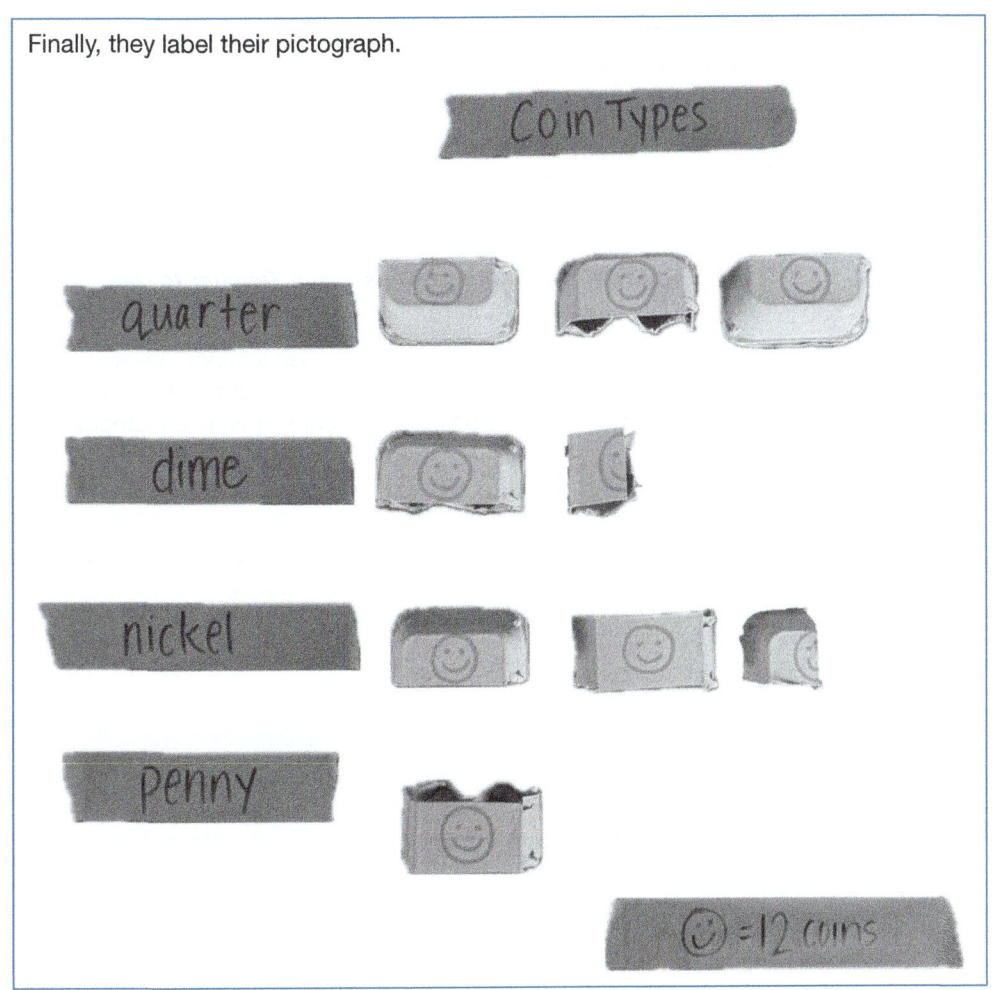

Engage in Math Discourse (Make the Mathematics Visible):

Observation: As students are finding relationships, encourage them to add information to the whole-group chart as they discover it. Students need not complete all the columns at the same time. As they work, encourage them to notice important relationships and push on early and partial understandings.

- How did you decide how many coins of each type to get?
- Which coin did you get the most of? the fewest of? Why?
- Do you have a prediction about how many coins you will need for this type? How are you predicting?
- I see you have three containers for quarters. How many quarters will you need? How did you figure that out?

Interview and **Show Me:** After students have completed their pictographs, bring the intervention group together for a discussion. Suggested prompts:

- (Indicate a group's pictograph) How can we figure out how many nickels this group had?

- (Indicate a pictograph that has a "half image more" relationship between two categories) I notice that in this pictograph, there are three "smiley faces" of quarters, but the nickels have two smiley faces plus one half of a smiley face. Did this group need to get the same number of quarters and nickels? Which one did they need more of? How could we figure that out?

|  | ***Bridging Prompt (Prompt for Classroom Teacher to Use During New Lesson Content):***
• Show students an example of a half symbol. Ask students who participated in the intervention group, "How can you figure out what a half symbol represents on a pictograph?" |

References

Achieve the Core. (2013). *Progressions documents for the Common Core State Standards in Mathematics*. Retrieved from https://achievethecore.org/page/254/progressions-documents-for-the-common-core-state-standards-for-mathematics

Aguirre, J. M., Mayfield-Ingram, K., & Martin, D. B. (2013). *The impact of identity in K–8 mathematics: Rethinking equity-based practices*. National Council of Teachers of Mathematics.

Asbell-Clarke, J. (2023). *Reaching and teaching neurodivergent learners in STEM*. Routledge.

Aspen Institute National Commission on Social, Emotional, and Academic Development. (2019). *From a nation at risk, to a nation at hope: Recommendations from the National Commission on Social, Emotional, and Academic Development*. Aspen Commission.

Bahr, D. L., Whiting, E. F., & Charlton, C. T. (2023). Enhancing access to general education mathematics instruction through special education preteaching: Special education–general education teacher collaboration. *Preventing School Failure: Alternative Education for Children and Youth*, 1–8.

Bay-Williams, J. M., SanGiovanni, J. J., Martinie, S. M., & Suh, J. M. (2022). *Figuring out fluency—addition and subtraction with fractions and decimals: A classroom companion*. Corwin Press.

Bechara, A., Damasio, H., Damasio, A. R. (2000). Emotion, decision making and the orbitofrontal cortex. *Cerebral Cortex, 10*(3), 295–307.

Berger, J. G. (2004). Dancing on the threshold of meaning: Recognizing and understanding the growing edge. *Journal of Transformative Education, 2*(4), 336–351.

Bismark, S., & Prosser, S. (2024). *Concrete-representational-abstract (CRA) approach in an algebra I inclusion class*. Paper presentation at the ICME Conference in the TSG on Students with Special Learning Needs. Sydney, Australia.

Black, P., & Wiliam, D. (1998). Assessment and classroom learning. *Assessment in Education: Principles, Policy & Practice, 5*(1), 7–74.

Blanton, L. P., Pugach, M. C., & Boveda, M. (2018). Interrogating the intersections between general and special education in the history of teacher education reform. *Journal of Teacher Education, 69*(4), 354–366.

Boaler, J., & Humphreys, C. (2005). *Connecting mathematical ideas: Middle school video cases to support teaching and learning*. Heinemann.

Bondy, E., & Ross, D. D. (2008). The teacher as warm demander. *Educational Leadership, 66*(1), 54–58.

Bonner, E. P. (2014). Investigating practices of highly successful mathematics teachers of traditionally underserved students. *Educational Studies in Mathematics, 86*, 377–399.

Brewster, B. J. M., & Miller, T. (2020). Missed opportunity in mathematics anxiety. *International Electronic Journal of Mathematics Education, 15*(3), em0600.

Brodesky, A. R., Fagan, E. R., & MacVicar, T. J. (2022). *Strengthening math intervention in the middle grades: A guide for leaders*. Education Development Center, Inc. https://www2.edc.org/accessmath/resources/documents/Math_Intervention_Guide_Leaders_2022_fin.pdf

Bruner, J. S. (1966). *Toward a theory of instruction*. Harvard University Press.

Bryk, A. S., & Schneider, B. (2002). Trust in schools: A resource for improvement. *School Leadership & Management, 22*(2), 169–187.

Carpenter, T. P., Fennema, E., Franke, M. L., Levi, L., & Empson, S. B. (2014). *Children's mathematics: Cognitively guided instruction* (2nd ed.). Heinemann.

Cassidy, C. (2024). Nearly 80% of Australian students say they "didn't fully try" in latest PISA tests. *The Guardian*. (January 11).

CAST. (2024). *Draft of UDL guidelines 3.0*. CAST.

Civil, M. (2007). Building on community knowledge: An avenue to equity in mathematics education. In N. Nasir & P. Cobb (Eds.), *Improving access to mathematics: Diversity and equity in the classroom*, 105–117. Teachers College Press.

Civil, M., & Andrade, R. (2003). Collaborative practice with parents: The role of the researcher as mediator. In A. Peter-Koop, V. Santos-Wagner, C. Breen, & A. Begg (Eds.), *Collaboration in teacher education: Examples from the context of mathematics education*, pp. 153–168. Springer.

Clarkson, P., Bishop, A. J., & Seah, W. T. (2010). Mathematics education and student values: The cultivation of mathematical wellbeing. In T. Lovat, R. Toomey, & N. Clement (Eds.), *International *research handbook on values education and student wellbeing* (pp. 111–135). Springer.

Clements, D. H., & Sarama, J. (2011). *Learning and teaching early math: The learning trajectories approach.* Routledge.

Clements, D., & Sarama, J. (2014). *Learning and teaching early mathematics: The learning trajectories approach.* Routledge.

Clements, D. H., & Sarama, J. (2019). *Learning and teaching early math: The learning trajectories approach* (2nd ed.). Routledge.

Clifton, D. O., & Harter, J. K. (2003). "Strengths investment." In K. S. Cameron, J. E. Dutton, & R. E. Quinn (Eds.), *Positive organizational scholarship* (pp. 111–121). Berrett-Koehler.

Clifton, D. O., & Nelson, P. (1992). *Soar with your strengths.* Dell.

Cohen, E., & Lotan, R. (2014). *Designing groupwork: Strategies for the heterogeneous classroom* (3rd ed.). Teachers College Press.

Confrey, J., Maloney, A., & Corley, D. (2014). *Learning trajectories: A framework for connecting standards with curriculum.* ZDM 46, 719–733. 10.1007/s11858-014-0598-7.

Cornelius-White, J. (2007). Learner-centered teacher–student relationships are effective: A meta-analysis. *Review of Educational Research, 77*(1), 113–143.

Council for Exceptional Children & CEEDAR Center. (2017). *High-leverage practices in special education: Foundations for student success.* Retrieved from https://highleveragepractices.org

Council of Chief State School Officers. (2018). *Key shifts in the Common Core State Standards.* Retrieved from https://ccsso.org

Daro, P., Mosher, F. A., & Corcoran, T. B. (2011). *Learning trajectories in mathematics: A foundation for standards, curriculum, assessment, and instruction.* Consortium for Policy Research in Education.

Delpit, L. (2013). *Multiplication is for white people: Raising expectations for other people's children.* The New Press.

Dixon, J. (2018). *Five ways we undermine efforts to increase student achievement (and what to do about it!).* Ignite Session at NCTM Annual Meeting. Washington, D.C.

Dougherty, B. (2008). Measure up: A quantitative view of early algebra. In J. Kaput, D. Carraher, & M. Blanton (Eds.), *Algebra in the early grades* (pp. 389–412). Erlbaum.

D'Souza, K. (2024). Biden budget plan includes $8 billion for learning recovery. *EdSource.*

Dweck, C. S. (2006). *Mindset: The new psychology of success.* Random House.

Dweck, C. S. (2008). Can we praise students for their intelligence and their performance? A cautionary tale. *The Chronicle of Higher Education, 54*(30), B6.

Evans, E., Bostic, J., & Yee, S. (2024). Productive problem-solving behaviors of students with learning disabilities. *Investigations in Mathematics Learning,* DOI: 10.1080/19477503.2024.2396720

Fennell, F., Kobett, B., & Wray, J. (2015). Classroom-based formative assessments: Guiding teaching and learning. In C. Suurtamm (Ed.) & A. McDuffie (Series Ed.), *Annual perspectives in mathematics education: Assessment to enhance teaching and learning* (pp. 51–62). National Council of Teachers of Mathematics.

Fennell, F., Kobett, B. M., & Wray, J. A. (2024). *The Formative 5 in action 2.0, grades K–12.* Corwin Press.

Fennema, E., Franke, M. L., Carpenter, T. P., & Carey, D. A. (1993). Using children's mathematical knowledge in instruction. *American Educational Research Journal, 30*(3), 555–583.

Fernandes, M. A., Wammes, J. D., & Meade, M. E. (2018). The surprisingly powerful influence of drawing on memory. *Current Directions in Psychological Science 27*(6), 302–308.

Findell, B., Swafford, J., & Kilpatrick, J. (Eds.). (2001). *Adding it up: Helping children learn mathematics.* National Academies Press.

Fischer, K. W. (2009). Mind, brain, and education: Building a scientific groundwork for learning and teaching. *Mind, Brain, and Education, 3*(1), 3–16.

Fuchs, L. S., Newman-Gonchar, R., Schumacher, R., Dougherty, B., Bucka, N., Karp, K. S., Woodward, J., Clarke, B., Jordan, N. C., Gersten, R., Jayanthi, M., Keating, B., & Morgan, S. (2021). *Assisting students struggling with mathematics: Intervention in the elementary grades* (WWC 2021006). National Center for Education Evaluation and Regional Assistance (NCEE), Institute of Education Sciences, U.S. Department of Education. https://whatworks.ed.gov/

Harbour, K. E., Livers, S. D., McDaniel, S. C., Gleason, J., & Barth, J. M. (2022). Professional development to support elementary mathematics and co-teaching practices: Collaborations between general and special education. *Mathematics Teacher Education and Development, 24*(2), 33–56.

Hardy, A. M., Douglas, G. P., MacLean, K. B., Mason, K. K., & Powell, S. R. (2025). How keywords impact word-problem performance: When "more" does not mean to add. *Learning Disability Quarterly* 1-11. doi/10.1177/07319487251322473

Hattie, J. (2009). *Visible learning*. Routledge.

Hattie, J., Fisher, D., Frey, N., Gojak, L. M., Moore, S. D., & Mellman, W. (2016). *Visible learning for mathematics, grades K–12: What works best to optimize student learning*. Corwin Press.

Heddens, J. W. (1964). *Today's mathematics*. Science Research Associates.

Heddens, J. W., & Speer, W. (2009). *Today's mathematics: Concepts, methods, and classroom activities*. (12th ed.) Wiley.

Henningsen, M., & Stein, M. K. (1997). Mathematical tasks and student cognition: Classroom-based factors that support and inhibit high-level mathematical thinking and reasoning. *Journal for Research in Mathematics Education, 28*(5), 524–549.

Hill, J. Kern, M. van Driel, J., & Seah, W. T. (2022). The importance of positive classroom relationships for diverse students' well-being in mathematics education (pp. 76–89). In D. Burghes & J. Hunger, *Mathematics education for sustainable economic growth and job creation*. Routledge.

Hord, S. M., & Xin, Y. P. (2001). Creating a supportive environment for special education inclusion. *Preventing School Failure, 46*(2), 73–78.

Howard, G. R. (2016). *We can't teach what we don't know: White teachers, multiracial schools*. Teachers College Press.

Huinker, D., Marshall, A. M., Rigelman, N., Yeh, C., & Barnes, D. (2020). *Catalyzing change in early childhood and elementary mathematics: Initiating critical conversations*. National Council of Teachers of Mathematics.

Hurtado, C. K. (2018). Community engagement: School and family partnerships to support mutual learning. *Allies for Education, 2*(1), 1–6. https://journals.library.csuci.edu/ojs/index.php/afe

IDEA—Individuals with Disabilities Education Act Amendments of 1997. (1997). Public Law 105-17.

Ishimaru, A. M., Barajas-Lopez, F., & Bang, M. (2015). Centering family knowledge to develop children's empowered mathematics identities. *Journal of Family Diversity in Education, 1*(4), 1–21. https://doi.org/10.53956/jfde.2015.63

Jacobs, V. R., Martin, H. A., Ambrose, R. C., & Philipp, R. A. (2014). Warning signs! *Teaching Children Mathematics, 21*(2), 107–113.

Jitendra, A., Dougherty, B., Sanchez, V., Harwell, M. R., & Harbour, S. (2023). Building conceptual understanding multiplicative reasoning content in third graders struggling to learn mathematics: A feasibility study. *Learning Disabilities Research & Practice, 38*(4), 285–295.

Karp, K. (2013). *The invisible 10%—Preparing teachers to teach mathematics to students with special needs* [Keynote presentation of the Judith Jacobs Lecture]. Annual Meeting of the Association of Mathematics Teacher Educators. Orlando, FL.

Karp, K. S., Bush, S. B., & Dougherty, B. J. (2019). Avoiding the ineffective keyword strategy. *Teaching Children Mathematics, 25*(7), 428–435.

Karp, K. S., Dougherty, B. J., & Bush, S. B. (2021). *The math pact: Achieving instructional cohesion within and across grades—Elementary*. Corwin Press.

Kestel, E., & Forganz, H. (2024). *Supporting years 7–9 students significantly behind in mathematics: Insights from an umbrella review*. Paper presentation at ICME Conference in the TSG on Students with Special Needs, Sydney, Australia.

King, S. (2022). The education context for twice-exceptional students: An overview of issues in special and gifted education. *Neurobiology of Learning and Memory, 193*, 107659.

Kobett, B., Fennell, F., Karp, K., Harrison, D., & Swartz, B. (2021). *Classroom-ready rich math tasks for grades 2–3: Engaging students in doing math*. Corwin Press.

Kobett, B., & Karp, K. (2020). *Strengths-based teaching and learning in mathematics: Five teaching turnarounds for grades K–6*. Corwin Press.

Koenig, E. A., & Buckley, J. A. (2018). Fostering mathematical resilience: The power of belonging, self-efficacy, and growth mindset. *Teaching Children Mathematics, 25*(5), 294–302.

Korbey, H. (2024). Reading supports abound in schools but effective math help much harder to find. *The 74: America's Education News Source*. https://www.the74million.org/

La Paro, K. M., & Pianta, R. C. (2000). Predicting children's competence in the early school years: A meta-analytic review. *Review of Educational Research, 70*(4), 443–484.

Lambert, R. (2024). *Rethinking disability and mathematics: A UDL math classroom guide for grades K–8*. Corwin Press.

Lambert, R., McNiff, A., Schuck, R., Imm, K., & Zimmerman, S. (2023). "UDL is a way of thinking"; Theorizing UDL teacher knowledge, beliefs, and practices. *Frontiers in Education*, 8, 1145293.

Larson, M. R., Fennell, F., Adams, T. L., Dixon, J. K., Kobett, B. M., & Wray, J. A. (2012). *Common Core mathematics in a PLC at work: Grades 3–5*. Solution Tree Press; NCTM.

Lewis, K. (2018). Difference, not deficit: Assessing issues of access in mathematics for students with disabilities. In *Access and equity: Promoting high quality mathematics in grades 3–5*. The National Council of Teachers of Mathematics.

Lewis, K. E. (2014). Difference not deficit: Reconceptualizing mathematical learning disabilities. *Journal for Research in Mathematics Education*, 45(3), 351–396.

Liljedahl, P. (2020). *Building thinking classrooms in mathematics: Grades K–12*. Corwin Press.

Lobado, J., Clarke, D., & Ellis, A. B. (2005). Initiating and eliciting in teaching: A reformation of telling. *Journal for Research in Mathematics Education*, 36(2), 101–136.

Ludwig, J., & Guryan, J. (2023). *Overcoming pandemic-induced learning loss*. Aspen Economic Strategy Group. November 8.

Marrongelle, K. A., & Foltz, L. (2014). Goals and goal-setting in mathematics. *Mathematics Teacher*, 108(2), 126–131.

Matthews, J., Gray, D. L., Lachaud, Q., McElveen, T., Chen, X. Y., Victor, T., ... & Cha, E. (2021). Belonging-centered instruction: An observational approach toward establishing inclusive mathematics classrooms. OSF Preprints. https://osf.io/n7bv2

Munk, J. H., Gibb, G. S., & Caldarella, P. (2010). Collaborative preteaching of students at risk for academic failure. *Intervention in School and Clinic*, 45(3), 177–185. https://doi.org/10.1177/1053451209349534

NAEP. (2019). *Highlights from the 2019 grade 4 and 8 assessment*. https://www.nationsreportcard.gov/highlights/mathematics/2019/

NAEP. (2022). *Highlights from the 2022 grade 4 and 8 assessment*. https://www.nationsreportcard.gov/highlights/mathematics/2022/

NAEP. (2023). *Long term trend (LTT) performance*. https://www.nationsreportcard.gov/highlights/ltt/2023/

NAEP. (2024). *Highlights from the 2022 grade 4 and 8 assessment*. https://www.nationsreportcard.gov/highlights/mathematics/2024/

National Center for Education Statistics. (2022). *Characteristics of 2020–21 Public and Private K–12 Schools in the United States: Results From the National Teacher and Principal Survey* (NCES 2022-111). U.S. Department of Education.

National Council of Teachers of Mathematics. (2014). *Principles to actions: Ensuring mathematics success for all*. NCTM.

National Council of Teachers of Mathematics. (2000). *Principles and standards for school mathematics*. NCTM.

National Council of Teachers of Mathematics. (2023). *Ability labels: Disrupting "high," "medium," and "low" in mathematics education*. Retrieved from NCTM.

National Council of Teachers of Mathematics and the Council on Exceptional Children. (2024). *Joint position statement on teaching mathematics to students with learning disabilities*. NCTM and CEC.

National Council of Teachers of Mathematics. (2025, January 29). Statement from the National Council of Teachers of Mathematics on the Release of 2024 NAEP Mathematics Scores, NCTM.

NGA Center (National Governors Association Center for Best Practices) & CCSSO (Council of Chief State School Officers). (2010). *Common core state standards*. Retrieved from corestandards.org

Osher, D., Cantor, P., Berg, J., Steyer, L., & Rose, T. (2020). Drivers of human development: How relationships and context shape learning and development. *Applied Developmental Science*, 24(1), 6–36.

Paivio, A. (2013). *Imagery and verbal processes*. Psychology Press.

Paivio, A., & Csapo, K. (1973). Picture superiority in free recall: Imagery or dual coding? *Cognitive Psychology*, 5, 176–206.

Panicucci, J. (2007). Cornerstones of adventure education. In D. Prouty, J. Panicucci, & R. Collinson (Eds.), *Adventure education: Theory and applications* (pp. 33–48). Human Kinetics.

Pianta, R. C., & Stuhlman, M. W. (2004). Teacher–child relationships and children's success in the first years of school. *School Psychology Review*, 33(3), 444–458.

Pilotti, D. M. (2025). *Developing mathematical thinking and reasoning for students with learning disabilities through professional learning communities*. [Doctoral dissertation, University of South Carolina].

Pilotti, D. M., Harbour, K. E., Miller, B. T., & Larkin, E. K. (2023). Creating cohesion and collaboration in mathematics classrooms: Implementing interdepartmental professional learning communities to support students with disabilities. *Education Sciences, 14*(1), 50.

Powell, S. R., Namkung, J. M., & Lin, X. (2022). An investigation of using keywords to solve word problems. *The Elementary School Journal, 122*(3), 452–473.

Reinhart, Steven. (2000, April). Never say anything a kid can say! *Mathematics Teaching in the Middle School, 5*(8), 478–483.

Robinson, Carly D., & Susanna Loeb. (2021). *High-impact tutoring: State of the research and priorities for future learning.* (EdWorkingPaper: 21-384). Retrieved from Annenberg Institute at Brown University: https://doi.org/10.26300/qf76-rj21

Ryan, R. M., Reeve, J., Kaplan, H., Matos, L., & Cheon, S. H. (2023). Education as flourishing: Self-determination theory in schools as they are and as they might be. *The Oxford handbook of self-determination theory* (pp. 591–618). Oxford University Press.

Scandurra, J. (2024). *Using visualization to unlock number sense.* Unpublished paper.

Schoenfeld, A. H. (2011). *How we think: A theory of goal-oriented decision making and its educational applications.* Routledge.

Schwartz, S. (2023, August 9). 7 states now require math support for struggling students. Here's what's in the new laws. *Education Week, 43*(2).

Sethi, J., & Scales, P. C. (2020). Developmental relationships and school success: How teachers, parents, and friends affect educational outcomes and what actions students say matter most. *Contemporary Educational Psychology, 63*, N.PAG. doi:10.1016/j.cedpsych.2020.101904

Shumow, L., & Schmidt, J. A. (2016). Empowering students' strengths: A path to greater equity in mathematics education. *Mathematics Teaching in the Middle School, 21*(8), 476–483.

Siegler, R., Carpenter, T., Fennell, F., Geary, D., Lewis, J., Okamoto, Y., Thompson, L., & Wray, J. (2010). *Developing effective fractions instruction for kindergarten through 8th grade: A practice guide* (NCEE #2010-4039). National Center for Education Evaluation and Regional Assistance, Institute of Education Sciences, U.S. Department of Education. Retrieved from whatworks.ed.gov/ publications/ practiceguides

Skemp, R. R. (1978). Relational understanding and instrumental understanding. *Arithmetic Teacher, 26*(3), 9–15.

Smith, M. S., & Stein, M. K. (1998). Selecting and creating mathematical tasks: From research to practice. *Mathematics Teaching in the Middle School, 3*(5), 344–349.

Sparks, S. (2022a, October 24). How do schools come back from "catastrophic" drops in math, reading on NAEP? *Education Week.*

Sparks, S. (2022b, October 24). Explaining that steep drop in math scores on NAEP: 5 takeaways. *Education Week.*

Star, J. R., & Verschaffel, L. (2016). Building a strong foundation in mathematics: Enhancing children's early learning through culturally diverse perspectives and multiple mathematical practices. *Early Childhood Education Journal, 44*(6), 517–527.

Stelitano, L., Diliberti, M. K., & Doan, S. (2021a). *Teachers' lesson modifications for students with disabilities: Findings from the 2020 American Instructional Resources Survey. Data Note: Insights from the American Educator Panels.* Research Report. RR-A134-6. RAND Corporation.

Stelitano, L., Mulhern, C., Feistel, K., & Gomez-Bendaña, H. (2021b). *How are teachers educating students with disabilities during the pandemic? Data note: Insights from the American Educator Panels.* Research Report. RR-A1121-1. RAND Corporation.

Suh, J. M., Wills, T., Kirschner, S., Wearly, A., Vora, M. E., & Roscioli, K. (2022). Developing asset-based instruction through learning trajectory-based curricular design. *Conference Papers—Psychology of Mathematics & Education of North America* (pp. 215–223).

Sun, J., & Tan, S. (2019). Promoting mathematical discourse in mathematics classrooms. *Mathematics Education Research Journal, 31*(1), 79–96.

The White House. (2024, January 17). *Fact sheet: Biden–Harris administration announces improving student achievement agenda in 2024.* Retrieved from the Digital Public Library of America, http://catalog.gpo.gov/F/?func=direct&doc_number=001413572&format=999.

Thomas, J., Bicard, S., & Simmons, K. D. (2024). The effects of concrete and virtual manipulatives on solving algebraic equations in students with disabilities. *Journal of Special Education Technology.* https://doi.org/10.1177/01626434241263055

Thompson, I. (1999). Written methods of calculation. In I. Thompson (Ed.), *Issues in teaching numeracy in primary schools* (pp. 169–183). Open University Press.

Tomlinson, C. A. (1999). Mapping a route toward differentiated instruction. *Educational Leadership, 57*, 12–17.

Tulving, E., & Schacter, D. L. (1990). Priming and human memory systems. *Science 247*, 301–306.

United States Department of Education, Office of Special Education and Rehabilitative Services. (2000, July). *A guide to the Individualized Education Program*. Washington, DC: U.S. Department of Education. Retrieved from http://www.ed.gov/offices/OSERS

United States Supreme Court. (2017). 15-827 *Endrew F., et al v. Douglas County School District* RE-1.

Usiskin, Z. (2007). The case of the University of Chicago school mathematics project—secondary component. In C. R. Hirsch (Ed.), *Perspectives on the design and development of school mathematics curricula* (pp. 173–182). NCTM.

Van de Walle, J. A., Karp, K. S., & Bay-Williams, J. M. (2023). *Elementary and middle school mathematics: Teaching developmentally* (11th ed.). Pearson.

van Garderen, D., Lannin, J. K., & Kamuru, J. (2020). Intertwining special education and mathematics education perspectives to design and intervention to improve student understanding of symbolic numerical magnitude. *Journal of Mathematical Behavior, 59*, 100782.

Virginia Department of Education. (2023). *All in tutoring*, VDOE.

Virginia Standards of Learning for Mathematics. (2023). https://www.doe.virginia.gov/teaching-learning-assessment/instruction/mathematics/standards-of-learning-for-mathematics

Wang, T. (2023). Teachers as the agent of change for student mental health: The role of teacher care and teacher support in Chinese students' well-being. *Frontiers in Psychology, 14*, 1283515.

Ware, F. (2006a). Warm demander pedagogy. *Urban Education, 41*(4), 427–456.

Ware, F. (2006b). The warm demander: Teacher–student relationships in the diverse classroom. *Journal of Advanced Academics, 18*(2), 267–279.

Wentzel, K. R. (1997). Student motivation in middle school: The role of perceived pedagogical caring. *Journal of Educational Psychology, 89*(3), 411–419.

Wyer, R. S. (2007). Principles of mental representation. *Social Psychology: Handbook of Basic Principles, 2*, 285–307.

Yonelinas, A. P., Aly, M, Wang, W. C., & Koen, J. D. (2010). Recollection and familiarity: Examining controversial assumptions and new directions. *Hippocampus, 20*, 1178–1194.

Young, J. D., Demirdöğen, B., & Lewis, S. E. (2023). Students' sense of belonging in introductory chemistry: Identifying four dimensions of belonging via grounded theory. *International Journal of Science and Mathematics Education, 22*, 1515–1535.

Zavala, M., & Aguirre, J. M. (2024). *Cultivating mathematical hearts: Culturally responsive mathematics teaching in elementary classrooms*. Corwin Press.

Zhang, Q., Sun, J., & Yeung, W. (2022). Effects of using picture books in mathematics teaching and learning: A systematic literature review from 2000–2022. *Review of Education, 11*(1), e3383.

Index

Addition
    fractions: Fraction Frenzy task, 241–249; Fraction Sums task, 295–303, 296 (figure), 298–300 (figure); Fraction Zap task, 290–294, 291–293 (figure)
    whole numbers: Adding It Up task, 105–113, 106 (figure), 107–108 (figure); Picture This task, 149–157; Targeted Sum task, 128-133; Things That Come in Groups task, 170–179
AIMS Education Foundation, 377, 378
Algebraic thinking
    equal sign: Balancing Act task, 158–165
    functional thinking: Magic Pot Patterns task, 234–240
Arrays as a representation
    area connection: Garden Spaces task, 180–186, 181 (figure), 182 (figure), 184 (figure), 185 (figure)
    commutative property connection: Rearrange It task, 197, 198 (figure), 200, 200 (figure)
    distributive property connection: Speed Estimator task, 220–221, 222 (figure), 224 (figure), 225 (figure), 227–228 (figure), 230–232 (figure)
    equation connection: Build It Bingo task, 187–196, 189 (figure), 191–194 (figure); Speed Estimator task, 220–221, 222 (figure), 224 (figure), 225 (figure), 227–228 (figure), 230–232 (figure)
    multiplication and division facts: Multiplication Chain task, 65–67, 209–219, 210–211 (figure), 212–213, 213–215 (figure), 216, 216–217 (figure), 218–219 (figure)
Assessment. *See* Formative assessment; Summative assessment
Assisting Students Struggling With Mathematics: Intervention In The Elementary Grades (*IES Practice Guide*; Fuchs et al., 2021), xviii, 8, 10, 14, 15
    recommendations, 9–10
Associative property of multiplication
    Rearrange It task, 197–208, 198 (figure), 200–201 (figures), 203 (figure)

Catalyzing Change in Early Childhood and Elementary Mathematics (Huinker), 8
Children's literature in mathematics interventions, 16, 76, Appendix A
Collaboration, 11, 12 (figure), 14, 16–17, 19 (figure), 23 (figure), 26, 32, 49, 53, 68, 69

Common hiccups, 19, 27, 55, 66, 92, 159, 374, 376, 401
Commutative property of multiplication
    Build It Bingo task, 187–196, 189 (figure), 191–194 (figure); Rearrange It task, 197–208 (figures)
Concrete, semiconcrete, and abstract (CSA), xviii, 9, 19 (figure), 45, 55, 56 (figure), 69, 76
    multiple representations, 14–16, 19 (figure)
    priming interventions, 19 (figure), 76
Connections, 8, 9, 23 (figure), 53, 57–59
    assessment and instruction, 49–51
    home, 32, 32–33 (figure), 70, Appendix B
Culturally responsive mathematics teaching, 31–32

Data collection using real-world objects: Tip and Plot task, 388–395, 390–391 (figure), 393–394 (figure)
Decimals
    compare and order: Valuable Digits task, 140–148
    notation: Come on Down! task 88–94, 89 (figure), 94 (figure)
    represent equivalent fractions and decimals: Sport Stats task, 284–289, 286 (figure), 287 (figure)
Digital resources, Appendix E
Division
    creating representations: Picture This task, 149–157
    facts: Multiplication Chain task 209–219, 210–211 (figure), 212–213, 213–215 (figure), 216, 216–217 (figure), 218–219 (figure)
    fraction and whole number: Fraction Races task, 313–319, 314–315 (figure), 317 (figure)
    missing factor: Build It Bingo task, 187–196, 189 (figure), 191–194 (figure)
    problem solving: Escape Room task, 166–169, 168 (figure)
    scale in graphing: Picture Perfect task, 396–408, 396 (figure), 398 (figure), 403–404 (figure), 406–407 (figure)

Equal groups, 152, 168
    multiplication representation: Multiplication Chain task, 209–219, 210–211 (figure), 212–213, 213–215 (figure), 216, 216–217 (figure), 218–219 (figure); Things That Come in Groups task, 170–179, 171–172 (figure)
    properties of multiplication: Rearrange It task, 197, 198 (figure), 200, 200 (figure)
    scale: Picture Perfect task, 396–408, 396 (figure), 398 (figure), 403–404 (figure), 406–407 (figure)

Equal signs
    Balancing Act task, 158–165, 159–161 (figure), 160 (figure), 163
Equivalent fractions, 252, 257, 279, 281–282, 298
    comparing: Cake Time task, 250–259
    renaming, 277, 296
    visual models: Fraction Equivalence Through Quilts and Fringes task, 279, 280–283 (figures); Fraction Zap task, 290–294, 291–293 (figure); Sports Stats task, 284–289, 286 (figure), 287 (figure); To Be or Not to Be: Equivalent Fractions task, 270–275, 272–274 (figure)
Escape Room task, 166–169, 168 (figure)

Feedback, 37, 37 (figure)
    communicating progress, 67, 67 (figure)
    strengths based, 27
Formative assessment
    classroom-based, 8, 35–36
    connection to instruction, 35, 49-51
    CSA component, 39 (figure), 40-41 (figure), 45
    interview tools, 41–44
    observation tools, 38–41
    Show Me technique, 45–48
Fractions. *See also* Equivalent fractions
    addition/subtraction: Fraction Sums task, 295–303, 296 (figure), 298–300 (figure); Fraction Zap task, 290–294, 291–293 (figure)
    changing to decimals, 284–289, 286 (figure), 287 (figure)
    comparing: Cake Time task, 250–259; Line 'Em Up task, 260–269, 262–268 (figure)
    decompositions: Fraction Frenzy task, 241–248, 242–248 (figure)
    Developing Effective Fractions Instruction for Kindergarten Through 8th Grade (IES Practice Guide; Siegler), 10
    division: Fraction Races task, 313–319, 314–315 (figure), 317 (figure)
    equivalence: Fraction Equivalence Through Quilts and Fringes task, 279, 280–283 (figures); Fraction Zap task, 290–294, 291–293 (figure); Sports Stats task, 284–289, 286 (figure), 287 (figure); To Be or Not to Be: Equivalent Fractions task, 270–275, 272–274 (figure); 276–283, 278 (figure), 280–282 (figure)
    multiplication: Farmland task 308–312, 309–311 (figures); Fundraiser task 304–312, 305–307 (figure), 309–311 (figure)
    representational models: number line, 260–269, 262 (figure), 264 (figure), 266 (figure), 268 (figure)
    size or magnitude: Cake Time task 250–259, 251–255 (figure)
    whole as a unit: "Make a whole" fractions, 292, 296, 296 (figure), 298, 300

Geometry
    three-dimensional shapes: attributes, Shape Shifting task, 331–336, 333 (figure)
    two-dimensional shapes: attributes, Shape Shifting task, 331–336, 333 (figure); comparing shapes: Alike and Different task, 325–330, 326 (figure); composing shapes: Shapes Build It! task, 337–342, 339 (figure), sorting, Shape Sort task 320– 324, 322 (figure), 323 (figure); What's My Rule task, 343–351, 345–349 (figure)
Grade-level content and planning, xi, xiii, xiv, 5, 7, 36, 49, 55–56, 59, 68

High-leverage practices (HLP), 11
Home connections, 32, 32–33 (figure), Appendix B

Individualized Education Program (IEP), 7–8, 17, 36, 49, 50, 67
Individuals with Disabilities Education Act (IDEA) of 1997, 6
Interventions, mathematics instruction, 12–16, 12 (table), 15 (figure)
    at a glance, Appendix D
    collaboration, 16–17
    digital resources, Appendix E
    mathematics teaching, culturally responsive, 31–32
    multiple representations, 14–16, 15 (figure)
    proactive approach, 17–18, 19 (figure)
    See also Formative assessment; Proactive intervention
Interview Planning Tool, 42, 43 (figure)
    formative assessment technique, 42
    Individual Student Prompt, 44, 44 (figure)

Learning-trajectory-based instruction, 5, 6, 18, 20, 54

Math anxiety, 33
Math sketches, xviii, 27
   fraction multiplication: Fundraiser task, 304–312, 305–307 (figure), 309–311 (figure)
   Fraction Zap task, 290–294, 291–293 (figure)
   Picture This task, 149–157, 150–152 (figure), 155–156 (figure)
   Things That Come in Groups task 170–179, 171–172 (figure)
Measurement
   area: Garden Spaces task, 180–186, 181 (figure), 182 (figure), 184 (figure), 185 (figure)
   comparing units of measure: Comparing With Measurement task, 369–372, 370–371 (figure)
   converting length measures; Field Day task, 377–381, 380 (figure)
   fractional units: Farmland task 308–312, 309–311 (figures); Fraction Sums task, 295–303, 296 (figure), 298–300 (figure)
   home connection, 32, Appendix B
   length: Measure Up! task 365–368, 366 (figure), 370; Tip and Plot task, 388–395, 390–391 (figure), 393–394 (figure)
   perimeter: On the Edge task, 373–376, 374 (figure), 375 (figure)
   planning discussion for intervention, 61
   problem solving: Escape Room task 166–169
   time. *See* Telling time
Mental residue, 4
Missing-factor equation, 187–196, 189 (figure), 191–194 (figure)
Money
   coin values: Coin Tower challenge 363–364
   counting coins, 352–358, 353 (figure), 354 (figure)
   problem solving: Smart About Money Task, 359–364, 364 (figure)
Multiplication
   building fluency, 65–67
   Chain task, 209–219, 210–211 (figure), 212–213, 213–215 (figure), 216, 216–217 (figure), 218–219 (figure)
   creating representations: Picture This task, 149–157
   distributive property: Speed Estimator task, 220–221, 222 (figure), 224 (figure), 225 (figure), 227–228 (figure), 230–232 (figure)
   equal-groups model, 210

   fractions: Fraction Multiplication: Farmland task 308–312, 309–311 (figures); Fraction Multiplication: Fundraiser task, 304–312, 305–307 (figure), 309–311 (figure)
   identity property, 276
   missing factor: Build It Bingo task, 187–196, 189 (figure), 191–194 (figure); Picture Perfect task, 396–408, 396 (figure), 398 (figure), 403–404 (figure), 406–407 (figure)
   scale on a pictograph: Picture Perfect task 396–408, 396 (figure), 398 (figure), 403–404 (figure), 406–407 (figure)
Multitiered system of support (MTSS), xi, ii, 6, 36, 69

National Assessment of Educational Progress (NAEP), 3–4, 3 (table), 54
   long-term trend data (LTT), 4
National Center for Education Statistics, 3
National Council of Teachers of Mathematics, xiv, 11–12
   NCTM/CEC joint position statement, xix, 6, 8, 14

Observation tools for assessment, 63
   Individual Student Mathematics Strengths Log, 40–41
   intervention activities, 38
   Small-Group Implementation and Recording Tool, 38, 38–40 (figure)

Patterns: Magic Pot Patterns task, 234–240, 235–237 (figure), 239 (figure)
Perimeter
   On the Edge task, 373–376, 374 (figure), 375 (figure)
Person-first language, xix
Pictographs
   Picture perfect task 396–408, 396 (figure), 398 (figure), 403–404 (figure), 406–407 (figure)
Place value
   addition: Adding It Up task 105–113 106–111 (figure); Targeted Sum task, 128–133
   comparing numbers: Valuable Digits task, 140–148, 142–145 (figure)
   decimal notation, notation: Come on Down! task, 88–94, 89 (figure), 94 (figure)
   multiplication: Rearrange It task, 197–208, 198 (figure), 200, 200 (figure)
   regrouping: Trading Places task, 95–102, 96–100 (figure)

renaming/rounding and three-digit numbers: Mystery Number Riddles task, 80–87, 81–82 (figure), 83–84 (figure)
subtraction: Targeted Differences task, 134–139, 135 (figure), 137 (figure); What's the Difference task, 114–127, 115–126 (figure)

Planning discussion for intervention, 61

Planning model, grade-level classroom
conversation, 60–61
four weeks before new lesson, 55
general and special education teams, 55, 56 (figure)
intervention planning tool, 59
new content lesson, 58
one week before new lesson, 57
repetition, 58
three weeks before new lesson, 55, 56 (figure), 58
two weeks before the new lesson, 57

Positive mathematical identities, xiii, 11, 19 (figure), 22, 31, 42

Priming approach, xi, xiv, xv, xvii, xviii (figure), 14, 22
intervention planning tool, 59, 59–60 (figure), 69–70
planning of, 54–61
strengths-based instruction, 8, 70–72

Proactive intervention
bridging to grade level content, 19 (figure), 45, 57–59
consistency and predictability, 7
family involvement, 32, 33 (figure), Appendix B
instructional materials, 69
mathematical language, 9
mathematics intervention program, 54
priming, 7–8
program size, 68–69
representations, 9
systematic instruction, 9
teacher-preparation programs, 5
fluency activities, 10–12, 10 (table)
tutoring, 14

Problem-solving, xiv, xxi, 24, 25, 62 (figure), 76
activity, 166–167
context based problems, 29
Escape Room task, 166–169, 168 (figure)
math discourse, 168–169
using word problems, 166
variation, 167–168, 168 (table)

Professional learning community (PLC), xiv, xv, 7, 16

Regrouping. *See* Place value
Relationship building, 29–30
Representations
semiconcrete models, 396–408, 396 (figure), 398 (figure), 403–404 (figure), 406–407 (figure)
student work, 27, 28 (figure)

Research base, 8–12, 14
Response to intervention (RtI), xi, 16, 67, 69
Rounding numbers, 85, 87
addition within 1,000, 128–133, 130–131 (figure)
subtraction within 1,000, 134–139, 135 (figure)

Scale
graphing: Picture Perfect task, 396–408, 396 (figure), 398 (figure), 403–404 (figure), 406–407 (figure)

Semiconcrete representations, 396–408, 396 (figure), 398 (figure), 403–404 (figure), 406–407 (figure)

Sense of belonging, 29–31
Show Me assessment technique, 45, 46–47 (figure)
Staffing, 68
Standards for Mathematical Practice (SMPs), xiv, 7
Strengths-based approaches, xiii, xx, 8, 18, 21–22, 53
and formative assessment, 62–64, 62 (figure), 63 (figure)
language, xix
Progress Monitoring Tool, 63, 63 (figure)

Strengths-spotting process
collaborative mathematics learning, 26
feedback, 27
growth and effort, 24–25, 25 (figure)
leverage, 24
naming, 23
priming approach, 22
rapport and relationships, 29–34
self-reflection and goal-setting, 26
students' strengths, 23, 23 (figure)
translating, 24
valuing diverse problem solving approaches, 25

Subtraction problems
creating representations: Picture This task, 149–157
problem solving: Escape Room task 166–169, 168 (figure)
regrouping: Targeted Differences task, 134–139, 135 (figure), 137 (figure); What's the Difference task 114–127, 115–126 (figure)
renaming/rounding and three-digit numbers: Mystery Number Riddles task, 80–87, 81–82 (figure), 83–84 (figure)

Summative assessment, 20, 35–36, 54

Tape diagram, 272, 305, 305 (figure), 306, 309
Task selection in assessment, 48
Telling time
analog clock, 382–387, 384 (figure), 383–386 (figure)
to the hour, half hour, or minute: Broken clock task, 382–387, 383–386 (figure)

Think-aloud process, 15
"Think multiplication" Build It Bingo task, 187–196, 189 (figure), 191–194 (figure)
Tutoring model, 3, 4, 6, 14, 22, 35, 36, 53, 66
Two-dimensional shapes. *See* Geometry

Universal Design for Learning (UDL), 11–12, 45
  Rethinking Disability and Mathematics; A UDL Math Classroom Guide for Grades K–8 (Lambert), xv

U.S. Supreme Court Endrew F. et al. vs. Douglas County Colorado School District RE-1 (2017), 17

Venn diagram: What's My Rule task 343–351, 345–349 (figure)

"Warm demander" pedagogy, 30–31, 33
Whole-group chart, 151–152, 151 (figure), 173, 243–244, 246–248 (figure), 251–252, 252 (figure), 322, 326, 360

Supporting Teachers, Empowering Students

Peter Liljedahl

Fourteen optimal practices for thinking that create an ideal setting for deep mathematics learning to occur.
Grades K–12

Peter Liljedahl, Maegan Giroux, Kyle Webb

Delve deeper into the implementation of the fourteen practices from Building Thinking Classrooms in Mathematics by focusing on the practice through the lens of tasks.
Grades K–5, 6–12

Rachel Lambert

Discover UDL for math, a way to design math classrooms that equips all students for meaningful and joyful math learning.
Grades K–8

Jo Boaler, Cathy Williams

Introduce data science to your students across disciplines with real-world stories and teacher testimonials to transform your classroom experience.
Grades K-8

Kimberly Rimbey, Katie Basham, Chryste Berda

Focus on making mathematics meaningful through multiple strategies and representations to help foster a love for mathematics in your students.
Grades K–6

Sherri Martinie, Jessica Lane, Janet Stramel, Jolene Goodheart Peterson, Julie Thiele

Build critical intrapersonal and interpersonal skills *through* mathematics to help all students grow the life-skills they'll carry forever.
Grades K–12

To order your copies, visit corwin.com/math

CORWIN Mathematics

Our research-based and high-quality content is written by trusted experts and provides clear pathways to helping all students gain access to rigorous mathematics learning; to learn to truly think, reason, collaborate, and fluently discuss mathematics; to form positive and strengths-based mathematical identities; and to see and use mathematics as a tool to effect change in their lives and communities.

Sue Chapman, Holly Burwell, Mary Mitchell

Focus on developing eight essential habits that support mathematical competence and confidence in students through a personalized, practice-based professional learning experience.

Grades K–5

Pamela Weber Harris

Guide students through domains of mathematical reasoning, from counting and adding strategies to more complex proportional and functional reasoning—without resorting to algorithms.

Grades K–12

John J. SanGiovanni, Dennis McDonald

Enhance your students' understanding and engagement in geometry, measurement, and data while also fostering a deeper connection between math and the real world.

Grades K–5

Vanessa "The Math Guru" Vakharia

Equip students to develop the skills they need to truly believe anything is possible, even a better relationship with math!

Grades K–12

Liesl McConchie

Reexamine what it means to have a positive math identity—and learn to use brain-based tools in a humorous and friendly way to build on a positive math identity for your students.

Grades K–12

Irina Lyublinskaya, Xiaoxue Du

Integrate AI literacy into K–12 classrooms, blending theory, practical lesson plans, and ethical considerations to empower students as critical thinkers.

Grades K–12

To order your copies, visit corwin.com/math

CORWIN

To help every educator help every student

We believe that every single student deserves a great education

We believe that knowing our impact is both a privilege and a responsibility

We believe that a fair, stable, and thriving society is built on education

In compliance with GPSR, should you have any concern about the safety of this product, please advise: International Associates Auditing and Certification Limited, The Black Church, St Mary's Place, Dublin 7, D07 P4AX Ireland EUAR@ie.ia-net.com

Printed and bound by CPI Group (UK) Ltd, Croydon, CR0 4YY
10/04/2026
02087348-0001